A COURSE IN MATHEMATICAL ANALYSIS
Volume I: Foundations and Elementary Real Analysis

The three volumes of *A Course in Mathematical Analysis* provide a full and detailed account of all those elements of real and complex analysis that an undergraduate mathematics student can expect to encounter in the first two or three years of study. Containing hundreds of exercises, examples and applications, these books will become an invaluable resource for both students and instructors.

Volume I focuses on the analysis of real-valued functions of a real variable. Besides developing the basic theory it describes many applications, including a chapter on Fourier series. It also includes a Prologue in which the author introduces the axioms of set theory and uses them to construct the real number system. Volume II goes on to consider metric and topological spaces, and functions of several variables. Volume III covers complex analysis and the theory of measure and integration.

D. J. H. GARLING is Emeritus Reader in Mathematical Analysis at the University of Cambridge and Fellow of St John's College, Cambridge. He has fifty years' experience of teaching undergraduate students in most areas of pure mathematics, but particularly in analysis.

A COURSE IN
MATHEMATICAL ANALYSIS

Volume I
Foundations and
Elementary Real Analysis

D. J. H. GARLING

Emeritus Reader in Mathematical Analysis,
University of Cambridge, and
Fellow of St John's College, Cambridge

CAMBRIDGE
UNIVERSITY PRESS

CAMBRIDGE
UNIVERSITY PRESS

University Printing House, Cambridge CB2 8BS, United Kingdom

Cambridge University Press is part of the University of Cambridge.

It furthers the University's mission by disseminating knowledge in the pursuit of education, learning and research at the highest international levels of excellence.

www.cambridge.org
Information on this title: www.cambridge.org/9781107614185

First published 2013

A catalogue record for this publication is available from the British Library

Library of Congress Cataloguing in Publication data
Garling, D. J. H.
Foundations and elementary real analysis / D. J. H. Garling.
pages   cm. – (A course in mathematical analysis; volume 1)
Includes bibliographical references and index.
ISBN 978-1-107-03202-6 (hardback) – ISBN 978-1-107-61418-5 (paperback)
1. Mathematical analysis.   I. Title.
QA300.G276   2013
515–dc23   2012044420

ISBN 978-1-107-61418-5 Paperback

Contents

Contents

Volume II

Volume III

Introduction

This book is the first of three volumes of a full and detailed account of those elements of real and complex analysis that mathematical undergraduates may expect to meet in the first two years or so of the study of analysis. This volume is concerned with the analysis of real-valued functions of a real variable. Volume II considers metric and topological spaces, and functions of several variables, while Volume III is concerned with complex analysis, and with the theory of measure and integration.

Mathematical analysis depends in a fundamental way on the properties of the real numbers, and indeed much of analysis consists of working out their consequences. It is therefore essential to develop a full understanding of these properties. There are two ways of doing this. The traditional and appropriate way is to take the fundamental properties of the real numbers as axioms – the real numbers form an ordered field in which every non-empty subset which has an upper bound has a least upper bound – and to develop the theory – convergence, continuity, differentiation and integration – from these axioms. This programme is carried out in Part Two. This theory is meant to be used, and Part Two ends with an extensive collection of applications. **The reader is strongly recommended to follow this tradition, and to begin at the beginning of Part Two.**

It is however right to ask about the foundations on which these axioms, and the rest of mathematical analysis, are built. These foundations are considered in the Prologue. In the twentieth century, analysis was placed in a set-theoretic setting, and it is worth understanding what this involves. Chapter 1 contains an account of Zermelo–Fraenkel set theory, together with a brief discussion of the axiom of choice and its variants. The Zermelo–Fraenkel axioms lead naturally to the construction of the natural numbers. In Chapter 2 it is shown that there is then a steady progression through the integers and the rational numbers to the real numbers and the complex numbers. The problem with the natural numbers, the integers and the rational numbers is that they are very familiar; this part of the journey may appear to be spent proving the obvious. The construction of the real numbers is a quite different matter. There are many possible constructions, but we describe the first, given by Richard Dedekind. This has great virtue, since it involves both order and metric properties of the rational numbers and of the real numbers. **The reader is urged to defer a detailed reading of the Prologue until the occasion demands,** for example when it becomes clear how important the fundamental properties of the real numbers are, or when it is important to consider carefully the role of induction, recursion and the axiom of dependent choice.

The text includes plenty of exercises. Some are straightforward, some are searching, and some contain results needed later. Many concern applications, and all help develop an understanding of the theory: do them!

I have worked hard to remove errors, but undoubtedly some remain. Corrections and further comments can be found on a web page on my personal home page at `www.dpmms.cam.ac.uk`.

Part One

Prologue: The foundations of analysis

1

The axioms of set theory

It is probably sensible to read through this chapter fairly quickly, to find out the terminology and notation that we shall use, and then to return later to read it and think about it more carefully.

1.1 The need for axiomatic set theory

Mathematics is written in many languages, such as French, German, Russian, Chinese, and, as in the present case, English. Mathematics needs a particular precision, and within each of these languages, most of mathematics, and all the mathematics that we shall do, is written in the language of sets, using statements and arguments that are based on the grammar and logic of the predicate calculus. In this chapter we introduce the set theory that we shall use. This provides us with a framework in which to work; this framework includes a model for the natural numbers $(1, 2, 3, \ldots)$, together with tools to construct all the other number systems (rational, real and complex) and functions that are the subject of mathematical analysis.

The predicate calculus involves rules of grammar for writing 'well-formed formulae', and for providing mathematical arguments which use them. Well-formed formulae involve variables, and logical operations such as conjunction (P and Q), disjunction (P or Q (or both)), implication (P implies Q), negation (not P), and quantifiers 'there exists' and 'for all', together, in our case, with sets and the relation $\in$. We shall not describe the predicate calculus, which formalizes the everyday use of these logical operations (for example, 'P implies Q' if and only if '(not Q) implies (not P)'), but all our arguments and constructions will be based on it, and we shall give plenty of examples of well-formed formulae.[1]

---

[1] For a good account, see A. G. Hamilton, *Logic for Mathematicians*, Cambridge University Press, 1988.

Since the beginning of the study of set theory by Cantor in the 1870s and the introduction of Venn diagrams by Venn in 1881, the simple idea of a set has become commonplace, and young children happily manipulate sets such as {Catherine of Aragon, Ann Boleyn, Jane Seymour, Anne of Cleves, Kathryn Howard, Katherine Parr}, or more prosaically {Alice, Bob}, or the set of numbers $\{5, 13, 17, 29, 37, 41, 53, 61, 73, 89\}$. In mathematics, we consider sets of mathematical objects, such as the last of these examples. Can we not simply consider a mathematical object to be a collection of all those things which can be defined by a well-formed formula? Then a set would be something of the form 'the collection of those things a for which the well-formed formula $P(a)$ holds', where $P(x)$ is a well-formed formula with one free variable x, and conversely, each such formula would define a set. This approach is known as the *comprehension principle*. Unfortunately, it leads to contradictions. Consider the well-formed statement 'x does not belong to x'; according to the comprehension principle, there should be a set b which consists of those sets which do not belong to themselves. Does b belong to b? If it does, it fails the criterion for belonging to b, and so it does not belong to b. But if it does not belong to b, then it meets the criterion, and so it belongs to b. Thus, either way, we reach a contradiction.

This phenomenon was described by Bertrand Russell in 1901, and is known as *Russell's paradox*. It caused him a great deal of pain, as he described in his autobiography.[2] Concerning the events of May 1901, he wrote

> Cantor had a proof that there is no greatest number, and it seemed to me that the number of things in the world should be the greatest possible. Accordingly, I examined his proof with some minuteness, and endeavoured to apply it to the class of all things there are. This led me to consider those classes which are not members of themselves, and to ask whether the class of all such classes is or is not a member of itself. I found that either answer implied its contradictory.

He continued to consider the problem for several years. Describing the summers of 1903 and 1904, he wrote

> I was trying hard to solve the contradictions mentioned above. Every morning I would sit down before a blank sheet of paper. Throughout the day, with a brief interval for lunch, I would stare at the blank sheet. Often when evening came it was still empty.

Russell's paradox required a new approach to the theory of sets, which would provide a framework where Russell's paradox, and other paradoxes,

---

[2] *The Autobiography of Bertrand Russell*, George Allen and Unwin, 1967–69.

are avoided. In 1908, Zermelo introduced a system of axioms; these were modified in 1922 by Fraenkel and Skolem. The resulting system, known as the *Zermelo–Fraenkel axiom system* ZF, has stood the test of time, and it is the one that we shall describe and use.

1.2 The first few axioms of set theory

In Zermelo–Fraenkel set theory, the basic objects are all called *sets*, denoted by upper- or lower-case letters, and there is one relation, $\in$. Thus, if a and b are sets, then either $a \in b$, or this is not so, in which case we write $a \notin b$. (We use the symbol $/$ to mean 'not', in a similar way, for other relations.) If $a \in b$, we say that a *belongs to* b, or that a is a *member* or *element* or *point* of b, or, more simply, that a *is in* b.

The sets and the relation $\in$ are required to satisfy certain axioms, and we shall spend the rest of this chapter introducing and explaining them.

Axiom 1: The extension axiom

This states that two sets are equal if and only if they have the same elements. Thus the set with members 1, 2 and 3 and the set with members 1, 3, 2 and 1 are the same; the order in which they are listed is unimportant, as is the fact that repetition can occur. Set theory is all about membership, and about nothing else.

If a and b are sets, and every member of a is a member of b, then we say that a is a *subset* of b, or that b *contains* a, and write $a \subseteq b$ or $b \supseteq a$. Thus the extension axiom says that $a = b$ if and only if $a \subseteq b$ and $b \subseteq a$. If $a \subseteq b$ and $a \neq b$, we say that a is a *proper subset* of b, or that a is *properly contained in* b, and write $a \subset b$ or $b \supset a$.

Axiom 2: The empty set axiom

This states that there is a set with no members. The extension axiom then implies that there is only one such set: we denote it by $\emptyset$ and call it the *empty set*. It is easy to overlook the empty set: arguments involving it take on an idiosyncratic form. It also has a rather paradoxical nature, since it is a subset of every set a (if not, there is a member b of $\emptyset$ which is not in a; but $\emptyset$ has no members). Thus (looking ahead to some familiar sorts of sets) we can consider the set F of natural numbers n greater than 2 for which there exist natural numbers a, b and c with $a^n + b^n = c^n$, and we can consider the set Q of those complex quadratic polynomials of the form $z^2 + az + b$ for which the equation $z^2 + az + b = 0$ has no complex solutions. Then $F = Q$, since each is the empty set.

The next four axioms are concerned with creating new sets from old.

Axiom 3: The pairing axiom

This says that if a and b are sets then there exists a set whose members are a and b. The extension axiom again says that there is only one such set: we denote it by $\{a, b\}$. Note that $\{a, b\} = \{b, a\}$: we have an *unordered pair*. We can take $a = b$: then the set $\{a, a\}$ has only one element a. We write this set as $\{a\}$ and call it a *singleton set*.

We can use the pairing axiom to define *ordered pairs*. If a and b are sets, we define the *ordered pair* (a, b) to be the set $\{\{a\}, \{a, b\}\}$.

Proposition 1.2.1   *If (a, b) and (c, d) are ordered pairs and $(a, b) = (c, d)$, then $a = c$ and $b = d$.*

Proof   The proof makes repeated use of the extension axiom. First, suppose that $a = b$. Then $(a, b) = \{\{a\}\} = \{\{c\}, \{c, d\}\}$, and so $\{c, d\} = \{a\}$, and $a = c = d$. Thus $a = b = c = d$. Similarly, if $c = d$ then $a = b = c = d$.

Finally, suppose that $a \neq b$ and $c \neq d$. Since $\{a\} \in (c, d)$, either $\{a\} = \{c\}$ or $\{a\} = \{c, d\}$. But if $\{a\} = \{c, d\}$ then $c = a = d$, giving a contradiction. Thus $\{a\} = \{c\}$ and $a = c$. Since $\{a, b\} \in (c, d)$, either $\{a, b\} = \{c\}$ or $\{a, b\} = \{c, d\}$. But if $\{a, b\} = \{c\}$, then $a = c = b$, giving a contradiction. Thus $\{a, b\} = \{c, d\}$, and so $b = c$ or $b = d$. But if $b = c$ then $b = c = a$, giving a contradiction. Thus $b = d$. $\square$

If A is a set, then all its members are sets, and they, in turn, can have members.

Axiom 4: The union axiom

This says that there is a set whose elements are exactly the sets which are members of members of A. We denote this set by $\cup_{a \in A} a$ (here a is a variable, so we could as well write $\cup_{x \in A} x$) and call it the *union* of the members of A. The essential feature of this axiom is that the sets whose members make up the union must all be members of a single set; we cannot form the union of *all* sets since, as we shall see, there is no set to which all sets belong. If A and B are sets, we can consider the set $\cup_{C \in \{A, B\}} C$. This is the set whose elements are either in A or in B: we write this as $A \cup B$.

Axiom 5: The power set axiom

There is an essential difference between the statements $b \in A$ (b is a member of A) and $b \subseteq A$ (b is a subset of A). The power set axiom states that if A is a set, then there exists a set, the *power set* $P(A)$ of A, whose elements are the subsets of A. Thus $b \in P(A)$ if and only if $b \subseteq A$. For example, the elements of $P(\{a, b\})$ are $\emptyset$, $\{a\}$, $\{b\}$ and $\{a, b\}$, and the ordered pair $(a, b) = \{\{a\}, \{a, b\}\}$ is an element of $P(P(\{a, b\}))$.

Axiom 6: The separation axiom

This is particularly important, and is an axiom that is used all the time in mathematics. It states that if A is a set and $Q(x)$ is a well-formed formula, then there exists a subset of A whose elements are just those members a of A for which $Q(a)$ holds. By extensionality, there is only one such set; we denote it by $\{x \in A : Q(x)\}$. With this axiom in place, we can use the argument of Russell's paradox to show that there is no universal set to which every set belongs.

Theorem 1.2.2 *There is no set Ω such that if a is a set then $a \in \Omega$.*

Proof Suppose that such a set were to exist. Then the formula $x \notin x$ is a well-formed formula, and so there exists a set $b = \{x \in \Omega : x \notin x\}$. Does $b \in b$? If it does, it fails the criterion for membership, giving a contradiction. If it does not, then it meets the criterion, and so belongs to b, giving another contradiction. This exhausts all possibilities, and so no such universal set can exist. □

Let us give some more examples of the use of the separation axiom. Suppose that A and B are sets. The expression $x \in B$ is a well-formed formula, and so the set $\{x \in A : x \in B\}$ is a subset of A, the *intersection* of A and B, denoted by $A \cap B$. Note that $A \cap B = B \cap A = \{x \in B : x \in A\}$, since a set c is an element of either intersection if and only if it belongs to both A and B. We say that A and B are *disjoint* if $A \cap B = \emptyset$; A and B are disjoint if A and B have no member in common. Similarly, the expression $x \notin B$ is a well-formed formula, and so the set $\{x \in A : x \notin B\}$ is a subset of A, the *set difference* $A \setminus B$. $A \setminus B$ is also called the *relative complement* of B in A. It frequently happens that we consider a particular set A, say, and are only concerned with subsets of A. In this case, if $B \subseteq A$, then we denote $A \setminus B$ by $C(B)$, or B^c, and call it the *complement* of B.

We can extend the notion of intersection considerably. Suppose that A is a set. The expression 'for all $a \in A$, $x \in a$' is a well-formed formula with a a bound variable and x a free variable, and so we can form the set

$$\{x \in \cup_{a \in A} a : \text{for all } a \in A, x \in a\}.$$

This is the *intersection* $\bigcap_{a\in A}a$ of all the sets a that belong to A: $b \in \bigcap_{a\in A}a$ if and only if $b \in a$, for each $a \in A$. Here again a is a variable, and we could also write $\bigcap_{x\in A}x$. We must reconcile the two definitions of intersection that we have made: this is easy because $A \cap B = \bigcap_{x\in\{A,B\}}x$.

A word about notation here. Our aim will be to be accurate and clear without being pedantic. Suppose that A is a set. For each $a \in A$, we can form the intersection $\bigcap_{\alpha\in a}\alpha$. Using the separation axiom, we can then define the set I whose elements are exactly these intersections, and can then form the set $\bigcup_{i\in I}i$. In fact, we write this in the form

$$\bigcup_{a\in A}\left(\bigcap_{\alpha\in a}\alpha\right),$$

and use other similar expressions. In the same way, we shall use natural variations of the notation $\{x \in A : Q(x)\}$ to denote sets whose existence is ensured by the separation axiom; but in each case such a set is a subset of a given set, and it can be written, at greater length, in the form $\{x{\in}A : Q(x)\}$.

From now on, we shall define sets without appealing to the axioms to ensure that they are in fact sets. It is a useful exercise for the reader to consider, in each case, how suitable justification can be given.

It is unfortunately the case that the separation axiom is not strong enough for all purposes, and another axiom, the *replacement axiom*, is needed. We shall defer discussion of this and of the other axioms of ZF, until later. Let us first see what we can do with the axioms that we now have.

Exercises

Suppose that A, B, C, D are sets.

1.2.1 Show that $A \cup (B \cap C) = (A \cup B) \cap (A \cup C)$.

1.2.2 Show that $A \cap (B \cup C) = (A \cap B) \cup (A \cap C)$.

1.2.3 Show that $A \setminus (B \cup C) = (A \setminus B) \cap (A \setminus C)$.

1.2.4 Which of the following statements are necessarily true?
    (a) $P(A \cap B) = P(A) \cap P(B)$.
    (b) $P(A \cup B) = P(A) \cup P(B)$.

1.2.5 Define a set I such that $\bigcup_{i\in I}i = \bigcup_{a\in A}\left(\bigcap_{\alpha\in a}\alpha\right)$.

1.2.6 Does $\bigcup_{a\in A}\left(\bigcap_{\alpha\in a}\alpha\right)$ necessarily contain $\bigcap_{a\in A}\left(\bigcup_{\alpha\in a}\alpha\right)$? Is $\bigcup_{a\in A}\left(\bigcap_{\alpha\in a}\alpha\right)$ necessarily contained in $\bigcap_{a\in A}\left(\bigcup_{\alpha\in a}\alpha\right)$?

1.2.7 The *symmetric difference* $a\Delta b$ of two sets a and b is the set $(a\setminus b)\cup(b\setminus a)$. Establish the following:
    (a) $A\Delta B = (A \cup B) \setminus (A \cap B)$.

(b) $A\Delta B = B\Delta A$.
(c) $A\Delta(B\Delta C) = (A\Delta B)\Delta C$.
(d) $A\Delta\emptyset = A$.
(e) $A\Delta A = \emptyset$.

1.3 Relations and partial orders

The *Cartesian product* $A \times B$ of two sets A and B is the set of all ordered pairs (a, b) with $a \in A$ and $b \in B$. More formally,

$$A \times B = \{x \in P(P(A \cup B)) : \text{there exists } a \in A \text{ and there exists } b \in B$$

such that $x = \{\{a\}, \{a, b\}\}\}$.

(The term *Cartesian* honours René Descartes, who introduced coordinates to the plane, so that points in the plane are represented by ordered pairs of real numbers; the plane is thus represented as the Cartesian product of two copies of the set of real numbers.)

A *relation* on $A \times B$ is then simply a subset R of $A \times B$. It is customary to write aRb if $(a, b) \in R$. The set

$$\{a \in A : \text{ there exists } b \in B \text{ such that } (a, b) \in R\}$$

is then called the *domain* of R, and the set

$$\{b \in B : \text{ there exists } a \in A \text{ such that } (a, b) \in R\}$$

is called the *range* of R. A relation on $A \times A$ is called a *relation on A*.

Let us give some examples. First, if A is a set then

$$\in_A = \{(b, B) \in A \times P(A) : b \in B\}$$

is a relation on $A \times P(A)$. Recall that we introduced the relation $\in$ on the collection of all sets, which we have seen is not a set; $\in_A$ is the restriction to a set and its subsets.

Secondly, if A is a set then

$$\subseteq_A = \{(B, C) \in P(A) \times P(A) : B \subseteq C\}$$

is a relation on $P(A)$. This is an example of a partial order relation. An order $\leq$ on a set A is a *partial order* or *partial order relation* if
(i) if $a \leq b$ and $b \leq c$ then $a \leq c$ (transitivity), and
(ii) $a \leq b$ and $b \leq a$ if and only if $a = b$.

If $a \leq b$ then we say that a is *less than or equal* to b, or that b is *greater than or equal* to a, and we also write $b \geq a$.

Partial order relations play an important part in analysis. We make some definitions concerning partial orders here, and will consider them in more detail later.

Suppose that $\leq$ is a partial order on a set A, that $a \in A$ and that B is a subset of A.

- a is an *upper bound* of B if $b \leq a$ for all $b \in B$.
- a is a *lower bound* of B if $a \leq b$ for all $b \in B$.

An upper bound of B need not belong to B. If it does, it is the *greatest* element of B. B has at most one greatest element, but may have no greatest element. *Least* elements are defined in the same way.

- a ia a *maximal* element of B if $a \in B$, and if $b \in B$ and $a \leq b$ then $a = b$.
- a ia a *minimal* element of B if $a \in B$, and if $b \in B$ and $b \leq a$ then $a = b$.

A greatest element of B is a maximal element of B, but the converse need not hold.

- a is the *supremum*, or *least upper bound*, of B if a is an upper bound of B, and if c is an upper bound of B, then $a \leq c$. In other words, a is the least element of the set of upper bounds of B.
- a is the *infimum*, or *greatest lower bound*, of B if a is a lower bound of B, and if c is an lower bound of B, then $c \leq a$. In other words, a is the greatest element of the set of lower bounds of B.

B has at most one least upper bound, but may have no least upper bound. If a is the least upper bound of B then a may or may not be an element of B. If a is an element of B, then a is the least upper bound of B if and only if a is the greatest element of B.

If $a \leq b$ or $b \leq a$ then we say that a and b are *comparable*. In general, not all pairs are comparable. If, however, any two elements of A are comparable, then we say that the relation is a *total order*. As an example, the usual order on the set of natural numbers $\mathbf{N} = \{1, 2, 3, \ldots\}$ (which we shall consider in Section 2.1) is a total order.

The definition of the notion of partial order includes equality. There is a closely related notion which forbids equality. Suppose that $\leq$ is a partial order relation on a set A. Then the relation

$$\{(a, b) \in A \times A : a \leq b \text{ and } a \neq b\}$$

is a *strict partial order* on A. It is denoted by $<$ and satisfies
 (i) if $a < b$ and $b < c$ then $a < c$ (transitivity), and
 (ii) $a < a$ does not hold for any $a \in A$.
 Conversely, if $<$ is a strict partial order on A then the relation

$$\{(a, b) \in A \times A : a < b \text{ or } a = b\}$$

is a partial order.

Exercises

1.3.1 Which of the following statements are necessarily true?
 (a) $A \times (B \cup C) = (A \times B) \cup (A \times C)$.
 (b) $(A \times B) \cup (C \times D) = (A \cup C) \times (B \cup D)$.
 (c) $(A \times B) \cap (C \times D) = (A \cap C) \times (B \cap D)$.
1.3.2 Suppose that $\leq_1$ is a partial order on A_1 and that $\leq_2$ is a partial order on A_2. Show that the relation

$$\{(a_1, a_2), (b_1, b_2) \in (A_1 \times A_2) \times (A_1 \times A_2) : a_1 \leq_1 b_1 \text{ and } a_2 \leq_2 b_2\}$$

is a partial order on $A_1 \times A_2$.
1.3.3 Show that a subset of a partially ordered set can have at most one greatest element, and at most one supremum.
1.3.4 This question assumes knowledge of the set $\mathbf{N}$ of natural numbers, and of counting. Let $P(\mathbf{N})$ be given the partial order defined by inclusion, as above. Let $P_n(\mathbf{N})$ be the set of subsets of $\mathbf{N}$ with at most n elements.
 (a) What are the upper bounds of $P_n(\mathbf{N})$ in $P(\mathbf{N})$?
 (b) Does $P_n(\mathbf{N})$ have a supremum? If so, is it an element of $P_n(\mathbf{N})$?
 (c) What are the maximal elements of $P_n(\mathbf{N})$?
1.3.5 Suppose that a is a maximal element of a subset B of a totally ordered set A. Show that a is the greatest element of B.
1.3.6 Give an example of a subset of a totally ordered set which has a supremum but no greatest element.

1.4 Functions

The notion of *function* developed slowly from the time of Descartes and Leibniz until the end of the nineteenth century. Originally, a function was something that was given by an analytic formula, but confusion and dispute arose about what this meant, and confusion was also caused by the fact that two formulae could give the same values. Here we simply define a *function*, or, synonymously, a *mapping*, or a *map* (we shall use the terms interchangeably), from a set A to a set B to be a relation f on $A \times B$ which satisfies the condition

for each $a \in A$, there is a unique $b \in B$ such that $(a, b) \in f$.

In these circumstances, we write $b = f(a)$, so that $f = \{x \in A \times B : x = (a, f(a))\}$. The element $f(a)$ of B is called the *image* of a under f.

It is however helpful to consider a function as some sort of dynamic process (perhaps taking place in a black box): an element a of A is put in, and $f(a)$ comes out:

$$a \longrightarrow \boxed{\text{black box}} \longrightarrow f(a).$$

Thus we write $f : A \to B$ for a function from A to B. The set $\{x \in A \times B : x = (a, f(a))\}$ is then called the *graph* G_f of f. The set of all mappings from A to B is denoted by B^A; the reason for this notation may become clear later.

Let us consider some examples. First, suppose that $f : A \to B$ is a function. Then we can define a function $P(f) : P(A) \to P(B)$ by setting

$$P(f)(C) = \{x \in B : \text{ there exists } a \in C \text{ such that } f(a) = x\},$$

for C a subset of A. It is unfortunately standard practice to denote this function by f. This can be misleading; for example, it may happen that $\emptyset \in A$, and that $f(\emptyset)$, an element of B, is not the empty set. Then $f(\emptyset) \neq \emptyset$, whereas $P(f)(\emptyset) = \emptyset$. In spite of this defect, we shall follow standard practice; with caution and common sense, we can avoid the difficulty we have just described. Following standard practice, the subset $f(C)$ of B is also called the *image* of C under f. We can also define a function $f^{-1} : P(B) \to P(A)$ by setting

$$f^{-1}(D) = \{x \in A : f(x) \in D\},$$

for D a subset of B. This notation is also unfortunate, as we shall shortly see. The set $f^{-1}(D)$ is called the *inverse image* of D; if $b \in B$ then the set $f^{-1}(\{b\})$ is called the *inverse image* of b.

Suppose that A is a set. For $a \in A$, define $s(a) = \{a\}$; s is a mapping from A into $P(A)$. It is an example of an injective mapping. A mapping $f : A \to B$ is *injective*, or an *injection*, or *one-one*, if distinct elements of A have distinct images in B; in other words, if $f(a) = f(a')$ then $a = a'$.

Suppose that B is a subset of a set A. The *inclusion map* $j_B : B \to A$ is defined by setting $j_B(b) = b$, for $b \in B$. Thus $b \in B$, whereas $j_B(b)$ is an element of A. j_B is again injective. As a special case, when $B = A$ we have the *identity map* $i_A : A \to A$ defined by setting $i_A(a) = a$ for $a \in A$.

Let us consider a Cartesian product $A \times B$, where A and B are non-empty sets. For $(a, b) \in A \times B$, let $\pi_A((a, b)) = a$ and let $\pi_B((a, b)) = b$. Then π_A is a mapping from $A \times B$ to A, and π_B is a mapping from $A \times B$ to B; they are the *coordinate projections* of $A \times B$ onto A and B, respectively. The elements a and b are the *coordinates* of (a, b). The mappings π_A and π_B are examples of surjective mappings. A mapping $f : A \to B$ is *surjective*, or a *surjection* or *onto*, if $f(A) = B$; every element of B is the image of at least one element of A.

A mapping $f : A \to B$ is *bijective*, or a *bijection*, or a *one-one correspondence*, if it is both injective and surjective; every element b of B is the image under f of exactly one element of A. We denote this element by $f^{-1}(b)$; then f^{-1} is a bijective mapping of B onto A. We have thus used the term f^{-1} in two different senses: if $f : A \to B$ is a mapping, the mapping $f^{-1} : P(B) \to P(A)$ is always defined; the mapping $f^{-1} : B \to A$ is only defined when f is bijective. Once again, caution and common sense are called for.

Suppose that $(A, \leq_A)$ and $(B, \leq_B)$ are two partially ordered sets and that $f : A \to B$ is a mapping from A to B. The mapping f is said to be *increasing* if $f(a) \leq_B f(a')$ whenever $a \leq_A a'$, and to be *strictly increasing* if $f(a) <_B f(a')$ whenever $a <_A a'$. It is said to be *decreasing* if $f(a) \geq_B f(a')$ whenever $a \leq_A a'$, and to be *strictly decreasing* if $f(a) >_B f(a')$ whenever $a <_A a'$. It is said to be *monotonic* if it is either increasing or decreasing, and to be *strictly monotonic* if it is either strictly increasing or strictly decreasing.

Suppose that f is a mapping from A to B and that g is a mapping from B to C. We can then define the *composite mapping* $g \circ f$ from A to C by setting $(g \circ f)(a) = g(f(a))$, for $a \in A$. Note the order of the terms: first we use the mapping f and then the mapping g, but the terms in the composite mapping $g \circ f$ come in the opposite order.

As examples, if f is a bijection from A onto B then $f^{-1} \circ f : A \to A$ is the identity mapping i_A on A, and $f \circ f^{-1} : B \to B$ is the identity mapping i_B on B.

The composition of mappings is *associative*: if $f : A \to B$, $g : B \to C$ and $h : C \to D$ are mappings then

$$(h \circ (g \circ f))(a) = h((g \circ f)(a)) = h(g(f(a)))$$
$$= (h \circ g)(f(a)) = ((h \circ g) \circ f)(a),$$

so that $h \circ (g \circ f) = (h \circ g) \circ f$.

Suppose that $f : A \to B$ is a mapping, and that $f(A) \subseteq D \subseteq B$. Then $G_f \subseteq A \times D$, and we can consider f as a mapping from A into D. We usually denote this mapping by f, unless this is likely to cause confusion. Let us here denote the mapping from A to $f(A)$ by $\tilde{f}$. Then we have the factorization $f = j_{f(A)} \circ \tilde{f}$, where $\tilde{f}$ is surjective, and the inclusion mapping $j_{f(A)} : f(A) \to B$ is injective.

Next, consider a subset C of A. Then $G_f \cap (C \times B)$ is the graph of a mapping from C to B. This is the *restriction* $f_{|C}$ of f to C. If $c \in C$ then $f_{|C}(c) = f(c)$.

A bijective mapping $f : A \to A$ is called a *permutation* of A. As an example, suppose that a and b are elements of A. The mapping τ, or $\tau_{a,b}$, from A to A defined by

$$\tau(a) = b, \tau(b) = a, \tau(c) = c \text{ for all other } c \in A,$$

the mapping which *transposes* a and b, is a permutation of A. The set of permutations of A is denoted by Σ_A.

We can describe the composition properties of Σ_A in algebraic terms.

A *group* is a non-empty set, together with a mapping or *operation* $\circ$: $G \times G \to G$ which satisfies:

(i) composition is associative: that is, $(g \circ h) \circ j = g \circ (h \circ j)$ for $g, h, j \in G$;
(ii) there exists $e \in G$ such that $e \circ g = g \circ e = g$, for all $g \in G$;
(iii) for each $g \in G$ there exists $g^{-1} \in G$ such that $g \circ g^{-1} = g^{-1} \circ g = e$.
If
(iv) $gh = hg$ for all $g, h \in G$, then G is said to be *abelian*, or *commutative*.

Note that the element e is uniquely determined by (ii), for if e also satisfies (ii), then $e' = e' \circ e = e$. The element e is called the *identity element* of G, and is frequently denoted by e_G. Similarly, if $g \in G$ then the element g^{-1} is uniquely determined by (iii); for if $g \circ h = e$ then

$$h = e \circ h = (g^{-1} \circ g) \circ h = g^{-1} \circ (g \circ h) = g^{-1} \circ e = g^{-1}.$$

The element g^{-1} is called the *inverse* of g.

It then follows immediately from the earlier discussions that Σ_A is a group, when the group composition is taken to be the composition of functions and the identity map i_A is taken as the identity element.

Axiom 7: The replacement axiom

Let us end this section by stating the replacement axiom, since it has a function-like quality. A well-formed formula $Q(x, y)$ with free variables x and y is said to *determine a function* if whenever a is a set then there is at most one set b for which $Q(a, b)$ holds. If there is a set b for which $Q(a, b)$ holds, then we write $b = Q(a)$, and call b the *image* of a. The *replacement axiom* then states that if $Q(x, y)$ is a well-formed formula which determines a function, and if A is a set, then the collection of all images $Q(a)$, as a varies in A, is a set, which we denote by $Q(A)$. Thus

$$\{x \in A : \text{there exists } b \in Q(A) \text{ for which } Q(a, b) \text{ holds}\}$$

is a subset $D_A(Q)$ of A, and Q defines a surjection of $D_A(Q)$ onto $Q(A)$. We shall not make explicit use of this axiom.

Exercises

1.4.1 Suppose that $f : A \to B$, and that C, D are subsets of A and that E, F are subsets of B. Which of the following statements are necessarily true?
   (a) $f(C \cup D) = f(C) \cup f(D)$.
   (b) $f(C \cap D) = f(C) \cap f(D)$.
   (c) $f^{-1}(C \cup D) = f^{-1}(C) \cup f^{-1}(D)$.
   (d) $f^{-1}(C \cap D) = f^{-1}(C) \cap f^{-1}(D)$.

1.4.2 Suppose that $f : A \to B$ and $g : B \to C$ are mappings. What is the graph of $g \circ f$? Verify that $g \circ f$ is a mapping.

1.4.3 Suppose that f is a mapping from A to B, where A and B are non-empty sets. A mapping $l : B \to A$ is a *left inverse* of f if $l \circ f = i_A$, the identity on A. Show that if f has a left inverse, then f is injective, and that if f is injective, then f has a left inverse.

1.4.4 Suppose that f is a mapping from A to B, where A and B are non-empty sets. A mapping $r : B \to A$ is a *right inverse* of f if $f \circ r = i_B$, the identity on B. Show that if f has a right inverse, then f is surjective. Does a surjective mapping always have a right inverse? Think about this, and then read Section 1.9.

1.4.5 Suppose that $f : A \to B$ has a left inverse l and a right inverse r. Show that f is a bijection and that $l = r = f^{-1}$.

1.4.6 This question establishes basic facts about groups that we shall need later. A mapping θ from a group G to a group G' is a *homomorphism* if $\theta(g_1 \circ g_2) = \theta(g_1) \circ \theta(g_2)$ for all $g_1, g_2 \in G$. Suppose that $\theta : G \to G'$ is a homomorphism.
   (a) Show that $\theta(e) = e'$, where e is the identity element of G, e' the identity element of G'.
   (b) Show that $\theta(g^{-1}) = (\theta(g))^{-1}$ for all $g \in G$.
   (c) Show that if H is a subgroup of G then $\theta(H)$ is a subgroup of G'.
   (d) Show that if H' is a subgroup of G' then $\theta^{-1}(H')$ is a subgroup of G.

(e) A bijective homomorphism is called an *isomorphism*. Show that if θ is an isomorphism then $\theta^{-1} : G' \to G$ is also an isomorphism.

1.5 Equivalence relations

There is another sort of relation that we shall use later. The axiom of extensionality tells us that two sets are equal if and only if they have the same members. There are however many occasions when two different sets serve the same purpose, and we would like to identify them in some way. For example, we express a positive rational number as a fraction p/q, where (p, q) is an ordered pair of natural numbers. The rational number $1/2$ is the same as the rational number $3/6$, but the ordered pairs $(1, 2)$ and $(3, 6)$ are different. In this circumstance, we say that $(1, 2)$ and $(3, 6)$ are *equivalent*. This leads to the concept of an equivalence relation.

An *equivalence relation* on a set A is a relation on A (frequently, as here, denoted by $\sim$) which satisfies

 (i) if $a \sim b$ and $b \sim c$ then $a \sim c$ (transitivity);
 (ii) if $a \sim b$ then $b \sim a$ (symmetry);
 (iii) $a \sim a$ for all $a \in A$ (reflexivity).

As a trivial example, the relation $a = b$ is an equivalence relation. For a less trivial example, suppose that $f : A \to B$ is a mapping. Let $a \sim a'$ if and only if $f(a) = f(a')$. Then it is easy to check that $\sim$ is an equivalence relation on A. We shall see that any equivalence relation can be expressed in this way.

Suppose that $\sim$ is an equivalence relation on a set A and that $a \in A$. We define the *equivalence class* E_a to be the set $\{x \in A : a \sim x\}$. This is the traditional name, but an equivalence class is certainly a set. Note that $a \in E_a$, so that E_a is a non-empty set.

Proposition 1.5.1 *Suppose that $\sim$ is an equivalence relation on a set A. If $a \sim a'$ then $E_a = E_{a'}$, and if $a \not\sim a'$ then E_a and $E_{a'}$ are disjoint.*

Proof Suppose that $a \sim a'$. If $a' \sim c$ then $a \sim c$, by transitivity, and so $E_{a'} \subseteq E_a$. Further $a' \sim a$, by reflexivity, and so $E_a \subseteq E_{a'}$.

Suppose that $b \in E_a \cap E_{a'}$. Then $a \sim b$ and $a' \sim b$, so that $b \sim a'$, by reflexivity, and $a \sim a'$, by transitivity. Thus if $a \not\sim a'$, then $E_a \cap E_{a'} = \emptyset$. $\square$

We now say that a subset E of A is an *equivalence class* if there exists $a \in A$ such that $E = E_a$. We denote the set of equivalence classes by $A/\sim$. $A/\sim$ is a subset of $P(A)$.

Corollary 1.5.2 $A = \cup_{E \in A/\sim} E$ *is a union of disjoint equivalence classes.*

Proof We have seen that distinct equivalence classes are disjoint. Their union is A, since if $a \in A$ then $a \in E_a$. □

This leads to the following definition. Suppose that A is a set. A subset Π of $P(A)$ is a *partition* of A if

(i) each $E \in \Pi$ is non-empty;
(ii) $A = \cup_{E \in \Pi} E$;
(iii) distinct elements of Π are disjoint.

Thus if $\sim$ is an equivalence relation on A then the set $A/\sim$ of equivalence classes is a partition of A.

Let $\mathcal{E}_A$ denote the set of all equivalence relations on A, and let $\mathcal{P}_A$ denote the set of all partitions of A. We shall show that there is a natural bijection of $\mathcal{E}_A$ onto $\mathcal{P}_A$. If $\sim \in \mathcal{E}_A$, let $k(\sim) = A/\sim$. Then k is a mapping from $\mathcal{E}_A$ to $\mathcal{P}_A$, and it is easy to see that this is injective. Conversely, if Π is a partition of A, and we set $a \sim b$ if a and b are in the same element of Π, then it is easy to check that $\sim$ is an equivalence relation on A and that $A/\sim = \Pi$. Thus k is surjective, and the mapping from $\mathcal{P}_A$ to $\mathcal{E}_A$ which we have just defined is the inverse of k: k is a bijection of $\mathcal{E}_A$ onto $\mathcal{P}_A$.

Suppose now that $\sim$ is an equivalence relation on a set A, and that $A/\sim$ is the corresponding partition of A. We define a mapping $q : A \to A/\sim$ by setting $q(a) = E_a$. Then q is a surjection, and $a \sim a'$ if and only if $q(a) = q(a')$. The set $A/\sim$ is called the *quotient* of A by $\sim$, and the mapping $q : A \to A/\sim$ is called the *quotient* mapping.

Now suppose that $f : A \to B$ is a mapping. Define an equivalence relation $\sim$ by setting $a \sim a'$ if and only if $f(a) = f(a')$, and let $q : A \to A/\sim$ be the quotient mapping. If $E = E_a \in A/\sim$ and $a' \in E$, then $f(a) = f(a')$. We can therefore define $\tilde{f}(E) = f(a)$, and we obtain a well-defined mapping $\tilde{f}$ of $A/\sim$ onto $f(A)$. Suppose that $\tilde{f}(E) = \tilde{f}(E')$, that $a \in E$ and that $a' \in E'$. Then $f(a) = f(a')$, so that $a \sim a'$ and $E = E'$. Thus $\tilde{f}$ is one-one, and so $\tilde{f} : A/\sim \to f(A)$ is a bijection. We have therefore factorized f as $f = j_{f(A)} \circ \tilde{f} \circ q$, where q is a surjection, $\tilde{f}$ is a bijection and the inclusion mapping $j_{f(A)} : f(A) \to B$ is injective. Thus we have the following diagram of mappings:

$$
\begin{array}{ccc}
A & \xrightarrow{\;f\;} & B \\[1em]
q \downarrow & & \uparrow j_{f(A)} \\[1em]
A/\sim & \xrightarrow{\;\tilde{f}\;} & f(A)
\end{array}
$$

This diagram is *commutative*: the outcome of the direct journey from A to B is the same as the outcome of the longer journey going round the other three sides of the diagram.

Exercises

1.5.1 Suppose that σ is a permutation of a non-empty set A. A subset B of A is σ-*invariant* if $\sigma(B) = B$. If $a \in A$ let

$$O_a = \cap\{B \in P(A) : a \in B \text{ and } B \text{ is } \sigma\text{-invariant}\}.$$

(a) Show that O_a is σ-invariant.
(b) Suppose that $O_a \cap O_b \neq \emptyset$. Show that $O_a = O_b$. (*Hint*: Consider $O_a \setminus O_b$.)
(c) A subset O of A is an *orbit* of σ if there exists $a \in A$ such that $O = O_a$. Show that the set of orbits is a partition of A. What is the corresponding equivalence relation?

1.5.2 A *subgroup* H of a group G is a subset of G with the properties

(i) the identity of G belongs to H;
(ii) if $h \in H$ then $h^{-1} \in H$;
(iii) if h and h' are in H then $h \circ h' \in H$.

Thus H is a group with the operations inherited from G.
Suppose now that H is a subgroup of Σ_A. A subset B of A is H-*invariant* if $\sigma(B) = B$ for each $\sigma \in H$. Carry out a programme similar to that of the previous question.

1.6 Some theorems of set theory

Although we have only met some of the axioms of ZF, we are already in a position to prove some interesting and important results.

Theorem 1.6.1 (The Knaster–Tarski fixed-point theorem)   *Suppose that A is a set and that $f : P(A) \to P(A)$ is an increasing function; if $B \subseteq C \subseteq A$ then $f(B) \subseteq f(C)$. Then there exists $G \subseteq A$ such that $f(G) = G$.*

Proof Note that f is defined as a mapping from $P(A)$ to itself: it is not defined in terms of a mapping from A to itself. Thus $\emptyset \subseteq f(\emptyset)$ and $A \supseteq f(A)$; the inclusions change direction. The theorem states that equality holds at some intermediate subset.

We shall show that there exists a set G such that $G \subseteq f(G)$ and $f(G) \subseteq G$; the axiom of extensionality then ensures that $G = f(G)$.

Let $\mathcal{G} = \{B \in P(A) : B \subseteq f(B)\}$, and let $G = \cup_{B \in \mathcal{G}} B$. If $B \in \mathcal{G}$ then $B \subseteq G$, and so $f(B) \subseteq f(G)$. Thus $B \subseteq f(B) \subseteq f(G)$. Consequently

$G = \cup_{B \in \mathcal{G}} B \subseteq f(G)$, and so $G \in \mathcal{G}$. On the other hand, since $G \subseteq f(G)$ it follows that $f(G) \subseteq f(f(G))$, and so $f(G) \in \mathcal{G}$. Thus $f(G) \subseteq \cup_{B \in \mathcal{G}} B = G$. □

Theorem 1.6.2 (The Schröder–Bernstein theorem) *Suppose that A and B are sets, and that $f : A \to B$ and $g : B \to A$ are injective mappings. Then there exists a bijection $h : A \to B$.*

Proof The existence of f says that 'A is no bigger than B' and the existence of g says that 'B is no bigger than A'. The conclusion then is that if both hold then 'A and B are the same size'. We shall consider the problem of whether two sets are always comparable in size later (Theorem 1.9.2).

We consider the mappings $f : P(A) \to P(B)$ and $g : P(B) \to P(A)$ determined by f and g; they are clearly increasing maps. On the other hand the mapping $C_A : P(A) \to P(A)$ defined by $C_A(D) = A \backslash D$ is order reversing, as is the corresponding mapping $C_B : P(B) \to P(B)$. Thus the composite mapping $S = C_A \circ g \circ C_B \circ f$ is an increasing mapping from $P(A)$ into itself. The Knaster–Tarski fixed-point theorem then tells us that there exists $D \subseteq A$ such that $S(D) = D$; the restriction $f_{|D}$ of f to D is a bijection of D onto $f(D)$. Let $E = f(D)$, so that $C_B(f(D)) = B \setminus E$. Thus

$$A \setminus D = C_A(D) = C_A(S(D)) = C_A(C_A g C_B f(D))$$
$$= g(C_B f(D))) = g(B \setminus E).$$

Consequently the restriction $g_{|B \setminus E}$ of g to $B \setminus E$ is a bijection of $B \setminus E$ onto $A \setminus D$; let $k : A \setminus D \to B \setminus E$ be its inverse. We now set $h(a) = f_{|D}(a)$ for $a \in D$, and set $h(a) = k(a)$ for $a \in A \setminus D$; h clearly has the required properties. □

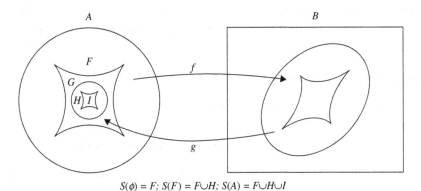

$S(\phi) = F; S(F) = F \cup H; S(A) = F \cup H \cup I$

Figure 1.6. The Schröder–Bernstein theorem.

The next result uses the argument of Russell's paradox.

Theorem 1.6.3 (Cantor's theorem)   *Suppose that f is a mapping from a set A to its power set $P(A)$. Then f is not surjective.*

Proof   Let $B = \{a \in A : a \notin f(a)\}$. We claim that B is not in the image of f. Suppose not, and suppose that $B = f(b)$. Does b belong to B? If it does, it fails the criterion for membership of B, giving a contradiction. If it does not, then it meets the criterion for membership of B, again giving a contradiction. This exhausts the possibilities, and so B is not in the image of f.   □

Corollary 1.6.4   *Suppose that A is a non-empty set and that $g : P(A) \to A$ is a mapping. Then g is not injective.*

Proof   The mapping $s : A \to P(A)$ defined by $s(a) = \{a\}$ is injective. If g were injective, then by the Schröder–Bernstein theorem there would be a bijection $h : A \to P(A)$, which contradicts the theorem.   □

1.7 The foundation axiom and the axiom of infinity

Suppose we start with the empty set. Repeatedly using the axioms that we have described so far to create new sets, we obtain an infinite collection of sets which satisfy these axioms. But each of these sets has only finitely many members. This may be satisfactory for certain areas of mathematics, such as finite group theory, or the mathematics of computer science, but in mathematical analysis we need to consider sets with infinitely many members. We now introduce two further axioms which enable us to do so.

Axiom 8: The foundation axiom

This states that if A is a non-empty set, then there exists an element a of A such that $a \cap A = \emptyset$: a and A have no element in common. As we shall see, this excludes the possibility of infinite regress. It also prevents us from going round in circles.

Proposition 1.7.1   *If a is a set then $a \notin a$.*

Proof   Consider the singleton set $\{a\}$. It has a member disjoint from $\{a\}$. But it only has one member, namely a, and so a and $\{a\}$ are disjoint. Since $a \in \{a\}$, $a \notin a$. Russell's paradox has completely disappeared.   □

Let us introduce a construction that will shortly be useful to us. If a is a set, we define a^+ to be the set $a \cup \{a\}$. The members of a^+ are the members of a, together with a. Thus $a \subseteq a^+$.

Corollary 1.7.2 *If a is a set then $a \neq a^+$.*

Proof For $a \in a^+$ and $a \notin a$. □

Here is another consequence of the foundation axiom.

Proposition 1.7.3 *If a and b are sets and $a \in b$ then $b \notin a$.*

Proof Consider the set $\{a, b\}$ with elements a and b. By the foundation axiom, either $a \cap \{a, b\} = \emptyset$ or $b \cap \{a, b\} = \emptyset$. But $a \in b \cap \{a, b\}$, and so $a \cap \{a, b\} = \emptyset$. Since $b \in \{a, b\}$, $b \notin a$. □

A set A is called a *successor set* if $\emptyset \in A$ and if $a^+ \in A$ whenever $a \in A$.

Axiom 9: The axiom of infinity

This states that there exists a set S which is a successor set.

Having postulated the existence of a successor set, we now show that there is a smallest one.

Theorem 1.7.4 *There exists a successor set $\mathcal{Z}^+$ such that if T is any successor set then $\mathcal{Z}^+ \subseteq T$.*

Proof Note that if A is a set, all of whose elements are successor sets, then it follows immediately from the definitions that the intersection $\cap_{B \in A} B$ is also a successor set. Suppose that S is a successor set. Let

$$\mathcal{Z}^+ = \cap \{B \in P(S) : B \text{ is a successor set}\}.$$

Then if T is a successor set, $T \cap S$ is a successor set, so that $\mathcal{Z}^+ \subseteq T \cap S \subseteq T$. □

The minimality of $\mathcal{Z}^+$ is very powerful, and leads to the principle of induction.

Let us use the foundation axiom to show that infinite regress is not allowed.

Proposition 1.7.5 *Suppose that $f : \mathcal{Z}^+ \to A$ is a mapping. Then there exists $n \in \mathcal{Z}^+$ such that $f(n^+) \notin f(n)$.*

Proof Consider the set $f(\mathcal{Z}^+)$. By the foundation axiom, there exists $n \in \mathcal{Z}^+$ such that no member of $f(n)$ is in $f(\mathcal{Z}^+)$. But $f(n^+) \in f(\mathcal{Z}^+)$, and so $f(n^+) \notin f(n)$. □

We now show that we can take the minimal successor set $\mathcal{Z}^+$ as a model for the natural numbers. Let us explain what this means. In 1888, Dedekind described an axiom system for the natural numbers $\mathbf{N} = (1, 2, 3, \ldots)$. Independently, Peano introduced them, in a pamphlet written in Latin. They are now known as *Peano's axioms*. Replacing 1 by 0, they also serve as axioms for the *non-negative integers* $\mathbf{Z}^+ = (0, 1, 2, \ldots)$; in this form, and in set-theoretic terms, they state the following. There is a set P and a mapping $s : P \to P$ (the *successor function*) such that

(P1) there is a distinguished element 0 of P;

(P2) if $n \in P$ then $s(n) \in P$ (this is included in the fact that s is a mapping from P to itself);

(P3) if $n \in P$ then $s(n) \neq 0$;

(P4) s is injective: if $m \in P$ and $n \in P$ and $s(m) = s(n)$ then $m = n$;

(P5) (the principle of induction) if $A \subseteq P$, if $0 \in A$ and if $s(A) \subseteq A$ then $A = P$.

We set $s(0) = 1$, $s(1) = 2$, and so on.

There are many ways of constructing a pair (P, s) which satisfies these axioms. Any pair (P, s) which does so is called a *model* for the non-negative integers $\mathbf{Z}^+$.

Theorem 1.7.6 *If $n \in \mathcal{Z}^+$, let $s(n) = n^+$. Then the pair $(\mathcal{Z}^+, s)$ is a model for $\mathbf{Z}^+$.*

Proof For (P1), we take the empty set $\emptyset$ to be the distinguished element. If $n \in \mathcal{Z}^+$ then $n^+ \in \mathcal{Z}^+$, so that (P2) holds. Since $n \in n^+$, $s(n) \neq 0$, so that (P3) holds. Suppose that $m^+ = n^+$, and that $m \neq n$. Then $m \in m^+ = n^+ = n \cup \{n\}$. Since $m \neq n$, $m \notin \{n\}$. Thus $m \in n$. Similarly, exchanging the roles of m and n, $n \in m$, contradicting Proposition 1.7.3. Thus (P4) holds. Finally, (P5) follows from Theorem 1.7.4.   □

As we have remarked, there are many other ways of constructing pairs (P, s) for which the Peano axioms hold. We need to show that any two are essentially the same, but we must wait until the results of the next section have been established before we can do this.

The principle of induction allows us to prove results relating to the non-negative integers. Suppose that $Q(x)$ is a well-formed formula and that we are interested in the subset T of P consisting of those n for which $Q(n)$ holds. Suppose that we can prove that $0 \in T$, and that we can also prove that if $Q(n)$ holds then it follows that $Q(s(n))$ holds. Then T satisfies the conditions of (P5), and so $T = P$; $P(n)$ holds for all $n \in P$. A proof which uses this procedure is known as a *proof by induction*. We shall give many such proofs. Here is one.

Proposition 1.7.7 *Suppose that (P, s) satisfy the Peano axioms, with distinguished element 0. Then $s(P) = P \setminus \{0\}$, and $s : P \to s(P)$ is a bijection.*

Proof The mapping $s : P \to s(P)$ is a bijection, by (P4), and $0 \notin s(P)$, by (P3). Let $A = \{0\} \cup s(P)$. Then $0 \in A$, and if $n \in A$ then $s(n) \in A$, so that by (P5), $A = P$. □

Exercises

1.7.1 Consider the two-point subset $\{0, 1\}$ of $\mathbf{Z}^+$. If A is a set, we denote the set of functions from A to $\{0, 1\}$ by 2^A. Suppose that $B \in P(A)$ and $x \in A$. Let $I_B(x) = 1$ if $x \in B$, and $I_B(x) = 0$ if $x \notin B$. I_B is the *indicator function* of B. Show that the mapping $B \to I_B : P(A) \to 2^A$ is a bijection.

1.8 Sequences, and recursion

In this section, we shall assume that (P, s) is a model for $\mathbf{Z}^+$, with distinguished element 0. We write P as $(0, 1, 2, \dots)$, where $1 = s(0)$, $2 = s(1)$, and so on. A function $f : P \to A$ is then called a *sequence*, or an *infinite sequence* in A, and is denoted by $(f_n)_{n \in P}$, or by $(f_n)_{n=0}^{\infty}$, or as $(f_0, f_1, f_2, \dots)$. This notation suggests another way of considering a function: the elements of P act as *labels* or *indices*. Since f need not be one-one, an element of A may have more than one label. Since f need not be surjective, some elements of A may have no labels.

It is important to distinguish the sequence $(f_n)_{n \in P}$ from its set of values

$$f(P) = \{x \in A : \text{there exists } n \in P \text{ such that } x = f_n\},$$

but some flexibility is needed. When we consider a *term* f_n of a sequence, we may consider f_n as the value of the sequence at n, but at the same time keep in mind its index or label n. For sequences, as for fashion, the label is as important as the object.

The principle of induction lets us prove results about sequences. *Recursion* allows us to construct sequences.

Theorem 1.8.1 (The recursion theorem) *Suppose that A is a non-empty set, that f is a mapping of A to itself and that $\bar{a} \in A$. Then there is a unique sequence $(a_n)_{n \in P}$ such that $a_0 = \bar{a}$ and $a_{s(n)} = f(a_n)$ for $n \in P$.*

Proof Recall that a sequence is a function from P to A, that a function is a relation satisfying certain conditions, and that a relation is a subset of $P \times A$. Let us consider the set of relations on $P \times A$. We say that a relation R is *recursive* if

(i) $0R\bar{a}$ and
(ii) if nRa then $s(n)Rf(a)$.

The set S of all recursive relations is non-empty, since $P \times A \in S$. Let $g = \bigcap_{R \in S} R$. We shall show that g is a function, that $g(0) = \bar{a}$ and that $g(s(n)) = f(n)$ for all $n \in P$. Thus $a_n = g(n)$ satisfies the conditions of the theorem.

Let

$$D(g) = \{n \in P : \text{there exists } a \in A \text{ with } (n, a) \in g\}$$

be the domain of g. We must show that $D(g) = P$; we prove this by induction. Since $(0, \bar{a}) \in R$ for all $R \in S$, $S(0, \bar{a}) \in g$. Thus $0 \in D(g)$. If $n \in D(g)$, there exists a such that $(n, a) \in g$, and so $(n, a) \in R$ for all $R \in S$. Then since each $R \in S$ is recursive, $(s(n), f(a)) \in R$ for all $R \in S$, and so $(s(n), f(a)) \in g$. Thus $s(n) \in D(g)$. By the induction principle, it follows that $D(g) = P$.

Next, we must show that if $n \in P$ then there exists exactly one $a \in A$ such that $(n, a) \in g$. Again, we prove this by induction. Let

$$U = \{n \in P : \text{if } (n, a) \in P \text{ and } (n, a') \in P \text{ then } a = a'\}.$$

First, we show that $0 \in U$. $(0, \bar{a}) \in g$. Suppose that $(0, a') \in g$ and that $a' \neq \bar{a}$. Let $g' = g \setminus \{(0, a')\}$. Then $(0, \bar{a}) \in g'$, since $a' \neq \bar{a}$. If $(n, a) \in g' \subseteq g$ then $(s(n), f(a)) \in g$, and $(s(n), f(a)) \neq (0, a')$, since $s(n) \neq 0$, so that $(s(n), f(a)) \in g'$. Thus $g' \in S$, and so $g \subseteq g'$, giving a contradiction.

Secondly, we show that if $n \in U$ then $s(n) \in U$. Suppose not. There exists a unique $a \in A$ such that $(n, a) \in g$, and so $(s(n), f(a)) \in g$. Since $s(n) \notin U$, there exists $a' \in A$ with $a' \neq f(a)$ such that $(s(n), a') \in g$. Let $g' = g \setminus \{(s(n), a')\}$. We shall show that $g' \in S$. As before, $(0, \bar{a}) \in g'$. Suppose that $(m, b) \in g' \subseteq g$. Then $(s(m), f(b)) \in g$. Thus if $(s(m), f(b)) \notin g'$ then $(s(m), f(b)) = (s(n), a')$. But then $m = n$ and $f(b) = a'$. Since $n \in U$, $(m, b) = (n, a)$, and so that $b = a$. Thus $f(a) = a'$, giving a contradiction. By the principle of induction, $U = P$, and so g is a function.

Finally, we show that g is unique. Once again, we prove this by induction. Suppose that g' is a function in S. Let $G = \{n \in P : g(n) = g'(n)\}$. Since $g'(0) = g(0) = \bar{a}$, $0 \in G$. Suppose that $n \in G$. Then $g'(s(n)) = f(g'(n)) = f(g(n)) = g(s(n))$, so that $s(n) \in G$. By the induction principle, $G = P$, so that $g = g'$. $\square$

If $n \in \mathbf{Z}^+$ and $\bar{a} \in A$, let $f^n(\bar{a}) = a_n$. Then f^n is a mapping of A into itself. We can therefore express the recursion theorem in the following way.

Theorem 1.8.2 *Suppose that A is a non-empty set and that f is a mapping of A to itself. For each $n \in \mathbf{Z}^+$ there exists a unique mapping $f^n : A \to A$ such that $f^0(a) = a$ for all $a \in A$ and such that $f^{s(n)}(a) = f(f^n(a))$ for all $n \in \mathbf{Z}^+$ and $a \in A$.*

The recursion principle can be extended to more complicated situations. See Exercise 1.1.1 below.

We can now show that any two models of $\mathbf{Z}^+$ have exactly the same properties.

Theorem 1.8.3 *Suppose that (P, s) and (P', s') satisfy the Peano axioms, with distinguished elements 0 and $0'$ respectively. Then there is a unique bijection $t : P \to P'$ with $t(0) = 0'$ and $s't(n) = ts(n)$ for each $n \in P$. Thus we have the diagram:*

$$
\begin{array}{ccccccccccc}
0 & \xrightarrow{s} & 1 & \xrightarrow{s} & 2 & \xrightarrow{s} & \cdots & \xrightarrow{s} & n & \xrightarrow{s} & s(n) & \xrightarrow{s} \cdots \\
t \downarrow & & t \downarrow & & t \downarrow & & & & t \downarrow & & t \downarrow \\
0' & \xrightarrow{s'} & 1' & \xrightarrow{s'} & 2' & \xrightarrow{s'} & \cdots & \xrightarrow{s'} & n' & \xrightarrow{s'} & s'(n') & \xrightarrow{s'} \cdots
\end{array}
$$

Proof Set $t(0) = 0'$, and apply recursion to the mapping s'. There is then a unique mapping $t : P \to P'$ such that $ts(n) = s'(t(n))$ for each $n \in P$. Similarly, there is a unique mapping $t' : P' \to P$ such that $t'(0') = 0$ and $t's'(n') = st'(n')$. We shall show by induction that $t't$ is the identity on P and that tt' is the identity on P', so that t is a bijection. Let $U = \{n \in P : t't(n) = n\}$. Since $t't(0) = t'(0') = 0$, $0 \in U$. Suppose that $n \in U$. Then

$$s(n) = st't(n) = t's't(n) = t'ts(n),$$

so that $s(n) \in U$. Thus $U = P$. Exchanging the roles of P and P', we also see that tt' is the identity on P'. □

From now on, we take the non-negative integers to be a set $\mathbf{Z}^+ = \{0, 1, 2, \ldots\}$, together with a map $s : \mathbf{Z}^+ \to \mathbf{Z}^+$, such that the pair $(\mathbf{Z}^+, s)$ satisfies the Peano axioms, and take the natural numbers $\mathbf{N} = \{1, 2, 3, \ldots\}$ to be the set $s(\mathbf{Z}^+)$. We could, for example, take $(\mathbf{Z}^+, s)$ to be the pair $(\mathcal{Z}^+, ^+)$. Properties of $\mathbf{Z}^+$ and $\mathbf{N}$ will however be derived from the Peano axioms, and not from any particular set-theoretical properties that the model might have.

Exercises

1.8.1 Suppose that $(A_n)_{n \in \mathbf{Z}^+}$ is a sequence of non-empty subsets of a set A, and that for each $n \in \mathbf{Z}^+$, f_n is a mapping from A_n into $A_{s(n)}$. Show that if $\bar{a} \in A_0$ then there exists a unique sequence $(a_n)_{n \in \mathbf{Z}^+}$ in A such that $a_n \in A_n$ for $n \in \mathbf{Z}^+$, $a_0 = \bar{A}$ and $a_{s(n)} = f_n(a_n)$ for $n \in \mathbf{Z}^+$. [*Hint*: Let

$$D = \{x \in \mathbf{Z}^+ \times A : x = (n, a) \text{ with } a \in A_n\},$$

and consider the mapping $\phi : D \rightarrow D$ defined by $\phi(n, a) = (s(n), f_n(a))$.

1.8.2 Suppose that A is a non-empty set and that $S : P(A) \rightarrow P(A)$ is an increasing function.

(a) Use recursion to show that there are sequences $(H_n)_{n \in \mathbf{Z}^+}$ and $(J_n)_{n \in \mathbf{Z}^+}$ in $P(A)$ such that $H_0 = \emptyset$ and $S(H_n) = H_{s(n)}$ for $n \in \mathbf{Z}^+$, and $J_0 = A$ and $S(J_n) = J_{s(n)}$ for $n \in \mathbf{Z}^+$.

(b) Show that $(H_n)_{n \in \mathbf{Z}^+}$ is an increasing sequence and that $(J_n)_{n \in \mathbf{Z}^+}$ is a decreasing sequence.

(c) Let $H = \bigcup_{n=0}^{\infty} H_n$ and $J = \bigcap_{n=0}^{\infty} J_n$. Show that if $G \in P(A)$ and $G = S(G)$ then $H \subseteq G \subseteq J$.

(d) Give examples where $H \neq S(H)$ and $J \neq S(J)$.

(e) Let S be the mapping defined in the proof of the Schröder–Bernstein theorem. Show that $S(H) = H$ and $S(J) = J$.

1.9 The axiom of choice

We have seen that a sequence, that is, a mapping from $\mathbf{Z}^+$ to a set B, can be considered as a way of labelling elements of B. We can extend this idea to other mappings; if f is a mapping from a set A to a set B, we can consider A as an *index set*, used to label those elements of B which are in the image $f(A)$. In this case, we denote the function by $(f_\alpha)_{\alpha \in A}$, and call it a *family* of elements of B, indexed by A. Once again, f need not be injective, and so there may be distinct α and α' for which $f_\alpha = f_{\alpha'}$.

Suppose now that $(B_\alpha)_{\alpha \in A}$ is a family of non-empty sets. The *Cartesian product* $\prod_{\alpha \in A}(B_\alpha)$ is then defined to be the set of all families $(c_\alpha)_{\alpha \in A}$ with values in $\bigcup_{\alpha \in A} B_\alpha$, such that $c_\alpha \in B_\alpha$ for each $\alpha \in A$. If $\beta \in A$, then the mapping π_β defined by $\pi_\beta(c) = c_\beta$ for $c = (c_\alpha)_{\alpha \in A}$ in $\prod_{\alpha \in A}(B_\alpha)$ is the *coordinate projection* of $\prod_{\alpha \in A}(B_\alpha)$ into B_β: $\pi_\beta(c) = c_\beta$ is the β-*th coordinate* of c.

The question then arises: does $\prod_{\alpha \in A}$ have any members? At first glance, it appears that it must; since each B_α is non-empty, there exists c_α in B_α, and we can take $(c_\alpha)_{\alpha \in A}$ as an element of $\prod_{\alpha \in A}$. The problem is that we must do this simultaneously, for all $\alpha \in A$. We require a further axiom to say that this is valid.

Axiom 10: The axiom of choice

This states that if $(B_\alpha)_{\alpha \in A}$ is a family of non-empty sets then $\prod_{\alpha \in A}(B_\alpha)$ is non-empty: there exists a function c, a *choice function*, from A to $\cup_{\alpha \in A} B_\alpha$, such that $c_\alpha = c(\alpha) \in B_\alpha$ for each $\alpha \in A$.

The axiom of choice has a particular position in axiomatic set theory, which we shall discuss further in the next section. On the one hand, the way that we have presented it makes it seem plausible. On the other hand, there is no procedure for producing a choice function, so that its use is highly non-constructive. Further, the axiom of choice leads to some conclusions that seem bizarre. A famous example is the Banach–Tarski paradox, which says that a solid ball B in three dimensions can be divided into a finite number of disjoint sets, which can be rearranged, by rotation and translation, into two disjoint copies of B.

Even when $(B_n)_{n \in \mathbf{Z}^+}$ is a sequence of non-empty sets, we require a version of the axiom of choice to ensure that there is a sequence $(c_n)_{n \in \mathbf{Z}^+}$ in $\cup_{n \in \mathbf{Z}^+}(B_n)$ with $c_n \in B_n$ for all $n \in \mathbf{Z}^+$. Restricting the axiom of choice to sequences, we obtain the *countable axiom of choice*; this certainly seems plausible, and we shall accept it, and use it, generally without comment.

Although recursion enables us to construct sequences, it requires the use of a given function f. Let us consider a more general situation. Suppose that A is a non-empty set, and that ϕ is a mapping from A into the set $P(A) \setminus \{\emptyset\}$ of non-empty sets of A. Suppose that $\bar{a} \in A$. Does there exist a sequence $(a_n)_{n \in P}$ such that $a_0 = \bar{a}$ and $a_{s(n)} \in \phi(a_n)$, for $n \in P$? At stage n, we choose $a_{s(n)}$ from the set $\phi(a_n)$. The *axiom of dependent choice* states that this is always possible. It is an easy consequence of the axiom of choice, and implies the countable axiom of choice, but is not equivalent to either of them. Again, we shall accept it, and use it, generally without comment.

In the general situation, though, we will state explicitly when we use the axiom of choice or use Zorn's lemma. Zorn's lemma is an axiom equivalent to the axiom of choice, and is particularly useful in analysis.

Zorn's lemma concerns partially ordered sets, and we need to make a further definition in order to formulate it. Suppose that $(A, \leq)$ is a partially ordered set. A subset C is a *chain* if it is totally ordered under the order inherited from the partial order on A; that is, if c and c' are elements of C then either $c \leq c'$ or $c' \leq c$.

Zorn's lemma then states that if $(A, \leq)$ is a partially ordered set in which each chain has an upper bound, then A has a maximal element.

Zorn's lemma implies the axiom of choice, and the axiom of choice implies Zorn's lemma. We shall prove the former statement here. The proof of the converse is long and technical; we give the details in Appendix A.

Theorem 1.9.1   *The axiom of choice is a consequence of Zorn's lemma.*

Proof Suppose that $(B_\alpha)_{\alpha \in A}$ is a non-empty family of non-empty sets. We consider the set E of all pairs (Δ, c), where Δ is a subset of A, and $c : \Delta \rightarrow \cup_{\delta \in \Delta} B_\delta$ is a choice function. E is certainly not empty: if $\Delta = \{\delta\}$, there exists $b \in B_\delta$, and we define $c(\delta) = b$. We give E a partial order by setting $(\Delta, c) \le (\Delta', c')$ if $\Delta \subseteq \Delta'$ and $c(\delta) = c'(\delta)$ for all $\delta \in \Delta$. (This way of ordering a set of ordered pairs (X, f), where X is a subset of a set Ω and f is a mapping from X to a set Y, is typical of the way that Zorn's lemma is used.) Suppose that C is a chain in E. Let

$$\Delta_C = \{\delta \in A : \delta \in \Delta \text{ for some } (\Delta, c) \text{ in } C\}.$$

If $\delta \in \Delta_C$ then $\delta \in \Delta$ for some (Δ, c) in C. Let $c_C(\delta) = c(\delta)$. Since C is a chain, if $\delta \in \Delta'$ for some other (Δ', c') in C then $c(\delta) = c'(\delta)$, so that c_C is well defined (it does not matter which pair we choose). Further, (Δ_C, c_C) is an upper bound for C.

We now apply Zorn's lemma to deduce that there is a maximal element (Δ_m, c_m) in E. We claim that $\Delta_m = A$, so that c_m is a choice function on A. Suppose not. Then there exists $\alpha \in A \setminus \Delta_m$ and there exists $b_\alpha \in B_\alpha$. Let $\tilde{\Delta} = \Delta_m \cup \{\alpha\}$, let $\tilde{c}(\delta) = c_m(\delta)$ for $\delta \in \Delta_m$, and let $\tilde{c}(\alpha) = b_\alpha$. Then $(\tilde{\Delta}, \tilde{c}) \in E$, $(\Delta_m, c_m) \le (\tilde{\Delta}, \tilde{c})$ and $(\Delta_m, c_m) \ne (\tilde{\Delta}, \tilde{c})$, contradicting the maximality of (Δ_m, c_m).   □

Let us give another application of Zorn's lemma, to obtain a result which complements the Schröder–Bernstein theorem.

Theorem 1.9.2   *Suppose that A and B are non-empty sets. Then either there exists an injective mapping $j : A \rightarrow B$ or there exists an injective mapping $k : B \rightarrow A$.*

Proof Let E be the set of ordered pairs (H, h), where H is a subset of A, and h is an injective mapping of H into B. We order E by setting $(H, h) \le (H', h')$ if $H \subseteq H'$ and $h'(a) = h(a)$ for $a \in H$. Suppose that C is a chain in E. As above, we set

$$H_C = \{a \in A : a \in H \text{ for some } (H, h) \text{ in } C\}.$$

If $a \in H_C$ then $a \in H$ for some (H, h) in C. Let $h_C(a) = c(a)$. Arguing as above, h_C is well-defined. We must check that it is injective. If a and a' are distinct elements of H_C, then $a \in H$ for some $(H, h) \in C$ and $a' \in H$ for some $(H', h') \in C$. Since C is a chain, either $H \subseteq H'$ or $H' \subseteq H$. Suppose that $H \subseteq H'$. Then $a \in H'$, and so $h_C(a) = h'(a) \neq h'(a') = h_C(a')$. A similar argument holds if $H' \subseteq H$.

We now apply Zorn's lemma to deduce that there is a maximal element (H_m, h_m) of E. If $H_m = A$, we are finished. Suppose that $H_m \neq A$. Then we claim that $h_m(H_m) = B$. For if not, there exist $\tilde{a} \in A \setminus H_m$ and $\tilde{b} \in B \setminus h_m(H_m)$. Let $\tilde{H} = H_m \cup \{a\}$ and define $\tilde{h} : \tilde{H} \to B$ by setting $\tilde{h}(a) = h_m(a)$ for $a \in H_m$ and $\tilde{h}(\tilde{a}) = \tilde{b}$. Then $(\tilde{H}, \tilde{h}) \in E$, $(H_m, h_m) \leq (\tilde{H}, \tilde{h})$ and $(H_m, h_m) \neq (\tilde{H}, \tilde{h})$, contradicting the maximality of (H_m, h_m). Thus h_m is a bijective mapping of H_m onto B, and we can take k to be the inverse mapping h_m^{-1}. $\qquad\square$

Exercises

1.10.1 Show that the axiom of choice implies the axiom of dependent choice.

1.10.2 Show that the axiom of dependent choice implies the countable axiom of choice.

1.10.3 Suppose that $(\phi_n)_{n=0}^{\infty}$ is a sequence of non-empty subsets of $\mathbf{Z}^+$. Use recursion to show that there exists a sequence $(f_n)_{n=0}^{\infty}$ in $\mathbf{Z}^+$ such that $f_0 \in \phi_0$ and $f_{n+1} \in \phi(f_n)$, for $n \in \mathbf{Z}^+$. Why is the axiom of dependent choice not needed?

1.10.4 Suppose that $(A, \leq_A)$ and $(B, \leq_B)$ are partially ordered sets. Define a relation $\leq_l$ on $A \times B$ by setting $(a, b) \leq_l (a', b')$ if either $a <_A a'$ or $a = a'$ and $b \leq b'$.

(a) Show that $\leq_l$ is a partial order on $A \times B$ (the *lexicographic order*).

(b) Show that if $\leq_A$ and $\leq_B$ are total orders then so is $\leq_l$.

1.10.5 Prove the following variant of Zorn's lemma. Suppose that $(A, \leq)$ satisfies the conditions of Zorn's lemma, and that C is a chain in A. Show that there is a maximal element m of A such that $c \leq m$ for all $c \in C$.

1.10 Concluding remarks

We have now described the set-theoretical foundations on which we shall build mathematical analysis. In the process, we have constructed a model for the non-negative integers, which satisfies the requirements of Peano's axioms.

How sound are these foundations? Are they consistent, or is it possible that they may lead to a contradiction? Are they adequate, or are there problems which we are unable to solve using them?

In order to discuss these questions, it is helpful to put them in a historical context. The idea of developing mathematics from a collection of axioms goes back to *Euclid's Elements* of the third century BC. Euclid gave five postulates, or axioms, from which he deduced geometric theorems. The fifth postulate, the *parallel postulate*, states, in the essentially equivalent form of Playfair's axiom, that given a straight line in the plane, and a point not on it, there exists a unique straight line in the plane which passes through the given point, and which does not meet the given line. This postulate raised particular interest, since it was felt that it should be possible to deduce it from the other postulates, and many unsuccessful attempts were made to do so. In the early part of the nineteenth century, Gauss, János Bolyai and Lobachevsky all developed the theory of non-Euclidean geometry (where the parallel postulate fails), but it was not until 1868 that Beltrami produced a model of a two-dimensional non-Euclidean geometry in the setting of three-dimensional Euclidean space, showing that the parallel postulate cannot be deduced from the other postulates. All this raised interest in the axioms, and in particular interest in their consistency. Hilbert studied this in detail and observed that if the postulates of Euclidean geometry are not consistent, then neither are Peano's axioms. He came to believe that there should be a consistent set of axioms for mathematics, from which all results could be deduced. In his famous address to the International Congress of Mathematicians in Paris in 1900, in which he set out his twenty-three important problems for the twentieth century, he talked of

> the conviction (which every mathematician shares, but which no-one has yet supported by a proof) that every definite mathematical problem must necessarily be susceptible of an exact settlement, either in the form of an actual answer to the problem posed or by the proof of the impossibility of solution and therewith the necessary failure of all attempts.

Here he clearly had in mind the necessary failure to prove the parallel postulate from the other postulates. Later on, he said

> This conviction of the solvability of every mathematical problem is a powerful incentive to the worker. We hear within us the perpetual call: There is the problem. Seek its solution. You can find it by pure reason, for in mathematics there is no *ignorabimus*.

This optimism was overturned by Gödel in a spectacular way in 1930 and 1931. First came his incompleteness theorem, which showed that within any logical theory (satisfying certain technical conditions, which reasonably can be expected to hold in a worthwhile theory) there are statements which cannot be proved, and whose negation cannot be proved. Not every mathematical problem is susceptible of an exact solution. Alas, *ignorabimus*! Next came his inconsistency theorem: if a proof of consistency can be given within the theory, then necessarily a proof of inconsistency can also be given. For a system to be consistent, it must be impossible to prove its consistency.

Where does this leave ZF? First, and we shall illustrate this in a moment, the axioms of ZF cannot be the axioms for all of mathematics, nor can we add to them to obtain a set of axioms for all mathematics. Secondly, they cannot be proved to be consistent. Nevertheless they have stood the test of time, and so provide us with a valuable starting point. It is interesting to speculate what would happen if an inconsistency were found. Mathematics would not collapse: mathematicians would continue their work, turning to their logician colleagues to produce a better set of axioms. The effect on the mathematical analysis that we shall be considering would be negligible.

What about the axiom of choice? In 1938, Gödel showed that if ZF is consistent, then so is the system obtained by adding the axiom of choice. In 1963, Cohen showed that there are models of ZF in which the axiom of choice does not hold. Thus the axiom of choice is independent of the axioms of ZF, and cannot be proved or disproved, starting from ZF. We can add the axiom of choice to obtain a stronger axiom system ZFC. Within this, there are further statements that cannot be proved or disproved, such as the continuum hypothesis (which states that if A is an uncountable subset of the set $\mathbf{R}$ of real numbers, then there exists a bijection $f : A \to \mathbf{R}$). In fact, we shall adopt the axiom of choice, but will use it as sparingly as possible.

2
Number systems

2.1 The non-negative integers and the natural numbers

In this chapter we study various number systems. We begin by developing the familiar properties of the non-negative integers $\mathbf{Z}^+ = (0, 1, 2, \ldots)$ and the natural numbers $\mathbf{N} = (1, 2, 3, \ldots)$, using the Peano axioms, induction and recursion.

We begin with addition. This is defined by repeatedly adding 1; we use recursion to formalize this. Suppose that $m \in \mathbf{Z}^+$. Considering the mapping $s : \mathbf{Z}^+ \to \mathbf{Z}^+$, setting $m_0 = m$, and using recursion, we see that there is a sequence $(m_n)_{n \in \mathbf{Z}^+}$ such that $m_0 = m$ and $m_{s(n)} = s(m_n)$. We call m_n the *sum* of m and n, and denote it by $m + n$. This $m + 0 = m$ and $m + 1 = s(m)$. The equation $m_{s(n)} = s(m_n)$ becomes

$$m + (n + 1) = (m + n) + 1. \qquad (*)$$

Here are the fundamental results about addition.

Theorem 2.1.1 *Suppose that $m, n, p \in \mathbf{Z}^+$.*

(i)	$m + n = n + m$	*(commutativity)*
(ii)	$(m + n) + p = m + (n + p)$	*(associativity)*
(iii)	if $m + n = p + n$ then $m = p$	*(cancellation)*
(iv)	if $m + n = 0$ then $m = n = 0$.	

Proof The proof uses induction many times over.

(i) We prove this in three steps. First we show that $m + 0 = 0 + m$ for all m. We use induction. Let $U = \{m \in \mathbf{Z}^+ : 0 + m = m + 0\}$. Then $0 \in U$, since $0 + 0 = 0 + 0$. Suppose that $m \in U$. Then $(m + 1) + 0 = m + 1$, and $0 + (m + 1) = (0 + m) + 1$, by $(*)$, and $(0 + m) + 1 = m + 1$. Thus $m + 1 \in U$, and so $U = \mathbf{Z}^+$, by induction.

Next, we show that

$$(m+1) + n = (m+n) + 1 \text{ for all } m, n \in \mathbf{Z}^+. \qquad (\dagger)$$

Again, we use induction. Let

$$V = \{n \in \mathbf{Z}^+ : (m+1) + n = (m+n) + 1 \text{ for all } m \in \mathbf{Z}^+\}.$$

Since $(m+1) + 0 = m + 1 = (m+0) + 1$, $0 \in V$. Suppose that $n \in V$. Then

$$
\begin{aligned}
(m+1) + (n+1) &= ((m+1) + n) + 1, & \text{by } (*), \\
&= ((m+n) + 1) + 1, & \text{since } n \in V, \\
&= (m + (n+1)) + 1, & \text{by } (*) \text{ again.}
\end{aligned}
$$

Thus $n+1 \in V$, and $V = \mathbf{Z}^+$, by induction.

Finally we establish (i), using induction once more. Let

$$W = \{n \in \mathbf{Z}^+ : m + n = n + m \text{ for all } m \in \mathbf{Z}^+\}.$$

Then $0 \in W$, by the first step. Suppose that $n \in W$. Then

$$
\begin{aligned}
m + (n+1) &= (m+n) + 1, & \text{by } (*), \\
&= (n+m) + 1, & \text{since } n \in W, \\
&= (n+1) + m, & \text{by } (\dagger).
\end{aligned}
$$

Thus $n+1 \in W$, and $W = \mathbf{Z}^+$, by induction.

(ii) Induction once more. Let

$$X = \{p \in \mathbf{Z}^+ : (m+n) + p = m + (n+p) \text{ for all } m, n \in \mathbf{Z}^+\}.$$

Since $(m+n) + 0 = m + n = m + (n+0)$, $0 \in X$. Suppose that $p \in X$. Then

$$
\begin{aligned}
(m+n) + (p+1) &= ((m+n) + p) + 1, & \text{by } (*), \\
&= (m + (n+p)) + 1, & \text{since } p \in X, \\
&= m + ((n+p) + 1), & \text{by } (*), \\
&= m + (n + (p+1)), & \text{by } (*) \text{ again.}
\end{aligned}
$$

Thus $p+1 \in X$, and $X = \mathbf{Z}^+$, by induction.

(iii) A final use of induction. Let

$$Y = \{n \in \mathbf{Z}^+ : \text{if } m + n = p + n \text{ then } m = p, \text{ for all } m, p \in \mathbf{Z}^+\}.$$

Since $m + 0 = m$ and $p + 0 = p$, $0 \in Y$. Suppose that $n \in Y$ and that $m + (n + 1) = p + (n + 1)$. Then

$$(m + n) + 1 = m + (n + 1) = p + (n + 1) = (p + n) + 1, \quad \text{by (*)},$$

and so $m + n = p + n$, by the Peano axiom $(P4)$. Since $n \in Y$, it follows that $m = p$, and so $n + 1 \in Y$. Thus $Y = \mathbf{Z}^+$, by induction.

(iv) Suppose that $m + n = 0$ and that $n \neq 0$. Then $n \in \mathbf{N} = s(\mathbf{Z}^+)$, and so $n = p + 1$ for some $p \in \mathbf{Z}^+$. Then $0 = m + (p + 1) = (m + p) + 1$, by (ii). This contradicts the Peano axiom $(P3)$. Thus $m = 0$, and so $n = n + 0 = 0 + n = 0$.

□

As a result of (ii), we can write $(m + n) + p = m + (n + p) = m + n + p$, omitting the brackets.

By now, proof by induction should be familiar! In future, the details of many such proofs will be left to the reader.

We now define multiplication recursively. Suppose that $n \in \mathbf{Z}^+$. Using recursion, we see that there exists a sequence $(p_m)_{m \in \mathbf{Z}^+}$ such that $p_0 = 0$ and $p_{m+1} = p_m + n$. We then set $p_m = m.n$ (or mn, if this causes no confusion). The number mn is the *product* of m and n. Arguing as in Theorem 2.1.1, we obtain the following.

Theorem 2.1.2 *Suppose that $m, n, p \in \mathbf{Z}^+$.*

(i) $m.n = n.m$ *(commutativity)*;
(ii) $0.n = 0$ *and* $1.n = n$;
(iii) $(m.n).p = m.(n.p)$ *(associativity)*;
(iv) *if* $m.n = p.n$ *and* $n \neq 0$ *then* $m = p$ *(cancellation)*;
(v) *if* $m.n = 0$ *then* $m = 0$ *or* $n = 0$.

Proof The proofs, by induction, are left as exercises for the reader.   □

Again, we can write $(mn)p = m(np) = mnp$, omitting the brackets. We also connect addition and multiplication.

Theorem 2.1.3 *Suppose that $m, n, p \in \mathbf{Z}^+$. Then $m.(n + p) = (m.n) + (m.p)$ (the distributive law).*

Proof The proof, by induction, is again left as an exercise for the reader. □

We write $(m.n) + (m.p) = mn + mp$: multiplication is done before addition.

Corollary 2.1.4 *Suppose that $m, n \in \mathbf{Z}^+$.*
(*i*) *If $mn = n$ then $n = 0$ or $m = 1$.*
(*ii*) *If $mn = 1$ then $m = n = 1$.*

Proof We use results from Theorems 2.1.1 and 2.1.2, without comment. Decide which results are used at each stage of the arguments.

(i) If $n \neq 0$ then $m \neq 0$, and so there exists $k \in \mathbf{Z}^+$ such that $m = k + 1$. Then $0 + n = n = mn = (k+1)n = kn + n$, so that $kn = 0$, by cancellation. Since $n \neq 0$, $k = 0$ and $m = 1$.

(ii) $m \neq 0$ and $n \neq 0$, so that there exist $k, l \in \mathbf{Z}^+$ such that $m = k + 1$ and $n = l + 1$. Then $1 = mn = (k+1)(l+1) = kl + k + l + 1$, so that $kl + (k+l) = 0$. Thus $k + l = 0$ and $k = l = 0$. Thus $m = n = 1$. $\quad\square$

We now use addition to define an order relation on $\mathbf{Z}^+$. If $m, n \in \mathbf{Z}^+$ we set $m \leq n$ if there exists $t \in \mathbf{Z}^+$ such that $n = m + t$. Note that $0 \leq n$ for all $n \in \mathbf{Z}^+$, since $n = n + 0$. We set $m < n$ if $m \leq n$ and $m \neq n$. Thus $m < n$ if and only if there exists $u \in \mathbf{N}$ such that $n = m + u$.

Theorem 2.1.5 $\mathbf{Z}^+$ *is well-ordered by the relation $\leq$. That is:*
(*i*) *if $m \leq n$ and $n \leq p$ then $m \leq p$;*
(*ii*) *If $m, n \in \mathbf{Z}^+$ then either $m \leq n$ or $n \leq m$;*
(*iii*) *if $m \leq n$ and $n \leq m$ then $m = n$;*
(*iv*) *if A is a non-empty subset of $\mathbf{Z}^+$ then there exists $a \in A$ such that $a \leq a'$ for all $a' \in A$ (a is the least element of A, and so is the infimum of A; we denote it by $\inf A$).*

Proof

(i) if $m \leq n$ and $n \leq p$ then there exist t, u in $\mathbf{Z}^+$ such that $n = m + t$ and $p = n + u$. Then $p = (m + t) + u = m + (t + u)$, so that $m \leq p$.
(ii) We use induction. Suppose that $n \in \mathbf{Z}^+$. Let

$$U_n = \{m \in \mathbf{Z}^+ : m < n \text{ or } n \leq m\}.$$

Then $0 \in U_n$. Suppose that $m \in U_n$. We consider two cases. First, suppose that $m < n$. Then $n = m + u$ for some $u \in \mathbf{N}$. Thus $u = r + 1$ for some $r \in \mathbf{Z}^+$, and so $n = m + (r + 1) = (m + 1) + r$; $m + 1 \leq n$ and $m + 1 \in U_n$. Secondly, suppose that $n \leq m$. Then $m = n + t$, for some $t \in \mathbf{Z}^+$, and so $m + 1 = n + t + 1$, and $m + 1 \in U_n$. It therefore follows by induction that $U_n = \mathbf{Z}^+$.
(iii) If $m \leq n$ and $n \leq m$ then there exist $t, u \in \mathbf{Z}^+$ such that $n = m + t$ and $m = n + u$. Thus $n + 0 = n = n + (t + u)$, so that $t + u = 0$. By Theorem 2.1.1 (iv), it follows that $t = u = 0$, so that $m = n$.

(iv) Another proof by induction. Suppose that A does not have a least element. Let

$$V = \{m \in \mathbf{Z}^+ : m \le a \text{ for all } a \in A\}.$$

Note that $A \cap V = \emptyset$. $0 \in V$, since $0 \le n$ for all $n \in \mathbf{Z}^+$. Suppose that $m \in V$ and that $a \in A$. Since $m \notin A$, $m < a$. Thus $a = m + t$, where $t \in \mathbf{N}$. Thus $t = r + 1$ for some $r \in \mathbf{Z}^+$, so that

$$a = m + (r + 1) = (m + 1) + r,$$

and $m + 1 \le a$. Since this holds for all $a \in A$, $m + 1 \in V$. By induction, $V = \mathbf{Z}^+$. Since $A \cap V = \emptyset$, it follows that A is empty, giving a contradiction. □

The well-ordering property provides an alternative approach to induction. Suppose that $Q(x)$ is a well-formed formula, that $T = \{n \in \mathbf{Z}^+ : Q(n) \text{ is true}\}$ and that $F = \{n \in \mathbf{Z}^+ : Q(n) \text{ is false}\}$. Suppose that we know that $0 \in T$, and can show that if $Q(n)$ holds then $Q(n+1)$ holds. Then $F = \emptyset$. For if not, F has a least element f. Then $f \ne 0$, and so $f = n + 1$ for some $n \in \mathbf{Z}^+$. But then $n < f$, so that $n \notin F$. Thus $n \in T$, and so $f \in T$, giving a contradiction.

If $m \le n$ and $n = m + t$ then we write $m = n - t$: we shall remove the restriction $m \le n$ in Section 2.4. Similarly, if $n = mk$, with $k \ne 0$, we write $m = n/k$ and say that m *divides* n. We shall consider division further in Sections 2.5 and 2.6.

Exercises

2.1.1 *(The complete induction principle)* Suppose that $Q(x)$ is a well-formed formula, that $Q(0)$ holds, and that we can show that if $Q(m)$ holds for all $m \le n$ then $Q(n + 1)$ holds. Show that $Q(n)$ holds for all $n \in \mathbf{Z}^+$
   (a) by induction, and
   (b) by using the well-ordering of $\mathbf{Z}^+$.

2.1.2 Define m^n recursively by $m^0 = 1$ (note that $0^0 = 1$) and $m^{n+1} = m^n.m$. Show that $(m^n)^p = m^{np}$ and that $(mn)^p = m^p n^p$.

2.1.3 Show that $n < 2^n$ for all $n \in \mathbf{Z}^+$. A number $n \in \mathbf{N}$ is *even* if 2 divides n, and *odd* if not. Show that if $n \in \mathbf{N}$ then there exist $k \in \mathbf{Z}^+$ and $j \in \mathbf{N}$ such that j is odd and $n = 2^k j$. Show that k and j are uniquely determined by n.

2.1.4 Show how to define $n!$ so that $0! = 1$ and $(n + 1)! = (n!)(n + 1)$.

2.1.5 The *Fibonacci sequence* $(F_n)_{n \in \mathbf{Z}^+}$ is defined by $F_0 = 0$, $F_1 = 1$, $F_{n+2} = F_n + F_{n+1}$ for $n > 1$.

(a) Explain how this definition can be justified by recursion. The numbers that occur in the sequence are called *Fibonacci numbers*.

(b) Show by induction that 2 divides F_k if and only if 3 divides k, and that 3 divides F_k if and only if 4 divides k. When does 5 divide F_k?

(c) Show that $F_{n+k+1} = F_k F_n + F_{k+1} F_{n+1}$.

2.1.6 Show that 5 divides $2^{2n+2} + 3^{2n}$ for all $n \in \mathbf{Z}^+$.

2.1.7 Suppose that $(A_n)_{n \in \mathbf{Z}^+}$ is a sequence of non-empty totally ordered sets and that $A = \prod_{n \in \mathbf{Z}^+} A_n$. If $x, y \in A$ and $x \neq y$, let $k(x, y) = \inf\{n \in \mathbf{Z}^+ : x_n \neq y_n\}$. If $x, y \in A$, set $x \leq y$ if $x = y$ or $x_{k(x,y)} < y_{k(x,y)}$. Show that this is a total order on A (the *lexicographic order* on A).

2.2 Finite and infinite sets

We are all familiar with the basic properties of finite sets. Nevertheless, we need to deduce these properties from Peano's axioms. Since we shall be concerned with counting, we shall work with the natural numbers $\mathbf{N}$, rather than with $\mathbf{Z}^+$.

An *initial segment* I of $\mathbf{N}$ is a non-empty subset of $\mathbf{N}$ with the property that if $n \in I$ and $m \leq n$ then $m \in I$.

Proposition 2.2.1 *If I is an initial segment of I then either $I = \mathbf{N}$ or there exists $n \in \mathbf{N}$ such that $I = I_n = \{m \in \mathbf{N} : m \leq n\}$.*

Proof It follows immediately from the definition of an initial segment that if $m \notin I$ and $n \geq m$ then $n \notin I$. If $I \neq \mathbf{N}$, then $\mathbf{N} \setminus I$ is non-empty; let m_0 be its least element. Suppose, if possible, that $m_0 = 1$. If $n \in \mathbf{N}$, then $n \geq 1$, so that $n \notin I$ and $I = \emptyset$. Thus $m_0 > 1$, and so there exists $n \in \mathbf{N}$ such that $m_0 = n + 1$. Then $n \in I$, and so $I_n \subseteq I$. But if $p > n$ then $p \geq n + 1 = m_0$, and so $p \notin I$. Thus $I \subseteq I_n$. $\qquad \square$

So far, we have defined a sequence to be a mapping from $\mathbf{Z}^+$ to a set A. We now extend the definition, to include mappings from $\mathbf{N}$ to A. A mapping f from an initial segment I to a set A is also called a *sequence*. If $I = I_n$, it is called a *finite sequence in A of length n*, or an *n-tuple*, and is denoted by $(f_j)_{j=1}^n$ or $(f_1, \ldots, f_n)$.

We say that a set A is *finite* if either A is empty or there exists $n \in \mathbf{N}$ and a bijective mapping $c : I_n \to A$. Thus the finite sequence $(c_1, \ldots, c_n)$ lists the elements of A, without repetition. A set is *infinite* if it is not finite.

Proposition 2.2.2 *If $j : I_m \to I_n$ is an injective mapping then $m \leq n$.*

Proof The proof is by induction on m. The result is trivially true if $m = 1$. Suppose that it holds for m, and that $f : I_{m+1} \to I_n$ is injective. Then $m + 1 \geq 2$, so that $f(I_{m+1})$ contains at least two points, and so $n = k + 1$, for some $k \in \mathbf{N}$. Let $\tau : I_n \to I_n$ be the mapping that transposes $f(m + 1)$ and n and leaves the other elements of I_n fixed. Then $\tau \circ f : I_{m+1} \to I_n$ is injective, and $\tau(f(I_m)) \subseteq I_k$. By the inductive hypothesis, $m \leq k$, and so $m + 1 \leq k + 1 = n$. $\qquad\square$

Corollary 2.2.3 *If A is a non-empty finite set, there exists a unique $n \in \mathbf{N}$ for which there exists a bijection $c : I_n \to A$.*

Proof Suppose that $c : I_n \to A$ and $c' : I_{n'} \to A$ are bijections. Then $c^{-1} \circ c' : I_{n'} \to I_n$ is a bijection, and so $n' \leq n$. Similarly, $n \leq n'$. $\qquad\square$

The number n is the *size* or *cardinality* of A; it is written as $|A|$, or as $\#(A)$. We assign the empty set size 0.

Proposition 2.2.4 *Suppose that A is a finite set, and that $f : A \to B$ is a bijection. Then B is finite, and $|B| = |A|$.*

Proof For if $C : I_{|A|} \to A$ is a bijection, then the mapping $f \circ c : I_{|A|} \to B$ is a bijection. $\qquad\square$

Proposition 2.2.5 *If A is a non-empty subset of I_n then A has a greatest element.*

Proof Let $U = \{m \in \mathbf{N} : a \leq m$ for all $a \in A\}$ be the set of upper bounds of A. Then $n \in U$, so that $U \neq \emptyset$. Let b be the least element of U. If $b = 1$ then $A = \{b\}$, so that $b \in A$. Suppose that $b \notin A$. Then $b \neq 1$, and so $b = c + 1$ for some $c \in \mathbf{N}$. But then $c \in U$, contradicting the minimality of b. Thus $b \in A$, and b is the greatest element of A. $\qquad\square$

Corollary 2.2.6 *If A is a non-empty subset of I_n with greatest element n, then A is finite, and $|A| \leq n$, with equality if and only if $A = I_n$.*

Proof We prove this by complete induction on n. The result is certainly true if $n = 1$, since then $A = \{1\}$ and $|A| = 1$. Suppose that it is true for all $c \leq n$, and that A is a subset of $\mathbf{N}$ with greatest element $n + 1$. If A is the singleton $\{n + 1\}$ then the result certainly holds. Otherwise, let $A' = A \setminus \{n + 1\}$. Then $A' \neq \emptyset$, and so A' has a greatest element n' with $n' \leq n$. By the inductive hypothesis, A' is finite, and $k = |A'| \leq n'$, with equality only if $A' = I_{n'}$. Let $c' : I_k \to A'$ be a bijection. If $m \in I_{k+1}$, let $c(m) = c'(m)$ if $m \leq k$ and let $c(k + 1) = n + 1$. Then c is a bijection of I_{k+1} onto A, so that

$|A| = k + 1 \leq n' + 1 \leq n$. Finally, $k + 1 = n + 1$ only if $k = n$, in which case $A' = I_n$ and $A = I_{n+1}$. □

Corollary 2.2.7 *Suppose that B is a subset of a finite set A. Then B is finite, and $|B| \leq |A|$, with equality if and only if $B = A$.*

Proof If B is empty, then B is finite. If B is not empty then A is not empty, and there exist $n \in \mathbf{N}$ and a bijection $c : I_n \to A$. Then $c^{-1}(B)$ is a non-empty finite subset of I_n, and so there exists $m \in \mathbf{N}$, with $m \leq n$, and a bijection $d : I_m \to c^{-1}(B)$. Then $c \circ d$ is a bijection of I_m onto B. Thus B is finite, and $|B| = m \leq n = |A|$. Equality holds if and only if $c^{-1}(B) = I_n$, and this happens if and only if $B = c(I_n) = A$. □

Corollary 2.2.8 *Suppose that A is a non-empty finite set and that $f : A \to A$ is an injective mapping. Then f is bijective.*

Proof Let $c : I_{|A|} \to A$ be a bijection. Then $f \circ c : I_{|A|} \to f(A)$ is a bijection. Thus $|f(A)| = |A|$, and so $f(A) = A$. □

Dedekind defined a set A to be infinite if there is an injective map $j : A \to A$ which is not surjective; such sets are now called *Dedekind infinite*. For example, $\mathbf{N}$ is Dedekind infinite, since the mapping $n \to 2n : \mathbf{N} \to \mathbf{N}$ is injective, and is not surjective.

Corollary 2.2.9 *A Dedekind infinite set is infinite.*

Corollary 2.2.10 $\mathbf{N}$ *is infinite.*

There are many other basic properties of finite sets, including those listed in the exercises. Use only induction, recursion, Peano's axioms and the results derived from them to establish them.

Exercises

2.2.1 Suppose that A is a finite set, and that $f : A \to B$ is a surjection. Show that B is finite, and that $|B| \leq |A|$, with equality if and only if f is a bijection.

2.2.2 Suppose that A is an infinite set and that f is a mapping from A into itself. Show that there exists a non-empty proper subset B of A such that $f(B) \subseteq B$.

[*Hint*: consider the set

$$\{a \in A : \text{there exists } n \in \mathbf{N} \text{ such that } f^n(a) = a\}.]$$

Does the same hold for finite sets?

2.2.3 Show that if A is finite and f is a mapping from A to B then $f(A)$ is finite.

2.2.4 Show that if A and B are finite subsets of a set X then $A \cup B$ is finite, and show that $|A \cup B| + |A \cap B| = |A| + |B|$.

2.2.5 Suppose that $A_1, \ldots A_n$ are finite subsets of a set X. Use induction and the result of the previous exercise to prove the *inclusion-exclusion principle*:

$$|A_1 \cup \cdots \cup A_n|$$
$$= \sum_{k=1}^{n} \left((-1)^{k+1} \sum \{ |A_{j_1} \cap \cdots \cap A_{j_k}| : 1 \leq j_1 < \cdots < j_k \leq n \} \right).$$

2.2.6 *The pigeonhole principle.* Suppose that f is a mapping from a set A to a finite set B. Show that if A is finite and $|A| > |B|$ then f is not injective. Show that if A is infinite, then there exists $b \in B$ such that $f^{-1}(\{b\})$ is infinite.

2.2.7 A tennis club has more than one member. During a season, each member plays against none, some or all of the other members. Show that there are two members who play against the same number of other members.

2.2.8 Suppose that M and W are non-empty finite sets and that H is a relation on $M \times W$. If $m \in M$, let $h(m) = \{ w \in W : (m, w) \in H \}$ and if $A \subseteq M$ let $h(A) = \cup_{m \in A} h(m)$. Show that the following are equivalent:

(a) $|h(A)| \geq |A|$ for all $A \subseteq M$.

(b) There exists an injective mapping $\chi : M \to W$ such that $(m, \chi(m)) \in H$, for all $m \in M$.

[Hint: use induction on $|M|$. Consider two cases:

(i) $|h(A)| > |A|$ for every non-empty proper subset A of M;

(ii) there exists a non-empty proper subset A of M for which $|h(A)| = |A|$.]

This is *Hall's marriage theorem*; M is a set of men, W is a set of women, and $(m, w) \in H$ if m and w know and like each other.

2.2.9 Suppose that $(k_n)_{n \in \mathbf{Z}^+}$ is a decreasing sequence in $\mathbf{Z}^+$ – if $m \geq n$ then $k_m \leq k_n$. Show that $(k_n)_{n \in \mathbf{N}^+}$ is eventually constant: there exists $N \in \mathbf{N}^+$ such that if $m \geq N$ then $k_m = k_N$.

2.2.10 Suppose that $(A, \leq)$ is a non-empty totally ordered set for which each non-empty subset has a least element and a greatest element. If $a \in A$, let $U(a) = \{ b \in A : a < b \}$ be the set of strict upper bounds of $\{a\}$ in A. Let $s(a)$ be the least element of $U(a)$ if $U(a)$ is non-empty, and let $s(a) = a$ otherwise. Show by recursion that there is a surjective mapping $f : \mathbf{Z}^+ \to A$ such that $f(m) \leq f(n)$ if $m \leq n$. Show that A is finite.

2.2.11 Suppose that $(a_1, \ldots, a_n)$ is a finite sequence in $\mathbf{Z}^+$. Show that there are sequences $(s_1, \ldots, s_n)$ and $(p_1, \ldots, p_n)$ such that $s_1 = p_1 = a_1$ and $s_{j+1} = s_j + a_{j+1}$, $p_{j+1} = p_j.a_{j+1}$ for $1 \leq j < n$. We write

$$s_n = a_1 + \cdots + a_n \text{ or } s_n = \sum_{j=1}^{n} a_j, \quad p_n = a_1. \cdots .a_n \quad \text{or} \quad p_n = \prod_{j=1}^{n} a_j.$$

(This clearly will extend to other settings.) Suppose that σ is a permutation of I_n. Show that

$$\sum_{j=1}^{n} a_{\sigma(j)} = \sum_{j=1}^{n} a_j \quad \text{and} \quad \prod_{j=1}^{n} a_{\sigma(j)} = \prod_{j=1}^{n} a_j.$$

2.2.12 Show that $1^3 + 2^3 + \cdots + r^3 = (1 + 2 + \cdots + r)^2$, for all $r \in \mathbf{N}$.

2.2.13 Show that $1^3 + 3^3 + \cdots + (2n-1)^3 = n^2(2n^2 - 1)$ for all $n \in \mathbf{Z}^+$.

2.2.14 Show that any $n \in \mathbf{N}^+$ can be written as the sum of a strictly decreasing sequence of Fibonacci numbers. Is this representation unique?

2.2.15 Suppose that A is finite and that $(B_\alpha)_{\alpha \in A}$ is a family of finite sets. Show that the Cartesian product $\prod_{\alpha \in A} B_\alpha$ is finite and determine its size.

2.2.16 Suppose that A and B are finite. Show that B^A is finite, and determine its size.

2.2.17 Suppose that A is finite. Show that $P(A)$ is finite, and determine its size. By considering mappings $f : A \to \{0, 1\}$, relate this result to the previous one.

2.2.18 Let Σ_A be the set of permutations of a non-empty set A. Show that if A is finite, then Σ_A is finite; determine its size.

2.2.19 Suppose that A and B are finite. Let I be the set of injective mappings from A to B.

(a) Determine the size of I.

(b) Define an equivalence relation on I by setting $f \sim g$ if $f(A) = g(A)$. Determine the size of the equivalence classes.

(c) Let $\binom{n}{k}$ denote the size of the set of subsets of I_n of size k. Show that if $k \leq n$ then

$$\binom{n}{k} = \frac{n!}{(n-k)!k!}.$$

(d) Prove *de Moivre's formula*

$$\binom{n+1}{k} = \binom{n}{k} + \binom{n}{k-1}$$

and its generalization, *Vandermonde's formula,*

$$\binom{m+n}{k} = \sum_{j=0}^{k} \binom{m}{j}\binom{n}{k-j}.$$

(e) By considering the largest member of a subset of I_{n+1} of size $k+1$, show that

$$\binom{n+1}{k+1} = \binom{k}{k} + \binom{k+1}{k} + \cdots + \binom{n}{k}.$$

2.2.20 Suppose that $k_1, \ldots, k_r \in \mathbf{Z}^+$ and that $k_1 + \cdots + k_r = n$. Show that there are

$$\frac{n!}{k_1! \ldots k_r!}$$

r-tuples $(A_1, \ldots, A_r)$ of pairwise disjoint subsets of I_n, with $|A_j| = k_j$ for $1 \le j \le r$.

2.2.21 Show that if A is a non-empty finite set then the number of subsets of A of even size is the same as the number of subsets of A of odd size.

2.2.22 Suppose that $n, k \in \mathbf{N}$. Show that n can be written as $a_1 + \cdots + a_k$, with $a_i \in \mathbf{Z}^+$ for $1 \le i \le k$, in $\binom{n+k-1}{k-1}$ distinct ways. How many distinct ways are there of writing n as $b_1 + \cdots + b_k$, with $b_i \in \mathbf{N}$ for $1 \le i \le k$?

2.3 Countable sets

A set A is *countable* if it is finite or if there is a bijection $c : \mathbf{N} \to A$; otherwise it is *uncountable*. Thus a set is countable if it is empty or if there is a bijection from an initial segment of $\mathbf{N}$ onto A. The function c is called an *enumeration* of A. A set is *countably infinite* if it is infinite and countable.

Thus A is countably infinite if and only if the elements of A can be *listed*, or *enumerated*, as an infinite sequence $(c_1, c_2, \ldots)$, without repetition.

If A is countable (countably infinite) and $j : A \to B$ is a bijection, then B is countable (countably infinite).

Not every set is countable, since it is an immediate consequence of Theorem 1.6.3 that the set $P(\mathbf{N})$ of subsets of $\mathbf{N}$ is not countable. It was Cantor who first showed, in 1873, that there are different sizes of infinite set, showing that the set of real numbers is uncountable. We shall prove this in Section

3.6, where we shall also describe the consternation which Cantor's result produced. Meanwhile, let us concentrate on countable sets.

Theorem 2.3.1 *If A is a subset of $\mathbf{N}$ without a greatest element then there exists a unique strictly increasing function $f : \mathbf{N} \to \mathbf{N}$ (that is, $f(n) < f(n+1)$ for all $n \in \mathbf{N}$) such that $f(\mathbf{N}) = A$.*

Proof We construct the function recursively. If $n \in \mathbf{N}$ then $A_n = A \setminus I_n = \{m \in A : m > n\}$ is non-empty, by hypothesis. Let $g(n)$ be the least element of A_n. Then g is a mapping from $\mathbf{N}$ to $\mathbf{N}$, and $g(n) > n$ for all $n \in \mathbf{N}$. By recursion, there exists a mapping $f : \mathbf{N} \to \mathbf{N}$ such that $f(1) = g(1)$ and $f(n+1) = g(f(n))$, for all $n \in \mathbf{N}$. Since $g(f(n)) > f(n)$, f is strictly increasing; further, $f(\mathbf{N}) \subseteq A$.

Next we show that $f(\mathbf{N}) = A$. If not, let b be the least element of $A \setminus f(\mathbf{N})$. Then $1 \le f(1) < b$, so that the set $A \cap \{n \in \mathbf{N} : n < b\}$ is not empty. By Proposition 2.2.5, it has a greatest element c. Then $g(c) = b$. But $c \in A$ and $c < b$, so that $c \in f(\mathbf{N})$; if $c = f(k)$, then $b = f(k+1)$, giving the required contradiction.

It remains to show that f is unique. Suppose that $h : \mathbf{N} \to \mathbf{N}$ is a strictly increasing function such that $h(\mathbf{N}) = A$, and that $h \ne f$. Then there exists a least n such that $h(n) \ne f(n)$. Since $f(1) = h(1) = g$, where g is the least element of A, $n > 1$. Suppose that $h(n) > f(n)$. Then $h(n-1) = f(n-1) < f(n) < h(n)$. But $f(n) \in A$, and so $f(n) = h(m)$ for some $m \in \mathbf{N}$. Since h is strictly increasing, $n - 1 < m < n$, giving a contradiction. A similar argument applies if $h(n) < f(n)$. Hence f is unique. $\square$

The mapping f is called the *standard enumeration* of A.

Corollary 2.3.2 *Suppose that A is a non-empty subset of $\mathbf{N}$. If A has an upper bound in $\mathbf{N}$, then A is finite; otherwise, A is countably infinite.*

Proof If A has an upper bound, then it is finite, by Proposition 2.2.5 and Corollary 2.2.6. Otherwise, A does not have a greatest element, so that there is bijection $f : \mathbf{N} \to A$, and A is countably infinite. $\square$

Corollary 2.3.3 *A subset B of a countable set A is countable.*

Proof If B is finite, then B is countable. If B is infinite, then A is infinite, and there exists a bijection $g : \mathbf{N} \to A$. Then $g^{-1}(B)$ is infinite, and so does not have a greatest element. By the theorem, there exists a bijection $f : \mathbf{N} \to g^{-1}(B)$. Then $g \circ f : \mathbf{N} \to B$ is a bijection, so that B is countably infinite. $\square$

It is useful to have simple sufficient conditions for a set to be countable. The next proposition provides these.

Proposition 2.3.4 *Suppose that A is a set. The following are equivalent.*

(i) A is countable.
(ii) Either $A = \emptyset$ or there exists a surjective mapping $f : \mathbf{N} \to A$.
(iii) There exists an injective mapping $j : A \to \mathbf{N}$.

Proof Suppose that A is a countable non-empty set. If A is finite, there exists a bijection $f : I_{|A|} \to A$. Extend f to a surjection $f : \mathbf{N} \to A$ by setting $f(n) = f(1)$ for $n > |A|$. If A is countably infinite, there is a bijection of $\mathbf{N}$ onto A. Thus (i) implies (ii).

Suppose that (ii) holds. If A is empty, then the empty mapping is an injective mapping of A into $\mathbf{N}$. Otherwise, if $a \in A$ then $\{n \in \mathbf{N} : f(n) = a\}$ is non-empty; let $g(a)$ be its least element. Then $g : A \to \mathbf{N}$ is an injective mapping, and so (ii) implies (iii).

Finally, suppose that (iii) holds. If $A = \emptyset$, then A is finite, and so is countable. If $A \neq \emptyset$ and $j(A)$ is bounded above, then $j(A)$ is finite, and so A is finite. If $A \neq \emptyset$ and $j(A)$ is not bounded above, let $f : \mathbf{N} \to j(A)$ be the standard enumeration of $j(A)$. Then $j^{-1} \circ f$ is a bijection of $\mathbf{N}$ onto A, so that A is countable: (iii) implies (i). $\square$

In case (ii), each element of A is labelled, all the labels are used, but an element of A may have many labels. In case (iii), each element of A is given a separate label from $\mathbf{N}$, but all the labels need not be used.

When condition (ii) is used, it is important to remember that the empty set needs to be considered separately.

Corollary 2.3.5 *If $g : A \to B$, and A is countable, then $g(A)$ is countable.*

Proof If A is empty, then $g(A)$ is empty, and so is countable. Otherwise, there exists a surjective mapping f of $\mathbf{N}$ onto A. Then $g \circ f$ is a surjective mapping of $\mathbf{N}$ onto $g(A)$, so that $g(A)$ is countable. $\square$

Theorem 2.3.6 *The set $\mathbf{N} \times \mathbf{N}$ is countable.*

Proof Suppose that $(k, l) \in \mathbf{N}$. The mapping $f : \mathbf{N} \times \mathbf{N} \to \mathbf{N}$ defined by $f(k, l) = 2^{k-1}(2l - 1)$ is a bijection. (See Exercise 2.1.3.) $\square$

Corollary 2.3.7 *If A and B are countable sets then $A \times B$ is countable.*

Proof There exist injective mappings $j_A : A \to \mathbf{N}$ and $j_B : B \to \mathbf{N}$. If $(a, b) \in A \times B$, set $j((a, b)) = (j_A(a), j_B(b))$. Then $j : A \times B \to \mathbf{N} \times \mathbf{N}$

is injective, so that the mapping $f \circ j$ is injective. The result follows from Proposition 2.3.4. $\qquad\square$

Corollary 2.3.8 *If A is a countable set, and each $a \in A$ is countable, then $\cup_{a \in A} a$ is countable. (The countable union of countable sets is countable.)*

Proof First, let $B = \{a \in A : a \neq \emptyset\}$. Then $\cup_{a \in A} a = \cup_{a \in B} a$, and so we can suppose that each $a \in A$ is non-empty. Secondly, if A is empty then $\cup_{a \in A} a$ is empty, and so is countable. Thus we can suppose that the set A, and each of the sets $a \in A$, is non-empty. Using Proposition 2.3.4 (iii), there exists a surjection $c : \mathbf{N} \to A$, and for each $m \in \mathbf{N}$ there exists a surjection $f_m : \mathbf{N} \to c(m)$. (Note that here we use the countable axiom of choice; in many specific cases, this can be avoided.) Now if $(m, n) \in \mathbf{N} \times \mathbf{N}$, we set $g(m, n) = f_m(n)$: we use m to select an index $c(m)$ in A, and use n to select an element of $c(m)$. Then g is a surjection of $\mathbf{N} \times \mathbf{N}$ onto $\cup_{a \in A} a$, and so $\cup_{a \in A} a$ is countable, by Corollary 2.3.5. $\qquad\square$

If we assume the axiom of dependent choice, we can establish some properties of infinite sets.

Proposition 2.3.9 *Assuming the axiom of dependent choice, if A is an infinite set, then A contains a countably infinite subset.*

Proof Let $S(A)$ be the set of finite sequences in A. If $s = (a_0, \ldots, a_n) \in S(A)$, let $\phi(s) = \{(a_0, \ldots, a_n, y) : y \notin \{a_0, \ldots, a_n\}\}$. Then $\phi(s) \neq \emptyset$.

Let $\bar{a}$ be an element of A. Let $s_0 = (\bar{a})$. By the axiom of dependent choice, there exists a sequence $(s_n)_{n=0}^{\infty}$ in $S(A)$ such that $s_{n+1} \in \phi(s_n)$, for $n \in \mathbf{Z}^+$. Set $b_n = a_{n,n}$, where $s_n = (a_{n,0}, \ldots, a_{n,n})$. By the construction, $(b_n)_{n=0}^{\infty}$ is a sequence of distinct elements of A. $\qquad\square$

Corollary 2.3.10 *Assuming the axiom of dependent choice, if A is an infinite set then $P(A)$ is uncountable.*

Proof If C is a countably infinite subset of A, then $P(C) \subseteq P(A)$, and $P(C)$ is uncountable. $\qquad\square$

Corollary 2.3.11 *Assuming the axiom of dependent choice, an infinite set A is Dedekind infinite.*

Proof Let B be a countably infinite subset of A, and let $(b_1, b_2, \ldots)$ be a listing of the elements of B, without repetition. Let $f(b_j) = b_{2j}$ for $j \in \mathbf{N}$, and let $f(a) = a$ for $a \in A \setminus B$. Then f is an injective map of A into itself, and $A \setminus f(A) = \{b_1, b_3, b_5, \ldots\}$ is a countably infinite set. $\qquad\square$

Exercises

2.3.1 Show that a finite product of countable sets is countable. What about a countable product of finite sets?

2.3.2 *The countable pigeonhole principle.* Suppose that f is a mapping from an uncountable set A to a countable set B. Show that there exists $b \in B$ such that $f^{-1}(\{b\})$ is uncountable.

2.3.3 Suppose that A is a countably infinite set. Determine which of the following sets are countable and which are not.
   (a) The set of finite subsets of A.
   (b) The set of permutations of A.
   (c) The set of permutations σ of A for which σ^2 is the identity.
   (d) The set of permutations τ of A for which $\{a \in A : \tau(a) \neq a\}$ is finite.

2.3.4 Let J be the set of mappings $j : \mathbf{N} \to \mathbf{N}$ for which $j(m) \leq j(n)$ for $m \leq n$. Show that J is uncountable.

2.3.5 Let D be the set of mappings $d : \mathbf{N} \to \mathbf{N}$ for which $d(m) \geq d(n)$ for $m \leq n$. Show that D is countable.

2.3.6 Suppose that B is a disjoint set of subsets of $\mathbf{N}$: if $A, A' \in B$ and $A \neq A'$ then $A \cap A' = \emptyset$. Show that B is countable.

2.3.7 If $A \in P(\mathbf{Z}^+)$ and $n \in \mathbf{Z}^+$, let $f_A(n) = 2^n$ if $n \in A$ and let $f_A(n) = 0$ otherwise. Let $g_A(n) = \sum_{j=0}^{n} f_A(j)$, and let $G(A) = \{g_A(n) : n \in \mathbf{Z}^+\}$. Show that $\{G(A) : A \in P(\mathbf{Z}^+)\}$ is an uncountable subset of $P(\mathbf{Z}^+)$ with the property that $G(A) \cap G(A')$ is finite, if $A \neq A'$. [*Hint:* consider the binary expansion of $g_A(n)$.]

2.4 Sequences and subsequences

A strictly increasing function from $\mathbf{N}$ to $\mathbf{N}$ defines a sequence in $\mathbf{N}$. Such a sequence $(n_k)_{k=1}^{\infty}$ is called a *subsequence* of $\mathbf{N}$, and the set $\{n_k : k \in \mathbf{N}\}$ is called the *image* of the subsequence. Theorem 2.3.1 shows that there is a one-one correspondence between the infinite subsets of $\mathbf{N}$ and the subsequences of $\mathbf{N}$.

Proposition 2.4.1 *Suppose that $(m_k)_{k=1}^{\infty}$ and $(n_k)_{k=1}^{\infty}$ are subsequences of $\mathbf{N}$, with images A and B respectively. If $A \subseteq B$ then $m_k \geq n_k$ for all $k \in \mathbf{N}$.*

Proof We prove this by induction. First, $n_1 = \inf(A) \leq \inf(B) = m_1$. Suppose that $m_k \geq n_k$. If $m_k = n_k$ then

$$n_{k+1} = \inf\{a \in A : a > n_k\} \leq \inf\{b \in B : b > m_k\} = m_{k+1}.$$

If $m_k > n_k$ then $n_{k+1} = \inf\{a \in A : a > n_k\} \leq m_k < m_{k+1}$. □

Frequently we construct a sequence of subsequences, and use them to construct a further subsequence. This involves a *diagonal procedure*.

Theorem 2.4.2 (The diagonal procedure) *Suppose that $((n_k^{(j)})_{k=1}^\infty)_{j=1}^\infty$ is a sequence of subsequences of $\mathbf{N}$, that A_j is the image of $(n_k^{(j)})_{k=1}^\infty$, for $j \in \mathbf{N}$ and that $(A_j)_{j=1}^\infty$ is a decreasing sequence. Let $m_k = n_k^{(k)}$, for $k \in \mathbf{N}$. Then $(m_k)_{k=1}^\infty$ is a subsequence of $\mathbf{N}$, and $m_k \in A_l$ for $k \geq l$.*

Proof If $k \geq l$ then $m_k \in A_k \subseteq A_l$, so that $m_k \in A_l$. We must show that $(m_k)_{k=1}^\infty$ is strictly increasing. This follows from Proposition 2.4.1, since

$$m_k = n_k^{(k)} < n_{k+1}^{(k)} \leq n_{k+1}^{(k+1)} = m_{k+1}.$$ □

Suppose that $(a_n)_{n=1}^\infty$ is a sequence in a set A and that $(n_k)_{k=1}^\infty$ is a subsequence of $\mathbf{N}$. The composite $(a_{n_k})_{k=1}^\infty$ is called a *subsequence* of $(a_n)_{n=1}^\infty$. In fact, it would be more accurate to define the subsequence as the ordered pair $((a_n)_{n=1}^\infty, (n_k)_{k=1}^\infty)$, since the set $\{n_k : k \in \mathbf{N}\}$ is important. We call it the *support* of the subsequence, and denote it by $\mathrm{supp}\,(a_{n_k})_{k=1}^\infty$.

Let us give an important example.

Theorem 2.4.3 *Suppose that $(a_n)_{n=1}^\infty$ is a sequence in a totally ordered set A. Then there exists a subsequence $(a_{n_k})_{k=1}^\infty$ such that either*

(i) if $k < l$ then $a_{n_k} < a_{n_l}$ ($(a_{n_k})_{k=1}^\infty$ is strictly increasing), or
(ii) if $k < l$ then $a_{n_k} > a_{n_l}$ ($(a_{n_k})_{k=1}^\infty$ is strictly decreasing), or
(iii) if $k < l$ then $a_{n_k} = a_{n_l}$ ($(a_{n_k})_{k=1}^\infty$ is constant).

Proof Let us say that an index n is a *high point* if $a_n > a_m$ for all $m > n$. There are two possibilities. First, there are infinitely many high points $n_1 < n_2 < \cdots$. In this case, $(a_{n_k})_{k=1}^\infty$ is strictly decreasing. Secondly, there are only finitely many high points. In this case, there exists N such that if $n \geq N$ then n is not a high point, so that there exists a least $m > n$ with $a_m \geq a_n$. We can therefore recursively find a sequence $(n_1 < n_2 < \cdots)$ with $n_1 = N$ and $a_{n_{j+1}} \geq a_{n_j}$ for all j. Then either there exists k such that $a_{n_j} = a_{n_k}$ for all $j > k$, in which case we have a constant subsequence, or we can extract a further subsequence which is strictly increasing. □

Theorem 2.4.3 is a consequence of a much more general theorem. This has considerable theoretical importance, but we shall not use it later. It may therefore be omitted on a first reading. First we introduce some notation and terminology. Suppose that C is a finite set and that $f : A \to C$ is a surjective mapping. Then we call f a *colouring* of A. The elements of C are the *colours*; a has colour $f(a)$. The collection of sets $\{f^{-1}(\{c\}) : c \in C\}$ partitions A into sets of different colour.

If A is a set and $k \in \mathbf{N}$, we denote by $P_k(A)$ the set of all subsets of A of size k. We identify $P_1(A)$ with A. If $B \subseteq A$ then $P_k(B) \subseteq P_k(A)$.

Theorem 2.4.4 (Ramsey's theorem)   *Suppose that* $f : P_k(\mathbf{N}) \to C$ *is a colouring of* $P_k(\mathbf{N})$. *Then there exists an infinite subset* $M = \{n_1 < n_2 < \cdots\}$ *of* $\mathbf{N}$ *such that* $f(P_k(M))$ *is a singleton: all the subsets of* M *of size* k *have the same colour.*

Proof   The proof is by induction on k. The result is true if $k = 1$, since the finite collection of sets $\{f^{-1}(\{c\}) : c \in C\}$ is a partition of the infinite set $\mathbf{N}$, and so, by Exercise 2.2.5, one of the sets $f^{-1}(\{c\})$ must be infinite. We take this for M.

Suppose that the result is true for k, and that $f : P_{k+1}(\mathbf{N}) \to C$ is a colouring of $P_{k+1}(\mathbf{N})$. The sets in $P_{k+1}(\mathbf{N})$ have $k + 1$ elements, and, in order to use the inductive hypothesis, we need to relate them to sets with k elements. First, let $b_1 = 1$ and let $D_1 = \{n \in \mathbf{N} : n > b_1\}$. If $B \in P_k(D_1)$, let $g_1(B) = f(\{b_1\} \cup B)$; then g_1 is a colouring of $P_k(D_1)$. By the inductive hypothesis, there exist $c_1 \in C$ and an infinite subset E_1 of D_1 such that $g_1(B) = c_1$ for all $B \in P_k(E_1)$. Thus $f(A) = c_1$ for those A in $P_{k+1}(\{b_1\} \cup E_1)$ for which $b_1 \in A$. But of course there are many other subsets in $P_{k+1}(\{b_1\} \cup E_1)$. We therefore iterate the procedure.

We use recursion to show that there exists a sequence $(b_n, E_n, c_n)_{n=1}^{\infty}$, where $(b_n)_{n=1}^{\infty}$ is a strictly increasing sequence in $\mathbf{N}$, $(E_n)_{n=1}^{\infty}$ is a strictly decreasing sequence of infinite subsets of $\mathbf{N}$ and $(c_n)_{n=1}^{\infty}$ is a sequence of colours, with the following properties:

(i) $b_n < e$ for all $e \in E_n$;
(ii) b_{n+1} is the least element of E_n,
(iii) $f(\{b_n\} \cup A) = c_n$ for all $A \in P_k(E_n)$.

We have found (b_1, E_1, c_1). Suppose that we have found (b_j, E_j, c_j) which satisfy the conditions, for $1 \leq j \leq n$. Let b_{n+1} be the least element of E_n. Let $D_{n+1} = E_n \setminus \{b_{n+1}\}$. If $A \in P_k(D_{n+1})$, we set $g_{n+1}(A) = f(b_{n+1} \cup A)$. Then g_{n+1} is a colouring of $P_k(D_{n+1})$. By the inductive hypothesis, there exists an infinite subset E_{n+1} of D_{n+1} and $c_{n+1} \in C$ such that if $A \in E_{n+1}$ then $f(b_{n+1} \cup A) = g_{n+1}(A) = c_{n+1}$. This establishes the recursion.

Now consider the sequence $(c_n)_{n=1}^{\infty}$. If $c \in C$, let $A_c = \{n \in \mathbf{N} : c_n = c\}$. The finite collection $\{A_c : c \in C\}$ of subsets of $\mathbf{N}$ forms a partition of the infinite set $\mathbf{N}$, and so one of them, A_{c_0} say, must be infinite. We take this to be M. Finally, we show that if $A \in P_{k+1}(M)$ then $f(A) = c_0$, so that M satisfies the conclusions of the theorem. If $A \in P_{k+1}(M)$, we can write $A = \{b_n\} \cup B$, where b_n is the least element of A and $B = A \setminus \{b_n\}$. Then $b_n \in M$ and $B \in P_k(E_n)$, so that $f(B) = c_0$, by (iii). $\square$

Let us now see how Ramsey's theorem can used to prove Theorem 2.4.3. Consider $P_2(\mathbf{N})$. If $m < n$, colour the unordered pair $\{m, n\}$ red if $a_m < a_n$, yellow if $a_m > a_n$ and blue if $a_m = a_n$. Then there exists an infinite subset $M = \{n_1 < n_2 < \cdots\}$ such that the sets $\{n_j, n_k\}$ with $j \neq k$ all have the same colour. Thus the sets $\{n_j, n_{j+1}\}$ all have the same colour. If the colour is red, we have a strictly increasing subsequence; if yellow, a strictly decreasing subsequence; and if blue, a constant subsequence.

Exercises

2.4.1 Suppose that $(A_n)_{n=1}^\infty$ is a sequence of subsets of a set A. Show that there exists a subsequence $(A_{n_k})_{k=1}^\infty$ which is either constant, or strictly increasing, or strictly decreasing, or such that if $k \neq l$ then $A_{n_k} \not\subseteq A_{n_l}$ and $A_{n_l} \not\subseteq A_{n_k}$.

2.4.2 Suppose that $(g_n)_{n=1}^\infty$ is a sequence in a group G. Show that either there is a sequence $(g_{n_k})_{k=1}^\infty$ such that $g_{n_k} g_{n_l} = g_{n_l} g_{n_k}$ for $k, l \in \mathbf{N}$ or there is a sequence $(g_{n_k})_{k=1}^\infty$ such that $g_{n_k} g_{n_l} \neq g_{n_l} g_{n_k}$ if $k \neq l$.

2.5 The integers

Our next task will be to adjoin a set $-\mathbf{N}$ of negative numbers to $\mathbf{Z}^+$ to obtain the set $\mathbf{Z}$ of integers. There are many ways of doing this. We use a rather naïve one, which involves a certain amount of case-by-case checking. Another method appears in Exercise 2.7.5.

Define a mapping $n \to n^+$ from $\mathbf{Z}^+$ to $\mathbf{Z}^+ \times \mathbf{Z}^+$ by setting $n^+ = (0, n)$, and define a mapping $n \to n^-$ from $\mathbf{N}$ to $\mathbf{Z}^+ \times \mathbf{Z}^+$ by setting $n^- = (n, 0)$, and set

$$\mathbf{Z} = \{n^+ : n \in \mathbf{Z}^+\} \cup \{n^- : n \in \mathbf{N}\}.$$

We define addition in $\mathbf{N}$ by setting

$$n^+ + m^+ = (n + m)^+,$$

$$n^- + m^- = (n + m)^-, \text{ and}$$

$$n^+ + m^- = m^- + n^+ = \begin{cases} (n - m)^+ & \text{if } n \geq m, \\ (m - n)^- & \text{if } n < m. \end{cases}$$

Note that addition is commutative: if $p, q \in \mathbf{N}$ then $p + q = q + p$.

We now verify that addition is associative; we do this case by case. Certainly

$$(m^+ + n^+) + p^+ = (m + n + p)^+ = m^+ + (n^+ + p^+)$$
$$\text{and } (m^- + n^-) + p^- = (m + n + p)^- = m^- + (n^- + p^-).$$

Next,

$$(m^+ + n^+) + p^- = (m + n)^+ + p^- = \begin{cases} (m + n - p)^+ & \text{if } m + n \geq p, \\ (p - m - n)^- & \text{if } m + n < p, \end{cases}$$

while

$$m^+ + (n^+ + p^-) = \begin{cases} m^+ + (n - p)^+ = (m + n - p)^+ & \text{if } n \geq p, \\ m^+ + (p - n)^- = (m + n - p)^+ & \text{if } m + n \geq p > n, \\ m^+ + (p - n)^- = (p - m - n)^- & \text{if } m + n < p. \end{cases}$$

Thus $(m^+ + n^+) + p^- = m^+ + (n^+ + p^-)$. Using this, and the commutative property, we find that

$$(m^+ + p^-) + n^+ = n^+ + (m^+ + p^-) = (n^+ + m^+) + p^-$$
$$= (m^+ + n^+) + p^- = m^+ + (n^+ + p^-)$$
$$= (m^+ + n^+) + p^-,$$

and the other cases are dealt with in a similar way.

Note also that 0^+ acts as an identity: if $p \in \mathbf{N}$ then $p + 0^+ = 0^+ + p = p$, and if $n \in \mathbf{Z}^+$ then $n^+ + n^- = 0^+$.

Thus we have the following.

Theorem 2.5.1 $(\mathbf{Z}, +)$ *is an abelian group with identity element* 0^+, *generated by* 1^+. *The mapping* $\theta : \mathbf{Z}^+ \to \mathbf{Z}$ *defined by* $\theta(n) = n^+$ *is an injective mapping of* $\mathbf{Z}^+$ *into* $\mathbf{Z}$, *and* $\theta(n + m) = \theta(n) + \theta(m)$.

In particular, $-(n^+) = n^-$ and $-(n^-) = n^+$.

The set $\mathbf{Z}$ is the set of *integers*. We identify $\mathbf{Z}^+$ with $\theta(\mathbf{Z}^+)$, and $\mathbf{N}$ with $\theta(\mathbf{N})$. Thus $\mathbf{Z} = \mathbf{Z}^+ \cup (-\mathbf{Z}^+) = \mathbf{N} \cup \{0\} \cup (-\mathbf{N})$, and the latter is a disjoint union. If $n \in \mathbf{N}$, we say that n is *positive*; if $n \in \mathbf{Z}^+$, we say that n is *non-negative*; if $n \in -\mathbf{N}$ we say that n is *negative*, and if $n \in -\mathbf{N} \cup \{0\}$ we say that n is *non-positive*.

The fact that $(\mathbf{Z}, +)$ is a group is important; it leads to useful algebraic results.

Proposition 2.5.2 *Suppose that $(G, \circ)$ is a group and that $g \in G$. Then there exists a unique homomorphism ϕ of $(\mathbf{Z}, +)$ into G for which $\phi(1) = g$.*

Proof We define ϕ recursively on $\mathbf{Z}^+$. Define a mapping $r : G \to G$ by setting $r(h) = h \circ g$, for $h \in G$. By recursion, there exists a unique mapping $\phi : \mathbf{Z}^+ \to G$ such that $\phi(0) = e_G$, the identity in G, and $\phi(n+1) = r(\phi(n))$. Set $g^n = \phi(n)$. Then

$$g^{n+1} = \phi(n+1) = \phi(n) \circ g = g^n \circ g;$$

an easy induction shows that $g^{m+n} = g^m \circ g^n$, for $m, n \in \mathbf{Z}^+$. Now define $g^{-n} = (g^n)^{-1}$, for $-n \in \mathbf{N}^-$. It is again straightforward to check that $g^{a+b} = g^a \circ g^b$ for $a, b \in \mathbf{Z}$. In particular, $g^n \circ g^{-n} = g^{-n} \circ g^n = e$, so that g^{-n} is the inverse of g^n. Finally, uniqueness follows from the uniqueness of the recursion. $\square$

The image $\phi(G)$ is a subgroup of G. It is the smallest subgroup of G which contains g, and is denoted by $\mathrm{Gp}(g)$. If $\mathrm{Gp}(g) = G$, we say that G is a *cyclic group*, with *generator g*.

Proposition 2.5.3 *The additive group $(\mathbf{Z}, +)$ is a cyclic group, with generator 1.*

Proof Let $\mathrm{Gp}(1)$ be the subgroup of $\mathbf{Z}$ generated by 1. Then $0 \in \mathrm{Gp}(1)$. By induction, $n \in \mathrm{Gp}(1)$ for all $n \in \mathbf{N}$. But then $-n \in \mathrm{Gp}(1)$ for all $n \in \mathbf{N}$, and so $\mathrm{Gp}(1) = \mathbf{Z}$. $\square$

Next, we define an order on $\mathbf{Z}$. We set $k \leq j$ if $j - k \in \mathbf{Z}^+$. If $j - k \in \mathbf{Z}^+$ and $k - l \in \mathbf{Z}^+$ then $j - l = (j - k) + (k - l) \in \mathbf{Z}^+$; thus if $k \leq j$ and $l \leq k$ then $l \leq j$. If $k \not\leq j$ then $j - k \notin \mathbf{Z}^+$, so that $j - k \in \mathbf{N}^-$, and $k - j = -(j - k) \in \mathbf{N} \subseteq \mathbf{Z}^+$. Thus $j < k$. Consequently $\leq$ is a total order on $\mathbf{Z}$. Note that $j \leq k$ if and only if $j + l \leq k + l$, for any $j, k, l \in \mathbf{Z}$. We can arrange the integers in increasing order as a doubly infinite sequence of terms:

$$\ldots, -4, -3, -2, -1, 0, 1, 2, 3, 4, \ldots$$

The order and the group structure of $(\mathbf{Z}, +, \leq)$ are related. An *ordered group* is a group G, together with a total order on G with the property that if $g \leq g'$ and $h \in G$ then $h \circ g \leq h \circ g'$ and $g \circ h \leq g' \circ h$. We denote the set $\{g \in G : e \leq g\}$ by G^+. The preceding remarks show that $(\mathbf{Z}, +, \leq)$ is an ordered group. Further, the set $\mathbf{Z}^+$ is well-ordered, and $\mathbf{Z}$ has at least two elements. We now show that these properties characterize $\mathbf{Z}$.

Theorem 2.5.4 *Suppose that $(G, \circ, \leq)$ is an ordered group with at least two elements and that G^+ is well-ordered. Then there exists a unique order-preserving group isomorphism θ of $(\mathbf{Z}, +, \leq)$ onto $(G, \circ, \leq)$.*

Proof We do not assume that G is an abelian group, and so we write the group operation as multiplication. If $g \in G$, then either g or g^{-1} is in G^+ (if $g \notin G^+$ then $g \leq e$; composing with g^{-1}, $e = g \circ g^{-1} \leq g^{-1}$, so that $g^{-1} \in G^+$). Since G has at least two elements, the set $P = \{g \in G : e < g\}$ of strictly positive elements is not empty. Let 1_G be the least element of P. By Proposition 2.5.2, there exists a unique homomorphism $\theta : \mathbf{Z} \to G$ with $\theta(1) = 1_G$. An easy induction shows that $\theta(\mathbf{N}) \subseteq P$. Suppose that $j, k \in \mathbf{Z}$ and that $j < k$. Then $k - j \in \mathbf{N}$, so that $\theta(k - j) \in P$. Thus $e < \theta(k - j) = \theta(k) \circ (\theta(j))^{-1}$. Multiplying by $\theta(j)$, we see that $\theta(j) < \theta(k)$; θ is order-preserving. Since the order on $\mathbf{Z}$ is a total order, it follows that θ is injective.

Next we show that $\theta(\mathbf{Z}) = G$. If not, there exists $g \in G \setminus \theta(\mathbf{Z})$. Since $\theta(\mathbf{Z})$ is a subgroup of G, $g \neq 0$, and $-g \in G \setminus \theta(\mathbf{Z})$. As before, one of g and $-g$ is strictly positive, and so $P \setminus \theta(\mathbf{Z})$ is non-empty. Let g_0 be its least element. Since $1_G^{-1} = \theta(-1) \in \theta(\mathbf{Z})$ and $g_0 \notin \theta(\mathbf{Z})$, it follows that $1_G^{-1} \circ g_0 \notin \theta(\mathbf{Z})$. Since $e = \theta(0) \in \theta(\mathbf{Z})$, it follows $1_G^{-1} \circ g_0 \neq e$. Since $1_G^{-1} \circ g_0 < g_0$ and since g_0 is the least element of $P \setminus \theta(\mathbf{Z})$, it follows that $1_G^{-1} \circ g_0 < e$. Multiplying by 1_G, it follows that $g_0 < 1_G$. But $g_0 \in P$ and 1_G is the least element of P, and so we have a contradiction.

Uniqueness then follows from Proposition 2.5.2.   □

What about multiplication? We want to extend the multiplication defined on $\mathbf{Z}^+$, and to preserve the distributive law. Thus if $m, n \in \mathbf{Z}^+$ we require that

$$m.n + m.(-n) = m.(n + (-n)) = m.0 = 0 \text{ and}$$
$$n.m + (-n).m = (n + (-n)).m = 0.m = 0,$$

so that $m.(-n) = -(m.n) = -(n.m) = (-n).m$. In particular, we require that $0.(-n) = (-n).0 = 0$. Similarly we require that

$$(-m).n + (-m).(-n) = (-m).(n + (-n)) = (-m).0 = 0 \text{ and}$$
$$n.(-m) + (-m).(-n) = (n + (-n)).(-m) = 0.(-m) = 0,$$

so that $(-m).(-n) = m.n = n.m = (-n).(-m)$.

Summing up, we have the following multiplication table:

	0	$m \in \mathbf{N}$	$-m \in -\mathbf{N}$
0	0	0	0
$n \in \mathbf{N}$	0	nm	$-nm$
$-n \in -\mathbf{N}$	0	$-nm$	nm

With this multiplication, we have the following extension of Theorem 2.1.2 and Theorem 2.1.3.

Theorem 2.5.5 *Suppose that $j, k, l \in \mathbf{Z}$.*

(i) $j.k = k.j$ *(commutativity)*;

(ii) $0.j = 0,$ $1.j = j$ *and* $(-1).j = -j$;

(iii) $(j.k).l = j.(k.l)$ *(associativity)*;

(iv) *if* $j.k = l.k$ *and* $k \neq 0$ *then* $j = l$ *(cancellation)*;

(v) *if* $j.k = 0$ *then* $j = 0$ *or* $l = 0$.

(vi) $j.(k + l) = (j.k) + (j.l)$ *(the distributive law)*.

Proof The proof is again left as an exercise for the reader. $\square$

Again, we can write jk for $j.k$. Then $(jk)l = j(kl) = jkl$. We write $(jk) + (jl) = jk + jl$; multiplication is carried out before addition.

Exercises

2.5.1 Suppose that $x \in \mathbf{Z}$ and that $x \neq 0$. Show that $x^2 > 0$.

2.5.2 Show that $\mathbf{Z}$ is countable. Define an explicit bijection from $\mathbf{N}$ onto $\mathbf{Z}$.

2.6 Divisibility and factorization

We now consider divisibility in $\mathbf{N}$ and in $\mathbf{Z}$. If j and k are in $\mathbf{Z}$, we say that j *divides* k, and write $j|k$, if there exists $q \in \mathbf{Z}$ such that $k = qj$. It follows from Corollary 2.1.4 that the only elements of $\mathbf{Z}$ which divide every element of $\mathbf{Z}$ are 1 and -1: we call them the *units* of $\mathbf{Z}$.

In order to study divisibility, we first consider the additive group $(\mathbf{Z}, +)$, and ask the question: what are the subgroups of $(\mathbf{Z}, +)$? Suppose that $n \in \mathbf{N}$. By Proposition 2.5.2, there is a homomorphism $\theta : \mathbf{Z} \to \mathbf{Z}$ such that $\theta(1) = n$. Then

$$\theta(\mathbf{Z}) = \mathbf{Z}n = \{k \in \mathbf{Z} : k = jn \text{ for some } j \in \mathbf{Z}\} = \{k \in \mathbf{Z} : n|k\},$$

so that $\mathbf{Z}n$ is a subgroup of $(\mathbf{Z}, +)$ (note that $\mathbf{Z}0 = \{0\}$ and $\mathbf{Z}1 = \mathbf{Z}$). If $n \neq 0$ then n is the least positive element of $\mathbf{Z}n$, and so $\mathbf{Z}_m \neq \mathbf{Z}_n$ if $m \neq n$.

These subgroups are useful when considering division with remainder.

Proposition 2.6.1   *Suppose that $m, n \in \mathbf{N}$. There exist $q, r \in \mathbf{Z}^+$, with $0 \leq r < n$ such that $m = qn + r$.*

Proof   Let $L = \{j \in \mathbf{Z}n : 0 \leq j \leq m\}$. Since $0 \in L$, L is a non-empty finite set, and therefore it has a greatest element $l = qn$. Let $r = m - qn$, so that $r \geq 0$. Since $qn + n = (q+1)n \notin L$, $m < qn + n$, and so $r = m - nq < n$. $\square$

In fact, the subgroups $\mathbf{Z}n$ are the only subgroups of $\mathbf{Z}$.

Proposition 2.6.2   *If H is a subgroup of $(\mathbf{Z}, +)$ then $H = \mathbf{Z}n$ for some $n \in \mathbf{Z}^+$.*

Proof   If $H \neq \{0\} = \mathbf{Z}0$, then, since $h \in H$ if and only if $-h \in H$, the set $H \cap \mathbf{N}$ of positive elements of H is non-empty. Let n be its least member. Then $\mathbf{Z}n \subseteq H$. We shall show that $H = \mathbf{Z}n$. Suppose that $m \in H$ and that m is positive. By the previous proposition, we can write $m = qn + r$, where $0 \leq r < n$. But $qn \in H$, and so $r = m - qn \in H$. Since n is the least positive element of H and $r < n$, it follows that $r = 0$. Thus $m = qn \in \mathbf{Z}n$. If $m \in H$ and m is negative, then $-m \in H$, so that $-m \in \mathbf{Z}n$; consequently $m \in \mathbf{Z}n$. $\square$

Now let us return to divisibility. We restrict attention to $\mathbf{N}$. The relation $m|n$ is a partial order on $\mathbf{N}$, since if $m|n$ and $n|p$ then $m|p$, and since if $m|n$ and $n|m$ then $m = n$. A partially ordered set $(A, \leq)$ is called a *lattice* if whenever a and b are elements of A then the set $\{a, b\}$ has an infimum, denoted by $a \wedge b$, and a supremum, denoted by $a \vee b$.

Theorem 2.6.3   *(i) The partially ordered set $(\mathbf{N}, |)$ is a lattice.*
   (ii) If $m, n \in \mathbf{N}$ then there exist $k, l \in \mathbf{Z}$ such that $m \wedge n = km + ln$ (Bachet's theorem).
   (iii) $(m \wedge n)(m \vee n) = mn$.

The element $m \wedge n$ is called the *highest common factor* of m and n, and is traditionally written as (m, n) [risking confusion with the ordered pair (m, n)]; $m \vee n$ is called the *lowest common multiple* of m and n.
   Bachet's theorem is frequently called *Bézout's lemma*; Bachet established the result in 1624.

Proof   Suppose that $m, n \in \mathbf{N}$. Let

$$H = \{h \in \mathbf{Z} : h = um + vn, \text{ for some } u, v \in \mathbf{Z}\}.$$

Then $m = 1.m + 0.n \in H$ and $n = 0.m + 1.n \in H$. Since

$$(um + vn) + (u'm + v'n) = (u + u')m + (v + v')n \text{ and}$$
$$-(um + vn) = (-u)m + (-v)n,$$

H is a subgroup of $(\mathbf{Z}, +)$. Further, H is the smallest subgroup of $(\mathbf{Z}, +)$ containing m and n, since if K is a subgroup of $(\mathbf{Z}, +)$ which contains m and n then it contains all the elements $um + vn$, with $u, v \in \mathbf{Z}$. We call H the subgroup *generated by m and n*, and denote it by $\mathrm{Gp}(m, n)$. By Proposition 2.6.2 there exists $h \in \mathbf{Z}^+$ such that $H = \mathbf{Z}h$. Since $H \neq \{0\}$, $h > 0$. Then there exist $k, l \in \mathbf{Z}$ such that $h = km + ln$. Since $m, n \in H$, $h|m$ and $h|n$. Suppose that $h'|m$ and $h'|n$. Then $h'|(km + ln)$ and so $h'|h$. Thus h is the highest common factor of m and n.

Similarly $\mathbf{Z}m \cap \mathbf{Z}n$ is a subgroup of $(\mathbf{Z}, +)$, and $mn \in \mathbf{Z}m \cap \mathbf{Z}n$, so that $\mathbf{Z}m \cap \mathbf{Z}n \neq \{0\}$. Thus there exists $g \in \mathbf{N}$ such that $\mathbf{Z}m \cap \mathbf{Z}n = \mathbf{Z}g$. Since $g \in \mathbf{Z}m$, $m|g$, and similarly $n|g$. If $m|g'$ and $n|g'$ then $g' \in \mathbf{Z}m$ and $g' \in \mathbf{Z}n$, so that $g' \in \mathbf{Z}g$. Thus $g|g'$, and so g is the lowest common multiple of m and n.

We now show that $mn = hg$. Recall that $h = km + ln$. Since $m|g$, $mn|lng$, and similarly $mn|kmg$; Thus $mn|(km + ln)g$; that is, $mn|hg$. On the other hand, $m = sh$ and $n = th$ for some $s, t \in \mathbf{N}$. Then $m|sth$ and $n|sth$, so that sth is a common multiple of m and n; consequently, $g|sth$. Thus $hg|sth^2$. But $sth^2 = mn$, and so $hg|mn$. Consequently $mn = hg$. □

If the highest common factor of m and n is 1, we say that m and n are *coprime*, or *relatively prime*. Bachet's theorem has the following consequence.

Proposition 2.6.4 *If m and n are coprime, and $m|nr$, then $m|r$.*

Proof There exist $k, l \in \mathbf{Z}$ such that $1 = km + ln$, and so $r = kmr + lnr$. Since m divides each term on the right-hand side of this equation, it also divides r. □

Theorem 2.6.3 establishes the existence of the highest common factor of two numbers, but it does not tell us how to find them. For this, we use *Euclid's algorithm*; this was given in Euclid's *Elements*. This also enables us to determine the constants in Bachet's theorem.

It is convenient to work with $\mathbf{Z}^2 = \mathbf{Z} \times \mathbf{Z}$ with its product group structure: the identity element is $(0, 0)$, $(j, k) + (j', k') = (j + k, j' + k')$ and

$-(j,k) = (-j, -k)$. Any element (j, k) of $\mathbf{Z}^2$ can be written uniquely as $je_1 + ke_2$, where $e_1 = (1, 0)$ and $e_2 = (0, 1)$. Thus if $\theta : \mathbf{Z}^2 \to \mathbf{Z}^2$ is a homomorphism, then $\theta((j, k)) = j\theta(e_1) + k\theta(e_2)$. We can express θ in terms of matrices: if $\theta(e_1) = (\theta_{11}, \theta_{12})$ and $\theta(e_2) = (\theta_{21}, \theta_{22})$ then

$$\theta((j, k)) = (j, k) \begin{bmatrix} \theta_{11} & \theta_{12} \\ \theta_{21} & \theta_{22} \end{bmatrix} = (j\theta_{11} + k\theta_{21}, j\theta_{12} + k\theta_{22}).$$

Suppose that $m_0 > n_0 > 0$ and that we want to find $h_0 = m_0 \wedge n_0$. Thus we want to find h_0 such that $\mathrm{Gp}(m_0, n_0) = \mathbf{Z}h_0$.

We divide: by Proposition 2.6.1, there exist q_0 and r_0, with $0 \le r_0 < n_0$ such that $m_0 = q_0 n_0 + r_0$. We set $m_1 = n_0$ and $n_1 = r_0$. Thus

$$(m_1, n_1) = (m_0, n_0) \begin{bmatrix} 0 & 1 \\ 1 & -q_0 \end{bmatrix} = (m_0, n_0) M_1, \text{ say, and}$$

$$(m_0, n_0) = (m_1, n_1) \begin{bmatrix} q_0 & 1 \\ 1 & 0 \end{bmatrix} = (m_1, n_1) N_1, \text{ say}).$$

From these equations, it follows that m_1 and n_1 are in $\mathrm{Gp}(m_0, n_0)$, so that $\mathrm{Gp}(m_1, n_1) \subseteq \mathrm{Gp}(m_0, n_0)$, and that m_0 and n_0 are in $\mathrm{Gp}(m_1, n_1)$, so that $\mathrm{Gp}(m_0, n_0) \subseteq \mathrm{Gp}(m_1, n_1)$. Thus

$$\mathrm{Gp}(m_1, n_1) = \mathrm{Gp}(m_0, n_0) = \mathrm{Gp}(h_0).$$

If $n_0 | m_0$ then $n_1 = 0$ and $m_1 = h_0$. Otherwise, if $h_1 = m_1 \wedge n_1$, then $\mathrm{Gp}(h_1) = \mathrm{Gp}(m_1, n_1) = \mathrm{Gp}(h_0)$, so that $h_1 = h_0$; in this case we iterate the procedure. Since $0 \le n_j < n_{j-1}$, the procedure must stop after a finite number k of iterations. Then $m_k = h_{k-1} = \cdots = h_0$ and $n_k = 0$. Since we can write $(m_j, n_j) = (m_{j-1}, n_{j-1}) M_j$ for $1 \le j \le k$, it follows that

$$(m_j, n_j) = (m_0, n_0) M_1 \ldots M_j = (m_0, n_0) P_j,$$

$$\text{where } P_j = M_1 \ldots M_j = P_{j-1} M_j.$$

At each stage we can calculate the product $P_{j-1} M_j$, and so calculate P_j. In particular, $(h_0, 0) = (m_k, n_k) P_k$, so that if

$$P_k = \begin{bmatrix} p_{11} & p_{12} \\ p_{21} & p_{22} \end{bmatrix} \text{ then } h_0 = p_{11} m_0 + p_{21} n_0.$$

Let us give a numerical example. Let $m_0 = 1677$ and $n_0 = 1131$. Then

$$q_0 = 1, \quad r_0 = 546, \quad (m_1, n_1) = (1677, 1131) \begin{bmatrix} 0 & 1 \\ 1 & -1 \end{bmatrix} = (1131, 546)$$

$$q_1 = 2, \quad r_1 = 39, \quad (m_2, n_2) = (1131, 546) \begin{bmatrix} 0 & 1 \\ 1 & -2 \end{bmatrix} = (546, 39)$$

$$q_2 = 14, \quad r_2 = 0, \quad (m_3, n_3) = (546, 39) \begin{bmatrix} 0 & 1 \\ 1 & -14 \end{bmatrix} = (39, 0).$$

Thus the highest common factor of 1677 and 1131 is 39. Further

$$P_3 = \begin{bmatrix} 0 & 1 \\ 1 & -1 \end{bmatrix} \begin{bmatrix} 0 & 1 \\ 1 & -2 \end{bmatrix} \begin{bmatrix} 0 & 1 \\ 1 & -14 \end{bmatrix} = \begin{bmatrix} -2 & -29 \\ 3 & 43 \end{bmatrix},$$

so that $39 = -2.1677 + 3.1131$.

We now turn to factorization. Our aim is to factorize a number as a product of simpler numbers. An element p of $\mathbf{N}$ is a *prime*, or a *prime number*, if it is not a unit (that is, is not equal to 1), and if the only elements of $\mathbf{N}$ which divide it are 1 and p. Bachet's theorem provides an equivalent definition.

Proposition 2.6.5 *Suppose that $p \in \mathbf{N}$ and $p \neq 1$. The following are equivalent:*

(i) p is a prime;
(ii) if $p|mn$ then $p|m$ or $p|n$.

Proof Suppose that p is a prime, that $p|mn$ and that p does not divide m. Then the highest common factor of m and p is 1, and so by Bachet's theorem there exist $k, l \in \mathbf{Z}$ such that $1 = km + lp$. Thus $n = kmn + lpn$. Since p divides each of the terms on the right-hand side, p divides n.

If q is not a prime, then $q = mn$ for some m, n not equal to 1 or q. Then $q|mn$, but q does not divide either m or n, since m and n are smaller than q. □

Theorem 2.6.6 (The fundamental theorem of arithmetic) *If $n \in \mathbf{N}$ and $n > 1$ then n can be written uniquely as a product $p_1 \ldots p_k$ of primes, with $p_1 \leq p_2 \leq \cdots \leq p_k$.*

Proof First we use complete induction to show that n can be written as a product of primes. 2 is a prime, so $2 = p_1$ with $p_1 = 2$. Suppose that the result holds for m with $2 \leq m < n$. Let A be the set of divisors of n which are greater than 1. A is non-empty, since $n \in A$, and so A has a least element

p. p must be a prime, for otherwise $p = ab$, with $a, b > 1$; then $a \in A$ and $a < p$. Then $n = pq$ for some $q < n$. By the inductive hypothesis, we can write $q = p_1 \ldots p_k$ as a product of primes, with $p_1 \leq p_2 \leq \cdots \leq p_k$. Since $p_1 \in A$, $p \leq p_1$, and $n = pp_1 \ldots p_k$.

It is harder to show that the factorization is unique. Again we prove this by complete induction. It is certainly true when $n = 2$. Suppose that the result holds for m with $2 \leq m < n$. Let $n = p_1 \ldots p_k = q_1 \ldots q_l$ be two factorizations into primes, with $p_1 \leq p_2 \leq \cdots \leq p_k$ and $q_1 \leq q_2 \leq \cdots \leq q_l$. Let $s = p_2 \ldots p_k$ and $t = q_2 \ldots q_l$, so that $n = p_1 s = q_1 t$. First we show that $p_1 = q_1$. Suppose not, and suppose without loss of generality that $p_1 < q_1$. Since $p_1 | q_1 t$, and since p_1 does not divide q_1, $p_1 | t$, so that $t = p_1 u$, for some $u \in \mathbf{N}$. u has a factorization $u = r_1 \ldots r_m$ into primes, and so $t = p_1 r_1 \ldots r_m$ is a factorization into primes. Since $t < n$, the factorization is unique when the terms are rearranged in increasing order. Since $t = q_2 \ldots q_l$, with $q_2 \leq \cdots \leq q_l$, q_2 is the least of $p_1, r_1, \ldots, r_m$, and so $q_1 \leq q_2 \leq p_1$, giving a contradiction. Thus $p_1 = q_1$. Hence $s = p_2 \ldots p_k = t = q_2 \ldots q_l$. But $s < n$, and so the factorization of s is unique. Thus $k = l$ and $p_j = q_j$ for $2 \leq j \leq k$. $\square$

Corollary 2.6.7   *There are infinitely many primes.*

Proof Suppose, on the contrary that there are only finitely many primes $p_1, \ldots, p_k$. Let $n = p_1 \ldots p_k + 1$. Then p_j does not divide n, for $1 \leq j \leq k$, so that n has no prime divisors. $\square$

Exercises

2.6.1 Suppose that $(X, \leq)$ is a lattice. Show that $(a \wedge b) \wedge c = a \wedge (b \wedge c)$. Is $(a \wedge b) \vee c = a \wedge (b \vee c)$ always true?

2.6.2 Show that a maximal element of a lattice is the greatest element of L.

2.6.3 Show that the subgroups of a group, ordered by inclusion, form a lattice.

2.6.4 What is the highest common factor of the Fibonacci numbers F_{n+1} and F_n? How many steps does Euclid's algorithm take to evaluate it? What is the highest common factor of the Fibonacci numbers F_{n+2} and F_n?

2.6.5 Use Euclid's algorithm to find numbers m and n such that $81m - 100n = 1$.

2.6.6 Recall that two natural numbers a and b are *coprime* if their highest common factor is 1. Use Bachet's theorem to show that if a and b are coprime and a and c are coprime, then a and bc are coprime. Give another proof, using Theorem 2.6.6.

2.6.7 Show that given $k \in \mathbf{N}$ there exists $n \in \mathbf{N}$ such that $n + j$ is not a prime, for $1 \leq j \leq k$.

2.6.8 By considering numbers of the form $4p_1 \ldots p_k - 1$, show that there are infinitely many primes of the form $4t - 1$.

2.6.9 Show that there are infinitely many primes of the form $6t - 1$.

2.6.10 Suppose that p is a prime. Show that p divides $\binom{p}{r}$ for $1 \leq r < p$.

2.7 The field of rational numbers

In $\mathbf{Z}$, we can add, multiply, and subtract, but, as we have seen in the previous section, division is very limited, but also very interesting. In this section, we embed $\mathbf{Z}$ in a set $\mathbf{Q}$ of quotients, in which we can add, subtract, multiply and divide (but not by 0), according to the usual laws of algebra.

Let us make this last remark explicit. A *field* is a set F, together with two laws of composition, *addition* $(+)$ and *multiplication* $(\circ)$, with the following properties.

(i) $(F +)$ is an abelian group, with identity element 0.

(ii) Let $F^* = F \setminus \{0\}$. Then $(F^*, \circ)$ is an abelian group under multiplication, with identity element 1.

(iii) There is a distributive law:

$$a \circ (b + c) = (a \circ b) + (a \circ c), \text{ for } a, b, c \in F.$$

Note that $(b+c) \circ a = (b \circ a) + (c \circ a)$, by the commutativity of multiplication. Note also that $1 \in F^*$, so that $0 \neq 1$, and that $a \circ 0 = a \circ (0+0) = a \circ 0 + a \circ 0$, so that $a \circ 0 = 0$; Similarly $0 \circ a = 0$. We denote the additive inverse of a by $-a$, and the multiplicative inverse (if $a \neq 0$) by a^{-1}.

As an example, let $\mathbf{Z}_2$ consist of two elements 0 and 1. With the following laws of addition and multiplication

$$0+0 = 1+1 = 0; \quad 0+1 = 1+0 = 1; \quad 0 \circ 0 = 0 \circ 1 = 1 \circ 0 = 0; \quad 1 \circ 1 = 1,$$

$\mathbf{Z}_2$ becomes a field.

Proposition 2.7.1 *Suppose that F is a field and that $\phi : (\mathbf{Z}, +) \to (F, +)$ is the homomorphism of Proposition 2.5.2. Then $\phi(mn) = \phi(m)\phi(n)$ for $m, n \in \mathbf{Z}$.*

Proof Suppose that $n \in \mathbf{Z}$. If $m \in \mathbf{Z}$, let $\psi_n(m) = \phi(mn) - \phi(m)\phi(n)$. If $m_1, m_2 \in \mathbf{Z}$ then

$$\phi((m_1 + m_2)n) = \phi(m_1 n + m_2 n) = \phi(m_1 n) + \phi(m_2 n), \text{ and}$$

$$\phi(m_1 + m_2)\phi(n) = (\phi(m_1) + \phi(m_2))\phi(n) = \phi(m_1)\phi(n) + \phi(m_2)\phi(n),$$

so that $\psi_n(m_1 + m_2) = \psi_n(m_1) + \psi_n(m_2)$. Thus ψ_n is homomorphism of $(\mathbf{Z}, +)$ into $(F, +)$. But $\psi_n(1) = \phi(n) - n\phi(1) = 0$, and so $\psi_n(\mathbf{Z}^+) = \{0\}$. Thus $\phi(mn) = \phi(m)\phi(n)$. $\square$

A subset H of a field F is a *subfield* of F if H is a subgroup of the additive group $(F, +)$ and $H \cap F^*$ is a subgroup of the multiplicative group F^*. It then inherits the field structure from F.

A mapping θ from a field F to a field G is a *field homomorphism* if

- it is a homomorphism of the additive group $(F, +)$ into $(G, +)$, and
- $\theta(F^*) \subseteq G^*$ and $\theta_{|F^*}$ is a homomorphism of the multiplicative group $(F^*, \circ)$ into $(G^*, \circ)$.

In particular, if θ is a field homomorphism then $\theta(0_F) = 0_G$ and $\theta(1_F) = 1_G$.

Suppose that $\theta : F \to G$ is a field homomorphism, and that f and f' are distinct elements of F. Let $h = f - f'$. Then $h \neq 0_F$, and $\theta(h)\theta(h^{-1}) = 1_G$, Thus $\theta(f) - \theta(f') = \theta(h) \neq 0_G$, so that $\theta(f) \neq \theta(f')$. Consequently, θ is injective.

A surjective field homomorphism is called a *field isomorphism*.

Suppose that F is a field. A *polynomial over F of degree n* is an expression of the form $p(x) = a_n x^n + a_{n-1}x^{n-1} + \cdots a_1 x + a_0$, where the coefficients a_j are in F and $a_n \neq 0$. It is *monic* if $a_n = 1$. The polynomial p defines a *polynomial function* $p : F \to F$ defined by setting $p(r) = a_n r^n + a_{n-1}r^{n-1} + \cdots a_1 r + a_0$. An element r of F is a *root* of p if $p(r) = 0$.

We shall embed $\mathbf{Z}$ in a field $\mathbf{Q}$. We are all familiar with the notion of a fraction, and of the fact that different fractions, such as $2/3$ and $4/6$, represent the same number. Let us formalize this. Let $\mathbf{Z}^* = \mathbf{Z} \setminus \{0\}$ be the set of non-zero integers. We define a relation on $\mathbf{Z} \times \mathbf{Z}^*$ by setting $(p, q) \sim (r, s)$ if $ps = qr$.

Proposition 2.7.2 *The relation $(p, r) \sim (q, s)$ is an equivalence relation on $\mathbf{Z} \times \mathbf{Z}^*$.*

Proof It follows immediately from the definition that $(p, q) \sim (p, q)$ and that if $(p, q) \sim (r, s)$ then $(r, s) \sim (p, q)$. Suppose that $(p, q) \sim (r, s)$ and

$(r, s) \sim (t, u)$, so that $ps = qr$ and $ru = ts$. Thus

$$pusr = (ps)(ru) = (qr)(ts) = qtsr,$$

so that $(pu - qt)sr = 0$. Since $sr \neq 0$, $pu = qt$, and $(p, q) \sim (t, u)$. $\qquad\square$

We denote the set of equivalence classes by $\mathbf{Q}$, and denote the equivalence class $[(p, q)]$ by p/q, or $\frac{p}{q}$. The elements of $\mathbf{Q}$ are called *rational numbers*. If $r = p/q \in \mathbf{Q}$, we call p/q a *fraction*, *representing* r. Many different fractions represent r; for example, $2/3$ and $4/6$ represent the same element of $\mathbf{Q}$. It follows immediately from the definition of the equivalence relation on $\mathbf{Z} \times \mathbf{Z}^*$ that $j/k = j'/k'$ if and only if $jk' = j'k$. In particular, $j/k = (-j)/(-k)$, so that we can represent r as j/n, where $j \in \mathbf{Z}^*$ and $n \in \mathbf{N}$.

Let us consider the structure of the equivalence classes further.

Proposition 2.7.3 *(i) Suppose that $(m, n) \in \mathbf{N} \times \mathbf{N}$. Then there exists a unique $(m', n') \in [(m, n)]$ with m' and n' coprime. Then*

$$[(m, n)] = \{a \in \mathbf{N} \times \mathbf{N} : a = (km', kn') \text{ for some } k \in \mathbf{N}\}.$$

(ii) Suppose that $(-m, n) \in \mathbf{N} \times \mathbf{N}$. Then there exists a unique $(-m', n') \in [(-m, n)]$ with m' and n' coprime. Then

$$[(-m, n)] = \{a \in -\mathbf{N} \times \mathbf{N} : a = (-km', kn') \text{ for some } k \in \mathbf{N}\}.$$

Proof (i) Let h be the highest common factor of m and n, and let $m' = m/h$, $n' = n/h$. Then m' and n' are coprime, and $mn' = hm'n' = m'n$, so that $(m, n) \sim (m', n')$. If $(m'', n'') \in [(m, n)]$ then $m''n' = m'n''$, so that $m'|m''$, by Proposition 2.6.4. Let $m'' = km'$; then $km'n' = m''n' = m'n''$; dividing by m', we see that $n'' = kn'$. Thus $(m'', n'') = (km', kn')$. From this it follows that (m', n') is the only element of $[(m, n)]$ with m' and n' coprime.

The proof of (ii) is essentially the same as the proof of (i). $\qquad\square$

In other words, if $r \in \mathbf{Q}^*$, we can write r uniquely as $r = m/n$ or $r = (-m)/n$, with m and n coprime. In this case, we say that the fraction m/n is in *lowest terms*. As an example, a *dyadic number* or *dyadic rational number* is a rational number of the form $m/2^k$, where $m \in \mathbf{Z}$ and $k \in \mathbf{Z}^+$. If $k > 1$ then it is in lowest terms if and only if m is odd.

We now show how to define addition and multiplication in $\mathbf{Q}$, so that $\mathbf{Q}$ becomes a field. We give the details, though they are very straightforward. First we define addition. We define $p/q + r/s = (ps + qr)/qs$. If $(p, q) \sim (p', q')$

and $(r, s) \sim (r', s')$ then

$$(ps + qr)q's' = (pq')(ss') + (rs')(qq')$$
$$= (p'q)(ss') + (r's)(qq') = (p's' + q'r')qs$$

and so this is well-defined: it does not depend on the choice of representatives.

Proposition 2.7.4 $(\mathbf{Q}, +)$ *is an abelian group.*

Proof This is a matter of straightforward verification. Addition is associative, since

$$\left(\frac{p}{q} + \frac{r}{s}\right) + \frac{t}{u} = \frac{ps + qr}{qs} + \frac{t}{u} = \frac{psu + qru + qst}{qsu}$$

$$= \frac{p}{q} + \frac{ru + ts}{su} = \frac{p}{q} + \left(\frac{r}{s} + \frac{t}{u}\right),$$

and clearly $p/q + r/s = r/s + p/q$. The element $0/1$ is the identity, since $0/1 + p/q = p/q + 0/1 = p/q$ for all $(p, q) \in \mathbf{Z} \times \mathbf{Z}^*$. Similarly,

$$\frac{p}{q} + \frac{-p}{q} = \frac{pq - pq}{q^2} = \frac{0}{q^2} = \frac{0}{1},$$

so that $(-p)/q$ is the additive inverse of p/q. $\square$

Next we define multiplication. We define $(p/q)(r/s) = (pr)/(qs)$; once again, as the reader should verify, this does not depend on the choice of representatives. Let $\mathbf{Q}^* = \mathbf{Q} \setminus \{0/1\}$ be the set of non-zero rational numbers.

Proposition 2.7.5 $(\mathbf{Q}^*, .)$ *is an abelian group, with identity element* $1/1$. *The inverse of* p/q *is* q/p.

Proof The details are left as an easy exercise for the reader. $\square$

Theorem 2.7.6 $(\mathbf{Q}, +, .)$ *is a field.*

Proof It remains to prove the distributive law:

$$\frac{p}{q}\left(\frac{r}{s} + \frac{t}{u}\right) = \frac{p}{q}\left(\frac{ru + ts}{su}\right) = \frac{pru + pts}{qsu}$$

$$= \frac{pru}{qsu} + \frac{pts}{qsu} = \frac{pr}{qs} + \frac{pt}{qu} = \left(\frac{p}{q}\right)\left(\frac{r}{s}\right) + \left(\frac{p}{q}\right)\left(\frac{t}{u}\right).   \square$$

We now embed $\mathbf{Z}$ into $\mathbf{Q}$. If $n \in \mathbf{Z}$, let $\phi(n) = n/1$. It then follows immediately from the definitions that ϕ is an injective homomorphism of the additive group $(\mathbf{Z}, +)$ into the additive group $(\mathbf{Q}, +)$, and that $\phi(mn) = \phi(m)\phi(n)$ for $m, n \in \mathbf{Z}$. Summing up:

Theorem 2.7.7 *With addition and multiplication defined as above, $\mathbf{Q}$ is a field. $(\mathbf{Q}, +)$ has identity element $0/1$, and the multiplicative identity is $1 = 1/1$. The additive inverse of j/n is $(-j)/n$; and, if $m \in \mathbf{N}$, the multiplicative inverse of m/n is n/m and the multiplicative inverse of $(-m)/n$ is $(-n)/m$. There is an injective map $\phi : \mathbf{Z} \to \mathbf{Q}$ such that $\phi(0) = 0$, $\phi(1) = 1$, and $\phi(j + k) = \phi(j) + \phi(k)$, $\phi(jk) = \phi(j)\phi(k)$ for all $j, k \in \mathbf{Z}$.*

We identify $\mathbf{Z}$ with $\phi(\mathbf{Z})$, and consider $\mathbf{Z}$ as a subset of the field $\mathbf{Q}$. Thus we write n for $n/1$, so that 0 is the zero element of $\mathbf{Q}$, and 1 is the multiplicative inverse.

Exercises

2.7.1 Show that there is a field with four elements, and that there is no field with six elements.

2.7.2 Prove the *binomial theorem*: if F is a field, if $x, y \in F$ and if $n \in \mathbf{N}$ then

$$(x+y)^n = x^n + \binom{n}{1}x^{n-1}y + \cdots + \binom{n}{j}x^{n-j}y^j + \cdots + \binom{n}{n-1}xy^{n-1} + y^n.$$

2.7.3 Suppose that $r = m/n$ is a rational number in lowest terms, and that $0 < r < 1$. Show that there exists $k \in \mathbf{N}$ such that $1/(k+1) \le r < 1/k$. Show that if $r \ne 1/(k+1)$ and $r - 1/(k+1) = p/q$ in lowest terms, then $p < m$. Deduce that there exist $1 < n_1 < \cdots < n_t$ such that $r = 1/n_1 + \cdots + 1/n_t$.

2.7.4 We have adjoined additive inverses to $\mathbf{Z}^+$ to construct $\mathbf{Z}$, and we have adjoined multiplicative inverses to $\mathbf{Z}^*$ to construct $\mathbf{Q}$. These are special cases of a general construction to adjoin inverses. We need some definitions. A *monoid* is a set S with a binary associative operation $\circ : S \times S \to S$, together with an element e of S (the *identity element*) for which $s \circ e = e \circ s = e$, for all $s \in S$. S is *commutative*, or *abelian*, if $s \circ t = t \circ s$ for all $s, t \in S$. S has a *cancellation law* if whenever $s \circ u = t \circ u$ then $s = t$, and whenever $u \circ s = u \circ t$ then $s = t$. Suppose that S is a commutative monoid with a cancellation law.

(a) Define a relation on $S \times S$ by setting $(p, q) \sim (r, s)$ if $p \circ s = r \circ q$. Show that this is an equivalence relation on S. Let G be the set of equivalence classes.

(b) Suppose that $g = [(p, q)]$, $h = [(r, s)]$. Let $g + h = [(p \circ s, q \circ t)]$. Show that this is well-defined – it does not depend on the choice of representatives.

(c) Show that addition is associative and commutative.

(d) Show that $(G, +)$ is an abelian group, with identity $[(e, e)]$ and with $-[(p, q)] = [(q, p)]$.

(e) Let $\theta : S \to G$ be defined by $\theta(s) = [(s, e)]$. Show that θ is injective and that $\theta(s \circ t) = \theta(s) + \theta(t)$.

(f) Show that $G = \theta(S) - \theta(S)$.

2.7.5 Use the results of the previous question to provide another construction of $(\mathbf{Z}, +)$ from $\mathbf{Z}^+$.

2.7.6 There are circumstances (as in the construction of $\mathbf{Q}$), where in Exercise 2.7.4 it is natural to denote the composition in G multiplicatively. Do this, when $S = \mathbf{Z}^*[x]$ is the set of non-zero polynomials with integer coefficients, and where composition is the multiplication of polynomials:

$$\text{if } p = \sum_{i=0}^{m} a_i x^i \text{ and } q = \sum_{j=0}^{n} a_i x^i \text{ then } p \circ q = \sum_{k=0}^{m+n} c_k x^k,$$

where $c_k = \sum \{a_i b_j : i \geq 0, j \geq 0, i + j = k, \}$. What have you constructed?

2.8 Ordered fields

We introduce an order on $\mathbf{Q}$. We set $j/m \leq k/n$ if $jn \leq km$.

Proposition 2.8.1 (i) *The relation* $\leq$ *is a well-defined total order on* $\mathbf{Q}$.
   (ii) *If* $r \leq s$, *then* $r + t \leq s + t$ *for all* $t \in \mathbf{Q}$.
   (iii) *If* $r \leq s$, *then* $rt \leq st$ *for all* $t \in \mathbf{Q}$ *with* $t \geq 0$.
   (iv) *If* $m, n \in \mathbf{Z}$ *then* $m \leq n$ *in the order on* $\mathbf{Z}$ *if and only if* $m \leq n$ *in the order on* $\mathbf{Q}$.

Proof The straightforward verifications are left as an exercise for the reader. (Remember that m and n are positive.)   □

   A field with a total order that satisfies conditions (ii) and (iii) of Proposition 2.8.1 is called an *ordered field*. Note that if F is an ordered field, and $f \in F$, then $f^2 \geq 0$. For if $f \geq 0$, then $f^2 \geq 0$, and if $f < 0$ then $-f > 0$, so that $f^2 = (-f)^2 \geq 0$. In particular, $1 = 1^2 > 0$. If F is an ordered field, and $f \in F$, we say that f is *positive* if $f > 0$; we say that f is *non-negative* if

$f \geq 0$; we say that f is *negative* if $f < 0$, and we say that f is *non-positive* if $f \leq 0$.

An ordered field contains a copy of $\mathbf{Q}$ as a subfield. We prove this in two steps.

Proposition 2.8.2 *If F is an ordered field, there exists a unique injective map $\psi : \mathbf{Z} \to F$ such that $\psi(0) = 0_F$, $\psi(1) = 1_F$, $\psi(k+l) = \psi(k) + \psi(l)$. Further, $\psi(kl) = \psi(k)\psi(l)$ for all $k, l \in \mathbf{Z}$, and $\psi(k) \leq \psi(l)$ if $k \leq l$.*

Proof By Proposition 2.5.2, there exists a unique map $\psi : \mathbf{Z} \to F$ such that $\psi(0) = 0_F$, $\psi(1) = 1_F$ and $\psi(k+l) = \psi(k) + \psi(l)$, for $k, l \in \mathbf{Z}$.

A straightforward induction then shows that $\psi(ml) = \psi(m)\psi(l)$ for $m \in \mathbf{Z}^+$, $l \in \mathbf{Z}$. Since $\psi((-m)l) + \psi(ml) = \psi((-m)l + ml) = \psi(0) = 0$, $\psi((-m)l) = -\psi(ml) = -(\psi(m)\psi(l))$. Since $(\psi(m) + (-\psi(m)))\psi(l) = 0$, $-(\psi(m)\psi(l)) = (-\psi(m))\psi(l) = \psi(-m)\psi(l)$. Thus $\psi(-m)l) = \psi(-m)\psi(l)$, and $\psi(kl) = \psi(k)\psi(l)$ for all $k, l \in \mathbf{Z}$.

We show by induction that if $m \in \mathbf{N}$ then $\psi(m) > 0_F$. The result is true if $m = 1$, by the preceding remark. If it is true for m, then $\psi(m+1) = \psi(m) + \psi(1) = \psi(m) + 1_F > \psi(m) > 0_F$. Thus if $k \leq l$ then $l - k \in \mathbf{Z}^+$ and $\psi(l) - \psi(k) = \psi(l - k) \geq 0$: $\psi(k) \leq \psi(l)$. Further, ψ is injective, for if $k \neq l$ and $k < l$ then $\psi(l) - \psi(k) = \psi(l - k) > 0$, so that $\psi(l) \neq \psi(k)$: similarly, if $k > l$. $\square$

Theorem 2.8.3 *Suppose that F is an ordered field. Then there exists a unique injective field homomorphism $k : \mathbf{Q} \to F$. Further, k is order-preserving: if $r \leq s$ then $k(r) \leq k(s)$.*

Proof Let $\psi : \mathbf{Z} \to F$ be the unique mapping of the previous proposition. If $j \in \mathbf{Z}$, we define $k(j) = \psi(j)$, and if $r = j/n \in \mathbf{Q}$, we define $k(r) = \psi(j)(\psi(n))^{-1}$. Now $\psi(j)(\psi(n))^{-1} = \psi(j')(\psi(n'))^{-1}$ if and only if $\psi(jn') = \psi(j)\psi(n') = \psi(j')\psi(n) = \psi(j'n)$, and this happens if and only if $j/n = j'/n'$. Thus k is well defined, and is injective. It is a straightforward matter to verify that k satisfies the other requirements of the theorem. $\square$

This shows that every ordered field has a subfield isomorphic to $\mathbf{Q}$. $\mathbf{Q}$ itself has no proper subfield. For every subfield must contain 0 and 1, and so must contain (a copy of) $\mathbf{Z}$. Thus it must contain all elements of the form j/n, with $j \in \mathbf{Z}$ and $n \in \mathbf{N}$. Thus we have the following characterization of the rational numbers.

Corollary 2.8.4 *An ordered field F is isomorphic as a field to $\mathbf{Q}$ if and only if it has no proper subfields.*

We can therefore take any ordered field with no proper subfields as a model for the field $\mathbf{Q}$ of rational numbers.

Exercises

2.8.1 Suppose that A is a countable totally ordered subset with the intermediate property (if $a < b$ then there exists c with $a < c < b$) with no greatest or least element. Show that there is an order preserving bijection $j : A \to \mathbf{Q}$.

2.8.2 Give the details of the proof of Proposition 2.8.1.

2.8.3 (a) Suppose that a, b, v are elements of an ordered field F and that $a > v > b > 0$. Show that $ab < v(a + b - v)$.

(b) Suppose that $a_1, \ldots, a_k$ are positive elements of an ordered field. Let $v = (a_1 + \cdots + a_k)/k$. Use (a) and an inductive argument to show that $v^k \geq a_1 a_2 \ldots a_k$.

2.9 Dedekind cuts

The field of rational numbers is not adequate for our purpose. The ancient Greeks recognized the inadequacy of the rational numbers: the length of a diagonal of a square is not a rational multiple of the length of a side.

Proposition 2.9.1   *There is no rational number r with $r^2 = 2$.*

Proof Suppose that such an r exists; we can suppose that r is positive, and that $r = m/n$ in lowest terms. Then $m^2 = 2n^2$. Since 2 is prime, 2 divides m, and so $m = 2q$ for some $q \in \mathbf{N}$. Then $4q^2 = 2n^2$, and so $2q^2 = n^2$. This implies that 2 divides n, contradicting the fact that m and n are coprime. □

This result can be extended greatly.

Theorem 2.9.2   *If p is a monic polynomial with integer coefficients, then any $r \in \mathbf{Q}$ which is a root of p must be an integer.*

Proof If $r \neq 0$, let $r = j/q$, in lowest terms. Then

$$0 = q^{n-1}p(r) = \frac{j^n}{q} + [a_{n-1}j^{n-1} + a_{n-2}qj^{n-2} + \cdots + a_1 q^{n-2}j + a_0 q^{n-1}].$$

The term in square brackets is an integer, and so therefore is j^n/q. Since j and q are coprime, $q = 1$ and $r = j$, an integer. □

This result is due to Gauss.

Corollary 2.9.3 *If $a \in \mathbf{N}$ and $n \in \mathbf{N}$ then the polynomial $x^n - a$ has a rational root if and only if there exists $b \in \mathbf{N}$ such that $a = b^n$.*

Can we find a number system in which 2 has a square root, which avoids all other such anomalies, and which provides 'a purely arithmetic and perfectly rigorous foundation for the principles of infinitesimal analysis'? Richard Dedekind, whose phrase this is, was the first person to give a satisfactory answer. He found a solution to the problem on 24 November 1858, but did not publish his findings until 1872. His essential insight was that a number such as $\sqrt{2}$ or π could be characterized by the set of rational numbers greater than it, and by the set of rational numbers less than it. There are other ways of proceeding (see Exercise 3.4.3), but in many respects, Dedekind's approach remains the best way of defining the real numbers, and it is essentially the way that we shall follow. As we have seen, the rational numbers $\mathbf{Q}$ form an ordered field: the order relation and the algebra structure interact. In this section, following Dedekind, we use the order structure of $\mathbf{Q}$ to define the set of real numbers $\mathbf{R}$ as a totally ordered set. In the next section, we shall extend the algebraic operations of addition and multiplication from $\mathbf{Q}$ to $\mathbf{R}$.

Suppose that $(X, \leq)$ is a totally ordered set. A non-empty subset A of X is *bounded above* if it has an upper bound in X, is *bounded below* if it has an lower bound in X, and is *bounded* if it is bounded above and below. The totally ordered set $(X, \leq)$ is said to have the *supremum property* or *least upper bound property* if whenever A is a non-empty subset of X which has an upper bound then A has a supremum: there exists $\sup A \in X$ such that $\sup A$ is an upper bound for A and if b is any other upper bound, then $\sup A \leq b$. It is most important that $\sup A$ may or may not be an element of A. We shall require $\mathbf{R}$ to have the supremum property. This fundamental order property is the basis of almost all the analysis that we shall do.

Proposition 2.9.4 *A totally ordered set $(X, \leq)$ has the supremum property if and only if every non-empty subset B of X which has a lower bound has an infimum.*

Proof Suppose first that $(X, \leq)$ has the supremum property, and that B is a non-empty subset of X which is bounded below. Let L be the set of lower bounds of B. L is non-empty, and any element of B is an upper bound for L. Thus L has a supremum, s, say. We shall show that s is the infimum of B. If $b \in B$ then b is an upper bound for U, and so $b \geq s$. Thus s is a lower bound for B. If c is a lower bound for B, then $c \in L$, and so $c \leq s$; thus s is the greatest lower bound of B.

Conversely, suppose that the condition is satisfied and that A is a non-empty subset of X which has an upper bound. Then an exactly similar argument shows that the set U of upper bounds of A has an infimum t, and that t is the supremum of A □

We now show that we can embed the ordered field $\mathbf{Q}$ of rational numbers in an order-preserving way in a totally ordered set with the supremum property. This is the key construction.

Theorem 2.9.5   *There exists a totally ordered set $(\mathbf{R}, \leq)$ with the supremum property, together with an injective order-preserving map: $j : \mathbf{Q} \to \mathbf{R}$ such that*

(a) if $a, b \in \mathbf{R}$ and $a < b$ then there exists $s \in \mathbf{Q}$ such that $a < j(s) < b$, and

(b) $\mathbf{R}$ has neither a greatest element nor a least element.

Proof   We call a subset a of $\mathbf{Q}$ a *Dedekind cut* if it satisfies

(α) a is non-empty and bounded above,

(β) if $r \in a$ and $s < r$ then $s \in a$, and

(γ) a does not have a greatest element (if $r \in a$ there exists $t \in a$ with $t > r$).

(Dedekind, who considered the pair $\{a, \mathbf{Q} \setminus a\}$, used the word 'Schnitt', which can also be translated as 'section', 'slice', or 'intersection'.) As we shall see, conditions (α) and (β) say that a is a semi-infinite interval, and condition (γ) says that a is open.

Let $\mathbf{R}$ be the set of Dedekind cuts. We define an order on $\mathbf{R}$ by setting $a \leq b$ if $a \subseteq b$.

First, we show that this is a total order on $\mathbf{R}$. Suppose that $a, b \in \mathbf{R}$ and that b is not less than or equal to a. Thus b is not contained in a, so that there exists r in $b \setminus a$. If $s \geq r$, then $s \notin a$, since otherwise $r \in a$, by (β). Thus if $t \in a$, $t < r$, and so $t \in b$. Hence $a < b$.

Next, we show that $(\mathbf{R}, \leq)$ has the supremum property. Although this is the essential property of $\mathbf{R}$, the proof is quite straightforward. Suppose that A is a non-empty subset of $\mathbf{R}$ with upper bound u. Let us set $u_0 = \cup_{a \in A} a$.

First, we show that u_0 is a Dedekind cut. u_0 is non-empty, and $u_0 \leq u$, and so condition (α) is satisfied. Suppose that $r \in u_0$ and that $s < r$. Then $r \in a$ for some $a \in A$, and $s \in a$, by (β), so that $s \in u_0$. Thus condition (β) is satisfied. Further, there exists $t \in a$ with $t > r$, and then $t \in u_0$. Thus condition (γ) is also satisfied. Hence u_0 is a Dedekind cut.

Next, we show that u_0 is the supremum of A. If $a \in A$, then $a \subseteq u_0$, so that $u_0 \geq a$: u_0 is an upper bound for A. If d is an upper bound for A then

$d \geq a$ for all $a \in A$, so that $a \subseteq d$ for all $a \in A$, and so $u_0 = \cup_{a \in A} a \subseteq d$. Thus $d \geq u_0$; u_0 is the least upper bound of A.

Next, we define the mapping $j : \mathbf{Q} \to \mathbf{R}$. If $r \in \mathbf{Q}$, we set $j(r) = \{s \in \mathbf{Q} : s < r\}$. Let us show that $j(r)$ is a Dedekind cut. Since $\mathbf{Q}$ has no least element, $j(r) \neq \emptyset$, and r is an upper bound for $j(r)$, so that condition (α) is satisfied. Condition (β) is clearly satisfied. If $s \in j(r)$, let $t = (s + r)/2$. Then $s < t < r$, so that $t \in j(r)$. Thus condition (γ) is satisfied, and $j(r)$ is a Dedekind cut.

The mapping $j : \mathbf{Q} \to \mathbf{R}$ is clearly an order-preserving mapping from $\mathbf{Q}$ to $\mathbf{R}$, and j is injective, for if $r < s$ then $r \in j(s) \setminus j(r)$, so that $j(r) \neq j(s)$.

Let us now show that (a) holds. Suppose that $a, b \in \mathbf{R}$ and that $a < b$. Then there exists $r \in b \setminus a$. By condition (γ), there exists $s > r$ with $s \in b$. Then $s \in b \setminus j(s)$ and $r \in j(s) \setminus a$, so that $a < j(s) < b$.

Corollary 2.9.6 If $a \in \mathbf{R}$ then $a = \sup\{j(r) \in j(\mathbf{Q}) : j(r) < a\}$. In particular, if $t \in \mathbf{Q}$ then $j(t) = \sup\{j(r) : r < t\}$.

Now let us prove (b). Suppose that $a \in \mathbf{R}$. If $r \in a$, then $j(r) < a$, so that a is not a least element of $\mathbf{R}$. If s is an upper bound in $\mathbf{Q}$ for a, there exists $t \in \mathbf{Q}$ with $t > s$. Then $s \in j(t) \setminus a$, so that $a < j(t)$ and a is not a greatest element of $\mathbf{R}$. $\square$

We define the *real numbers* to be the pair $(\mathbf{R}, j)$. We shall usually identify $\mathbf{Q}$ with $j(\mathbf{Q})$. Thus $\mathbf{R}$ is a totally ordered set with the supremum property, with neither a greatest element nor a least element, which contains $\mathbf{Q}$ in an order-preserving way, and which has the property that if $a, b \in \mathbf{R}$ and $a < b$ then there exists $r \in \mathbf{Q}$ such that $a < r < b$. We shall deduce all the properties of $\mathbf{R}$ from this.

Exercises

2.9.1 Suppose that $n \in \mathbf{N}$ and that $(p/q)^2 = n$. Show that if $p - rq \neq 0$ then $((nq - rp)/(p - rq))^2 = n$. Use this to give another proof that if n has a rational square root then the square root is an integer.

2.9.2 Suppose that p is a polynomial of degree n with coefficients in a field F. Show that if c is a root of p in F, then $p(x) = (x - c)q(x)$, where q is a polynomial of degree $n - 1$. Show that p has at most n roots in F.

2.9.3 Show that there is an order-preserving bijection j of $\mathbf{Q}$ onto $\mathbf{Q} \setminus \{0\}$. [*Hint*: use the intermediate property, and define j recursively, using an enumeration of $\mathbf{Q}$.]

2.10 The real number field

Let $\mathbf{R}$ be the set of real numbers, and let $j : \mathbf{Q} \to \mathbf{R}$ be the inclusion mapping. So far, we have only established the order properties of $\mathbf{R}$. We now define the algebraic properties of $\mathbf{R}$. First, we define the addition of real numbers in such a way that $(\mathbf{R}, +)$ is an ordered abelian group and j is a group homomorphism.

If $x \in \mathbf{R}$ let $D(x) = \{r \in \mathbf{Q} : r < x\}$. By Corollary 2.9.6, $D(x)$ is a Dedekind cut.

Proposition 2.10.1   *Suppose that $x, y \in \mathbf{R}$. Then*

$$D(x)+D(y) = \{r+s : r \in D(x), s \in D(y)\} = \{r+s : r, s \in \mathbf{Q} : r < x, s < y\}$$

is a Dedekind cut.

Proof   Let us check the conditions.

(α) The set $D(x) + D(y)$ is not empty, and if r_x is an upper bound for $D(x)$ and r_y is an upper bound for $D(y)$ then $r_x + r_y$ is an upper bound for $D(x) + D(y)$.

(β) Suppose that $r \in D(x)$, that $s \in D(y)$ and that $t < r+s$. Let $u = t-s$, Then $u < r$, so that $u \in D(x)$. Hence $t = u + s \in D(x) + D(y)$.

(γ) Suppose that $w = r + s \in D(x) + D(y)$, with $r \in D(x)$, $s \in D(y)$. Then there exists $r' \in D(x)$ with $r < r'$. Then $w' = r' + s \in D(x) + D(y)$ and $w < w'$. $\square$

We define the real number $x + y$ to be the Dedekind cut $D(x) + D(y)$. Then $D(x + y) = D(x) + D(y)$.

Corollary 2.10.2   *If $x, y \in \mathbf{R}$ and $t \in \mathbf{Q}$, and if $j(t) < x + y$ then there exist $r, s \in \mathbf{Q}$ such that $j(r) < x$, $j(s) < y$ and $t = r + s$.*

Proof   For $t \in D(x + y) = D(x) + D(y)$. $\square$

Proposition 2.10.3   *Suppose that $x, y \in \mathbf{R}$ and that $r, s \in \mathbf{Q}$.*
  (i) $x + y = y + x$.
  (ii) $(x + y) + z = x + (y + z)$.
  (iii) $x + j(0) = x$.
  (iv) If $x \leq y$ then $x + z \leq y + z$.
  (v) $j(r) + j(s) = j(r + s)$.

Proof   (i)–(iv) are easy consequences of the definition, left as exercises for the reader.

(v) We must show that $D(j(r))+D(j(s)) = D(j(r+s))$. If $t \in D(j(r))$ and $u \in D(j(s))$, then $t < r$ and $u < s$, so that $t+u < r+s$ and $t+u \in D(j(r+s))$.

Thus $D(j(r)) + D(j(s)) \subseteq D(j(r+s))$. Conversely, if $t \in D(j(r+s))$ then $t < r+s$. As in Proposition 2.10.1, there exists $u \in \mathbf{Q}$ with $u < r$ such that $t = u+s$. Thus $t \in D(j(r))+D(j(s))$, so that $D(j(r+s)) \subseteq D(j(r))+D(j(s))$. Hence $D(j(r)) + D(j(s)) = D(j(r+s))$. □

We now need to define $-x$, for $x \in \mathbf{R}$, in such a way that $x + (-x) = 0$. If $-x$ exists, then

$$D(-x) = \{r \in \mathbf{Q} : j(r) < -x\} = \{r \in \mathbf{Q} : x < j(-r)\}.$$

We therefore define $M(x) = \{r \in \mathbf{Q} : x < j(-r)\}$.

Proposition 2.10.4 *If $x \in \mathbf{R}$ then $M(x)$ is a Dedekind cut.*

Proof Again, we check the conditions.

(α) There exists $s \in \mathbf{Q}$ such that $j(s) > x$. Let $r = -s$. Then $x < j(-r)$, so that $r \in M(x)$, and $M(x)$ is not empty. Similarly there exists $u \in \mathbf{Q}$ such that $j(u) < x$. Let $t = -u$. If $r \in M(x)$ then $j(-t) = j(u) < x < j(-r)$, so that $r < t$; t is an upper bound for $M(x)$.

(β) Suppose that $u \in M(x)$. If $t \in \mathbf{Q}$ and $t < u$ then $x < j(-u) < j(-t)$, so that $t \in M(x)$.

(γ) Suppose that $u \in M(x)$. There exists $s \in \mathbf{Q}$ such that $x < j(s) < -j(u)$, so that if $t = -s$ then $x < j(-t) < j(-u)$. Thus $t \in M(x)$ and $u < t$. □

We now define the real number $-x$ to be $M(x)$. Thus $M(x) = D(-x)$.

Theorem 2.10.5 *If $x \in \mathbf{R}$ then $x + (-x) = j(0)$.*

Proof First we show that $x + (-x) \leq j(0)$. If $r \in M(x)$ and $s \in D(x)$ then $j(s) < x < j(-r)$, so that $r + s < 0$, and $r + s \in D(j(0))$. Consequently, $x + (-x) \leq j(0)$.

Secondly, we show that $x+(-x) \geq j(0)$. Suppose that $t \in D(j(0))$, so that $t \in \mathbf{Q}$ and $t < 0$. There exists $r \in \mathbf{Q}$ such that $x + j(t) < j(r) < x$. Thus $x < j(r - t) = j(-(t - r))$, so that $t - r \in M(x)$. Since $j(r) < x$, $r \in D(x)$. Thus $t \in D(x) + M(x) = D(x) + D(-x) = D(x + (-x))$. Consequently $D(j(0)) \subseteq D(x + (-x))$, and so $x + (-x) \geq j(0)$. □

Thus $\mathbf{R}$ is an ordered abelian group under addition, with identity element $j(0)$, and the map $r \rightarrow j(r)$ is an order-preserving injective group homomorphism of $\mathbf{Q}$ into $\mathbf{R}$.

We now turn to multiplication. Here it is easiest first to define the product of two non-negative elements of $\mathbf{R}$, and then extend to the whole of $\mathbf{R}$, just as we did for **Z**. As we shall see, the programme is very similar to the programme for defining addition, and we shall therefore omit many of the details. If $x \in \mathbf{R}$ and $x > 0$, then the Dedekind cut $D(x)$ contains all the negative rational numbers, and we wish to avoid negative numbers. We therefore define a *positive Dedekind cut* to be a non-empty subset a^+ of $\{r \in \mathbf{Q} : r > 0\}$ which is bounded above, does not have a greatest element, and has the property that whenever $r \in a^+$, $s \in \mathbf{Q}$ and $0 < s < r$ then $s \in a^+$. If a^+ is a positive Dedekind cut, then $a = a^+ \cup \{r \in \mathbf{Q} : r \leq 0\}$ is a Dedekind cut.

If $x \in \mathbf{R}$ and $x > 0$, let $D^+(x) = \{r \in \mathbf{Q} : 0 < r < x\}$. Then $D^+(x)$ is a positive Dedekind cut, and $x = \sup(D^+(x))$. If x, y are positive real numbers, we define $D^+(x).D^+(y)$ to be $\{t \in \mathbf{Q} : t = rs \text{ for some } r \in D^+(x), s \in D^+(y)\}$.

Proposition 2.10.6   *Suppose that* x, y *are positive real numbers. Then* $D^+(x).D^+(y)$ *is a positive Dedekind cut.*

Proof   Just like the proof of Proposition 2.10.1.   □

Thus $D = D^+x.D^+y \cup \{r \in \mathbf{Q} : r \leq 0\}$ is a Dedekind cut. We set $xy = x.y = D$. Then $D(xy) = D = Dx.Dy$.

Corollary 2.10.7   *Suppose that* x, y *are positive real numbers, that* $t \in \mathbf{Q}$ *and that* $0 < j(t) < xy$. *Then there exist* $r, s \in \mathbf{Q}$ *such that* $0 < j(r) < x$, $0 < j(s) < y$ *and* $t = rs$.

Proof   For $t \in D^+(xy) = D^+(x)D^+(y)$.   □

Proposition 2.10.8   *Suppose that* x, y, z *are positive real numbers and that* r *and* s *are positive rational numbers.*

(i) $xy = yx$.
(ii) $(xy)z = x(yz)$.
(iii) $j(1).x = x$.
(iv) If $x \leq y$ then $xz \leq yz$.
(v) $x(y + z) = xy + xz$.
(vi) $j(rs) = j(r).j(s)$.

Proof   The proofs of (i)–(iv) follow from the definitions.
    (v) This is also easy, but here are the details. We need to show that

$$D^+(x).(D^+(y) + D^+(z)) = (D^+(x).D^+(y)) + (D^+(x).D^+(z)).$$

Clearly

$$D^+(x).(D^+(y) + D^+(z)) \subseteq (D^+(x).D^+(y)) + (D^+(x).D^+(z)).$$

Suppose that $rs + r't \in (D^+(x).D^+(y)) + (D^+(x).D^+(z))$. Let $r'' = \max(r, r')$. Then $r''(s + t) \in D^+(x).(D^+(y) + D^+(z))$ and $0 < rs + r't \le r''(s + t)$, so that $rs + r't \in D^+(x).(D^+(y) + D^+(z))$. Thus

$$(D^+(x).D^+(y)) + (D^+(x).D^+(z)) \subseteq D^+(x).(D^+(y) + D^+(z)).$$

(vi) Just like the proof of Proposition 2.10.3 (v). □

Suppose that $x \in \mathbf{R}$ and that $x > 0$. We want to show that x has a multiplicative inverse x^{-1}. Following the ideas behind the construction of additive inverses, we get

$$I(x) = \{r \in \mathbf{Q} : x < j(1/r)\} = \{x \in \mathbf{Q} : j(r)x < 1\}.$$

Proposition 2.10.9 *If $x \in \mathbf{R}$ and $x > 0$ then $I(x)$ is a Dedekind cut.*

Proof Just like the proof of Proposition 2.10.4. □

We now define x^{-1} to be $I(x)$. Thus $I(x) = D(x^{-1})$.

Theorem 2.10.10 *If $x \in \mathbf{R}$ then $x.x^{-1} = x^{-1}.x = j(1)$.*

Proof Just like the proof of Theorem 2.10.5. □

Thus $\{x \in \mathbf{R}^+ : x > 0\}$ is an abelian group under multiplication, and x^{-1} is the multiplicative inverse of x; we also write it as $1/x$.

We now extend multiplication to $\mathbf{R}$ and multiplicative inversion to $\mathbf{R}^* = \mathbf{R} \setminus \{0\}$. If $x, y \in \mathbf{R}^+$, we set $(-x)y = x(-y) = -(xy)$ and $(-x)(-y) = xy$, and if $x > 0$, we set $1/(-x) = -(1/x)$.

With these definitions, $\mathbf{R}$ becomes an ordered field with the supremum property. The mapping $j : \mathbf{Q} \to \mathbf{R}$ is an order-preserving field isomorphism of $\mathbf{Q}$ onto a subfield of $\mathbf{R}$, which we shall now identify with $\mathbf{Q}$. The elements of $j(\mathbf{Q})$ are rational numbers; the elements of $\mathbf{R} \setminus j(\mathbf{Q})$ are called *irrational numbers*.

We shall show in Theorem 3.3.1 that any ordered field with the supremum property is isomorphic as an ordered field to the ordered field $\mathbf{R}$ of real numbers, and that the isomorphism is unique.

After all this work, we should verify that we can use the real numbers $\mathbf{R}$ to solve the problem that we raised at the beginning of the previous section. In fact we can say more.

Theorem 2.10.11 *Suppose that y is a positive real number and that $n \in \mathbf{N}$. Then there exists a unique positive real number s such that $s^n = y$.*

We need the following lemma.

Lemma 2.10.12   *(i) Suppose that $0 < \epsilon < 1$ and that $n \in \mathbf{N}$. Then*

$$(1 - n\epsilon) \le (1 - \epsilon)^n < (1 + \epsilon)^n \le 1 + (2^n - 1)\epsilon.$$

(ii) Suppose that a and b are positive real numbers and that $n \in \mathbf{N}$. Then $a^n > b^n$ if and only if $a > b$.

Proof   (i) The proof is by induction on n: the result is true if $n = 1$. Suppose that it is true for n. Then

$$(1 - \epsilon)^{n+1} = (1 - \epsilon)(1 - \epsilon)^n \ge (1 - \epsilon)(1 - n\epsilon)$$
$$= 1 - (n + 1)\epsilon + n\epsilon^2 > 1 - (n + 1)\epsilon$$
$$\text{and } (1 + \epsilon)^{n+1} = (1 + \epsilon)(1 + \epsilon)^n \le (1 + \epsilon)(1 + (2^n - 1)\epsilon)$$
$$= 1 + (2^n - 1)\epsilon + \epsilon + (2^n - 1)\epsilon^2) < 1 + (2^{n+1} - 1)\epsilon.$$

(ii) This follows from the equation

$$a^n - b^n = (a - b)(a^{n-1} + a^{n-2}b + \cdots + ab^{n-2} + b^{n-1}). \qquad \square$$

Proof of Theorem 2.10.11   Let $B = \{x \in \mathbf{R} : x^n \le y\}$. Since $0 \in B$, B is non-empty. If $y \le 1$ then B is bounded above by 1. If $y > 1$ then $y^n > y$, so that if $x \in B$ then $x^n < y^n$ and $x < y$, by the lemma. Thus B is bounded above by y. Therefore B has a supremum s, say. We shall show that $s^n = y$. There are three possibilities; either $s^n < y$ or $s^n > y$ or $s^n = y$. We shall show that the first two of these cannot occur, so that $s^n = y$.

Suppose first that $s^n < y$. Choose $0 < \eta < (y - s^n)/(2^n - 1)y$. Note that $0 < \eta < 1$. By the lemma,

$$((1 + \eta)s)^n \le s^n + (2^n - 1)\eta s^n < s^n + (y - s^n) = y,$$

so that $(1 + \eta)s \in B$. Since $(1 + \eta)s > s$, this contradicts the fact that s is an upper bound for B.

Secondly, suppose that $s^n > y$. Choose $0 < \theta < (s^n - y)/ns^n$. Note that $0 < \theta < 1$. If $x \in B$ then

$$((1 - \theta)s)^n - x^n \ge (1 - n\theta)s^n - y = (s^n - y) - n\theta s^n > 0.$$

By the lemma, $(1 - \theta)s > x$, so that $(1 - \theta)s$ is an upper bound for B, contradicting the fact that s is the least upper bound of B.

Consequently, $s^n = y$. Finally, s is unique. For if $t^n = y$, then $s^n - t^n = 0$, and it follows from part (ii) of the lemma that $s = t$. $\qquad \square$

The number s is denoted by $y^{1/n}$: it is the *nth root* of y.

This proof is all very well, but it is very cumbersome. Surely there is a better proof! There certainly is, but it requires us to do a good deal of analysis first. In due course, we shall see that this result is an easy consequence of the intermediate value theorem (Theorem 6.4.1).

Part Two

Functions of a real variable

3

Convergent sequences

3.1 The real numbers

At the beginning of the nineteenth century, it became clear that mathematical analysis (the study of functions and of series) lacked a satisfactory firm foundation. In 1821, Augustin-Louis Cauchy published his *Cours d'Analyse*, which contained the first rigorous account of mathematical analysis. Cauchy however took the properties of the real numbers for granted. In 1858, when Richard Dedekind was preparing a course of lectures on the elements of the differential calculus at the Polytechnic School in Zürich, he 'felt more keenly than ever the lack of a really scientific foundation for arithmetic', and discovered the construction of the real number system that is described in the Prologue. In fact, he only published his results in 1872.[1] With hindsight, it has become clear that the properties of the real number system lie at the heart of all mathematical analysis, and that it is essential to obtain a full understanding of these properties in order to develop mathematical analysis.

In the Prologue, we have constructed Dedekind's model for the real numbers $\mathbf{R}$ and established some of its properties. It is however sensible to take the construction for granted, to write down the essential properties of $\mathbf{R}$, and to use these properties to develop the theory of mathematical analysis. This we shall do.

What are the essential properties of $\mathbf{R}$? First, $\mathbf{R}$ is a *field*: that is, addition, multiplication, subtraction and division have been defined to satisfy the usual conditions of arithmetic.

Secondly, there is a *total order* on $\mathbf{R}$: if $x, y \in \mathbf{R}$ then either $x \leq y$ or $y \leq x$, and both occur if and only if $x = y$; further if $x \leq y$ and $y \leq z$ then $x \leq z$. The order makes $\mathbf{R}$ an *ordered field*; if $x \leq y$ then $x + z \leq y + z$, and if $x \leq y$

---

[1] See Richard Dedekind, *Essays on the Theory of Numbers*, Dover, 1963.

and $z \geq 0$ then $xz \leq yz$. $\mathbf{R}$ contains a copy of the set of rational numbers $\mathbf{Q}$, and, within it, copies of the integers $\mathbf{Z}$ and the natural numbers $\mathbf{N}$.

The set $\mathbf{Q}$ of rational numbers is also an ordered field, but it is not adequate for analysis. The fundamental property of $\mathbf{R}$ that we shall use over and over again relates to the order structure of $\mathbf{R}$. If A is a non-empty subset of $\mathbf{R}$ and $b \in \mathbf{R}$, then b is an *upper bound* for A if $b \geq a$ for all $a \in A$. Then $\mathbf{R}$ has the *supremum property*: if A is a non empty set of $\mathbf{R}$ with an upper bound, then A has a *least upper bound*, or *supremum* $\sup(A)$: there exists an upper bound $c \in \mathbf{R}$ such that if b is any other upper bound of A then $c \leq b$. It is important that the supremum of a non-empty set may or may not belong to A. For example, if $\mathbf{R}^- = \{x \in \mathbf{R} : x < 0\}$ is the set of negative numbers, then $\sup(\mathbf{R}^-) = 0$, and $0 \notin \mathbf{R}^-$.

The supremum property is equivalent (Proposition 2.9.4) to the requirement that every non-empty subset A of $\mathbf{R}$ which is bounded below has a *greatest lower bound*, or *infimum*. For the set B of lower bounds of A is non-empty and bounded above by any element of A, and so B has a supremum, s, say. We show that s is the infimum of A. Suppose, if possible, that $a \in A$ and that $a < s$. Then a is not the least upper bound for B, and so there exists $b \in B$ with $a < b \leq s$. But then b is not a lower bound for A, giving a contradiction. Thus $a \geq s$, and so $s \in B$. If $c > s$ then c is not a lower bound for A, and so s is the infimum of A.

Here is a first application of the supremum property.

Theorem 3.1.1   *(i) Let $J = \{1/n : n \in \mathbf{N}\}$. Then 0 is the infimum of J.*

(ii) If $x \in \mathbf{R}$ and $x > 0$ then there exists $n \in \mathbf{N}$ with $n \geq x$.

(iii) If $x < y$ then there exists $r \in \mathbf{Q}$ such that $x < r < y$.

Proof   (i) 0 is a lower bound for J, and so $l = \inf J$ exists, and $l \geq 0$. Suppose, if possible, that $l > 0$. Then $2l > l$, and so $2l$ is not a lower bound for J. Thus there exists $n \in \mathbf{N}$ such that $1/n < 2l$. But then $1/2n < l$, giving a contradiction.

(ii) $1/x > 0$, so that $1/x$ is not a lower bound for J. There exists $n \in \mathbf{N}$ such that $1/n < 1/x$. Then $n > x$.

(iii) First, we prove this in the case where $0 \leq x < y$. By (i), there exists $n \in \mathbf{N}$ such that $1/n < y - x$. Let $A = \{k \in \mathbf{Z}^+ : k \leq nx\}$. A is non-empty ($0 \in A$) and finite, by (ii), and so it has a greatest element a. Then $a \leq nx < a + 1$, so that if we set $r = (a+1)/n$, then $x < r$. On the other hand $r = a/n + 1/n < x + (y - x) = y$.

If $x < 0 < y$, we can take $r = 0$; if $x < y \leq 0$, the result follows by considering $-y$ and $-x$. □

Statement (i) says that there are no infinitesimally small members of $\mathbf{R}^+$, and statement (ii), which is known as the *Archimedean property*, says that there are no infinitely large members.

Statement (iii) is an existence statement; when $x > 0$, it is desirable to give an explicit procedure for determining a rational r with $x < r < y$. There exists a least positive integer, q_0, say, such that $1/q_0 < y - x$, and there then exists a least integer, p_0, say, such that $x < p_0/q_0$. Then $x < p_0/q_0 < y$ and $r_0 = p_0/q_0$ is uniquely determined. Let us call r_0 the 'best' rational satisfying $x < r < y$.

Suppose that $x \in \mathbf{R}$. We set

$$\left.\begin{array}{l} x^+ = x \\ x^- = 0 \\ |x| = x^+ = x \end{array}\right\} \text{ if } x \geq 0, \text{ and } \left.\begin{array}{l} x^+ = 0 \\ x^- = -x \\ |x| = x^- = -x \end{array}\right\} \text{ if } x < 0.$$

Then $x = x^+ - x^-$ and $|x| = x^+ + x^- \geq 0$. $|x|$ is the *modulus*, or *absolute value*, of x; it measures the size of x. Note that if one of x, y is positive and the other negative then $|x + y| < |x| + |y|$; otherwise $|x + y| = |x| + |y|$; note also that $|x|.|y| = |xy|$. We set $d(x, y) = |x - y|$; $d(x, y)$ is the *distance* between x and y.

Proposition 3.1.2 *If $x, y, z \in \mathbf{R}$ then*

(i) $d(x, y) = d(y, x)$;
(ii) $d(x, y) = 0$ *if and only if* $x = y$;
(iii) $d(x, z) \leq d(x, y) + d(y, z)$ *(the* triangle inequality*).*

Proof (i) follows from the fact that $|x| = |-x|$, and (ii) from the fact that $|x| = 0$ if and only if $x = 0$. Finally,

$$|x - z| = |(x - y) + (y - z)| \leq |x - y| + |y - z|. \qquad □$$

The function $d : \mathbf{R} \times \mathbf{R} \rightarrow \mathbf{R}$ is a *metric* on $\mathbf{R}$. We shall study more general metrics in Volume II.

The problem of the existence of the square root of 2 arose as a problem in geometry. It is natural to think of the set $\mathbf{R}$ of real numbers geometrically, and to think of them as points on a line, the *real line*, arranged in order.

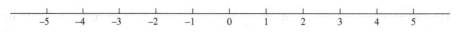

Figure 3.1. The real line.

If $a \in \mathbf{R}$, we can then consider the mapping $x \rightarrow x + a$ as a *shift*, or *translation*, shifting everything an amount a to the right, if $a \geq 0$, and an amount $|a|$ to the left, if $a < 0$. The mapping $x \rightarrow -x$ is a *reflection* about 0. If $a > 0$ then the mapping $x \rightarrow ax$ is a *dilation* or *scaling*, scaling x by a factor of a.

The totally ordered set $\mathbf{R}$ does not have a greatest or least element. It is sometimes convenient to add two points $+\infty$ and $-\infty$, to obtain the *extended real line* $\overline{\mathbf{R}}$. Thus $\overline{\mathbf{R}} = \{-\infty\} \cup \mathbf{R} \cup \{+\infty\}$. The order is extended by setting $-\infty < x < +\infty$ for every real number x. Then $\overline{\mathbf{R}}$ is *order complete* – every non-empty subset has an infimum and a supremum. If A is a non-empty subset of $\mathbf{R}$ then $\inf A = -\infty$ if and only if A does not have a lower bound in $\mathbf{R}$, and $\sup A = +\infty$ if and only if A does not have an upper bound in $\mathbf{R}$.

Some care must be taken in extending the algebraic operations from $\mathbf{R}$ to $\overline{\mathbf{R}}$. Common sense and prudence suggest the following rules.

$$\text{If } x \in \mathbf{R}, \text{ then } (+\infty) + x = x + (+\infty) = x - (-\infty) = +\infty,$$
$$(-\infty) + x = x + (-\infty) = x - (+\infty) = -\infty,$$
$$x/(+\infty) = x/(-\infty) = 0.$$

$$(+\infty) + (+\infty) = +\infty \text{ and } (-\infty) + (-\infty) = -\infty.$$

The sums $(+\infty) + (-\infty)$ and $(-\infty) + (+\infty)$ are not defined.

$$\text{If } x \in \mathbf{R}, \text{ and } x > 0 \text{ then } (+\infty).x = x.(+\infty) = +\infty,$$
$$(-\infty).x = x.(-\infty) = -\infty, \text{ and } x/0 = +\infty.$$

$$\text{If } x \in \mathbf{R}, \text{ and } x < 0 \text{ then } (+\infty).x = x.(+\infty) = -\infty,$$
$$(-\infty).x = x.(-\infty) = +\infty, \text{ and } x/0 = -\infty.$$

The products
$$0.(+\infty), \ 0.(-\infty), \ (+\infty).0 \text{ and } (-\infty).0$$
and the quotients
$$(+\infty)/(+\infty), \ (+\infty)/(-\infty), \ (-\infty)/(+\infty), \ (-\infty)/(+\infty),$$
$$0/0, \ (+\infty)/0 \text{ and } (-\infty)/0$$
are not defined.

Exercises

3.1.1 Show that the sum of a rational number and an irrational number is irrational. What about the product?

3.1.2 Suppose that r, r' are two rational numbers, with $r < r'$. Show that there exists an irrational number x with $r < x < r'$.

3.1.3 Suppose that A and B are non-empty subsets of $\mathbf{R}$ which are bounded below. Let $A + B = \{x \in \mathbf{R} : x = a + b \text{ for some } a \in A, b \in B\}$. Show that $A + B$ is bounded below and that $\inf(A + B) = \inf A + \inf B$. What about products?

3.1.4 Suppose that $(a_n)_{n=1}^\infty$ and $(b_n)_{n=1}^\infty$ are sequences in $\mathbf{R}$ such that the sets $\{a_n : n \in \mathbf{N}\}$ and $\{b_n : n \in \mathbf{N}\}$ are bounded above. Show that the set $\{a_n + b_n : n \in \mathbf{N}\}$ is bounded above. Is

$$\sup\{a_n + b_n : n \in \mathbf{N}\} = \sup\{a_n : n \in \mathbf{N}\} + \sup\{b_n : n \in \mathbf{N}\}?$$

3.1.5 Let $\mathbf{Q}(\sqrt{2})$ be the set of all real numbers of the form $r + s\sqrt{2}$, with $r, s \in \mathbf{Q}$. Show that $\mathbf{Q}(\sqrt{2})$ is a subfield of $\mathbf{R}$. Show that there are two total orderings of $\mathbf{Q}(\sqrt{2})$ under which it is an ordered field.

3.1.6 Suppose that $a_1, \ldots, a_n$ and $b_1, \ldots, b_n$ are real numbers. Establish *Lagrange's identity*:

$$\left(\sum_{1=1}^n a_i b_i\right)^2 + \sum_{\{(i,j):i<j\}} (a_i b_j - a_j b_i)^2 = \left(\sum_{i=1}^n a_i^2\right)\left(\sum_{i=1}^n b_i^2\right).$$

Deduce *Cauchy's inequality*:

$$\sum_{i=1}^n a_i b_i \le \left(\sum_{i=1}^n a_i^2\right)^{\frac{1}{2}} \left(\sum_{i=1}^n b_i^2\right)^{\frac{1}{2}},$$

with equality holding if and only if $a_i b_j = a_j b_i$ for $1 \le i, j \le n$.

3.1.7 If $\alpha \in \mathbf{R}$ and $k \in \mathbf{N}$, define the *binomial coefficient* to be

$$\binom{\alpha}{k} = \frac{\alpha(\alpha - 1) \ldots (\alpha - k + 1)}{k!}.$$

Prove de Moivre's formula

$$\binom{\alpha + 1}{k} = \binom{\alpha}{k} + \binom{\alpha}{k - 1}$$

and its generalization, Vandermonde's formula,

$$\binom{\alpha + \beta}{k} = \sum_{j=0}^{k} \binom{\alpha}{j}\binom{\beta}{k-j}.$$

[*Hint*: First suppose that $\beta \in \mathbf{N}$. Each side of the equation is a polynomial in α, and the two polynomials take the same values when $\alpha \in \mathbf{N}$. Now repeat the argument for β.]

3.2 Convergent sequences

Let us now look again at statement (i) of Theorem 3.1.1.

Theorem 3.2.1   *If $\epsilon > 0$ then there exists n_0 such that $0 < 1/n < \epsilon$ for $n \geq n_0$.*

Proof Since $l > 0$ and 0 is the infimum of $J = \{1/n : n \in \mathbf{N}\}$, there exists n_0 such that $1/n_0 < \epsilon$. If $n \geq n_0$, then $0 < 1/n < 1/n_0 < \epsilon$.   □

This suggests the following definition for more general sequences than $(1/n)_{n=1}^{\infty}$. (We consider sequences indexed either by $\mathbf{N}$ or by $\mathbf{Z}^+$, the set of non-negative integers – since we are concerned with what happens for large values of n, there is no real difference between the two cases, and we shall only state and prove results in one or other case.) A real-valued sequence $(a_n)_{n=0}^{\infty}$ *converges to 0 as n tends to ∞*, or *tends to 0 as n tends to ∞*, or is a *null sequence*, if whenever $\epsilon > 0$ there exists n_0 (which usually depends on ϵ) such that $|a_n| < \epsilon$ for $n \geq n_0$.

A couple of remarks are in order. First, the condition concerns the size $|a_n|$ of a_n, rather than a_n itself. We can write the condition as $-\epsilon < a_n < \epsilon$ for $n \geq n_0$; thus a_n has to satisfy two inequalities, and sometimes it is necessary to consider the two inequalities separately. Secondly, the sequence $(|a_n|)_{n=0}^{\infty}$ need not be decreasing. As an example, if we set $a_n = (-1)^n/n + 1/n^2$ then $(a_n)_{n \in \mathbf{N}}$ is a null sequence, although a_n is, after the first term, alternately negative and positive, and although $|a_n| - |a_{n+1}|$ is alternately negative and positive .

We can immediately generalize this definition. Suppose that $l \in \mathbf{R}$. We say that a real-valued sequence $(a_n)_{n=0}^{\infty}$ *converges to l*, or *tends to l*, as n *tends to ∞* if whenever $\epsilon > 0$ there exists n_0 (which usually depends on ϵ) such that $|a_n - l| < \epsilon$ for $n \geq n_0$. In other words, the sequence $(a_n - l)_{n=0}^{\infty}$ is a null sequence. Once again, we can split the definition into two: we require that $l - \epsilon < a_n < l + \epsilon$, for $n \geq n_0$.

If $(a_n)_{n=0}^{\infty}$ converges to l, we say that l is the *limit* of the sequence, and we write '$a_n \to l$ as $n \to \infty$', and write $l = \lim_{n\to\infty} a_n$. A warning: we can only write $l = \lim_{n\to\infty} a_n$ if we know that the sequence has a limit; and many sequences do not have a limit.

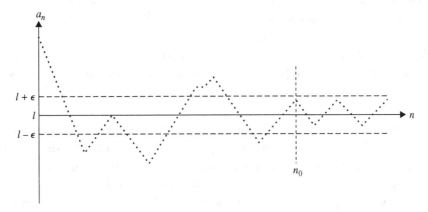

Figure 3.2. A convergent sequence.

When they exist, limits are unique.

Proposition 3.2.2 *If $a_n \to l$ as $n \to \infty$ and $a_n \to m$ as $n \to \infty$, then $l = m$.*

Proof Suppose not. Let $\epsilon = |l - m|/3$, so that $\epsilon > 0$. There exists n_0 such that $|a_n - l| < \epsilon$ for $n \geq n_0$, and there exists m_0 such that $|a_n - m| < \epsilon$ for $n \geq m_0$. Let $p_0 = \max(n_0, m_0)$. Then if $n \geq p_0$,

$$|l - m| \leq |a_n - l| + |a_n - m| < 2\epsilon = 2|l - m|/3,$$

giving a contradiction.   □

A subset B of $\mathbf{R}$ is *bounded* if it is bounded above and bounded below. A sequence $(a_n)_{n=0}^{\infty}$ is *bounded* if the set of values $\{a_n : n \in \mathbf{Z}^+\}$ is bounded.

Proposition 3.2.3 *If $a_n \to l$ as $n \to \infty$ then $(a_n)_{n=0}^{\infty}$ is bounded.*

Proof There exists n_0 such that $|a_n - l| < 1$ for $n \geq n_0$. Let $M = \max\{|a_1|, |a_2|, \ldots, |a_{n_0}|, |l| + 1\}$. If $n > n_0$ then $|a_n| \leq |a_n - l| + |l| \leq |l| + 1$, so that $|a_n| \leq M$ for all n.   □

Unbounded sequences can behave in many different ways: we pick out two where the behaviour is quite respectable. We say that $a_n \to +\infty$ as $n \to \infty$ if whenever $M \in \mathbf{R}^+$ there exists n_0 (which usually depends on M) such that

$a_n > M$ for $n \geq n_0$, and that $a_n \to -\infty$ as $n \to \infty$ if whenever $M \in \mathbf{R}^+$ there exists n_0 (which usually depends on M) such that $a_n < -M$ for $n \geq n_0$.

We now come to the most important result of this section.

Theorem 3.2.4 *Suppose that $(a_n)_{n=0}^{\infty}$ is an increasing sequence of real numbers. If $(a_n)_{n=0}^{\infty}$ is bounded, then $a_n \to \sup\{a_n : n \in \mathbf{Z}^+\}$ as $n \to \infty$; otherwise $a_n \to +\infty$ as $n \to \infty$.*

Suppose that $(a_n)_{n=0}^{\infty}$ is a decreasing sequence of real numbers. If $(a_n)_{n=0}^{\infty}$ is bounded, then $a_n \to \inf\{a_n : n \in \mathbf{Z}^+\}$ as $n \to \infty$; otherwise $a_n \to -\infty$ as $n \to \infty$.

Proof Suppose that $(a_n)_{n=0}^{\infty}$ is increasing and bounded. Let $l = \sup\{a_n : n \in \mathbf{Z}^+\}$. If $\epsilon > 0$ then $l - \epsilon$ is not an upper bound, and so there exists n_0 such that $a_{n_0} > l - \epsilon$. Since l is an upper bound, and since $(a_n)_{n \in \mathbf{N}}$ is increasing, $l - \epsilon < a_{n_0} \leq a_n \leq l$ for all $n \geq n_0$, so that $|a_n - l| < \epsilon$ for $n \geq n_0$.

Similarly if $(a_n)_{n \in \mathbf{N}}$ is increasing and unbounded and $M \in \mathbf{R}^+$, then there exists n_0 such that $a_{n_0} > M$; then $a_n \geq a_{n_0} > M$ for $n \geq n_0$.

Exactly similar arguments work for decreasing sequences.   □

Why is this so important, when the proof is so easy? First, it provides us with a rich supply of convergent sequences. Secondly, it is used in an essential way to prove further deep results. In fact, the results can be taken to provide another characterization of $\mathbf{R}$, as Exercise 3.2.16 shows.

The notion of convergence fits in well with the algebraic and order structure of $\mathbf{R}$, as the following collection of results shows.

Theorem 3.2.5 *Suppose that $(a_n)_{n=0}^{\infty}$ and $(b_n)_{n=0}^{\infty}$ are sequences of real numbers.*

(i) *If $a_n = l$ for all n, then $a_n \to l$ as $n \to \infty$.*

(ii) *If $(a_n)_{n=0}^{\infty}$ is a null sequence, and $(b_n)_{n=0}^{\infty}$ is bounded, then $(a_n b_n)_{n=0}^{\infty}$ is a null sequence.*

(iii) *If $a_n \to a$ and $b_n \to b$ as $n \to \infty$ then $a_n + b_n \to a + b$ as $n \to \infty$.*

(iv) *If $a_n \to a$ as $n \to \infty$ and $c \in \mathbf{R}$ then $ca_n \to ca$ as $n \to \infty$.*

(v) *If $a_n \to a$ and $b_n \to b$ as $n \to \infty$ then $a_n b_n \to ab$ as $n \to \infty$.*

(vi) *If $a_n \neq 0$ and $a \neq 0$ and $a_n \to a$ as $n \to \infty$ then $1/a_n \to 1/a$ as $n \to \infty$.*

(vii) *If $a_n \to a$ and $b_n \to b$ as $n \to \infty$ and $a_n \leq b_n$ for all n then $a \leq b$.*

(viii) *If $a_n \to a$ as $n \to \infty$ and if $(a_{n_k})_{k=0}^{\infty}$ is a subsequence, then $a_{n_k} \to a$ as $k \to \infty$.*

Proof The proofs are straightforward. We give details, but will subsequently leave similar proofs to the reader.

(i) is quite trivial: for any $\epsilon > 0$, take $n_0 = 0$.

(ii) There exists $M > 0$ such that $|b_n| \leq M$ for all n. Given $\epsilon > 0$, there exists n_0 such that $|a_n| < \epsilon/M$ for $n \geq n_0$. Then $|a_n b_n| < \epsilon$ for $n \geq n_0$.

(iii) Given $\epsilon > 0$, there exists n_0 such that $|a_n - a| < \epsilon/2$ for $n \geq n_0$, and there exists m_0 such that $|b_n - b| < \epsilon/2$ for $n \geq m_0$. Let $p_0 = \max(n_0, m_0)$. Then if $n \geq p_0$,

$$|(a_n + b_n) - (a + b)| \leq |a_n - a| + |b_n - b| < \epsilon.$$

(iv) Given $\epsilon > 0$, there exists n_0 such that $|a_n - a| < \epsilon/(|c| + 1)$ for $n \geq n_0$. Then $|c a_n - c a| < \epsilon$ for $n \geq n_0$.

(v) $a_n b_n - ab = (a_n - a)(b_n - b) + (a_n - a)b + a(b_n - b)$. The sequence $(b_n - b)_{n=0}^\infty$ is a bounded sequence, by Proposition 3.2.3; since $(a_n - a)$ is a null sequence, the sequence $((a_n - a)(b_n - b))_{n=0}^\infty$ is a null sequence, by (ii). The sequences $((a_n - a)b)_{n=0}^\infty$ and $(a(b_n - b))_{n=0}^\infty$ are also null sequences, by (iv), and so the result follows, using (iii).

(vi) There exists n_0 such that $|a_n - a| < |a|/2$ for $n \geq n_0$, so that if $n \geq n_0$ then $|a_n| \geq |a| - |a_n - a| \geq |a|/2$. Thus if $n \in \mathbf{Z}^+$ then

$$|1/a_n a| \leq \max(1/|a_0 a|, \ldots 1/|a_{n_0} a|, 2/|a|^2),$$

so that the sequence $(1/a_n a)_{n=0}^\infty$ is bounded. Since $1/a_n - 1/a = (a - a_n)/a_n a$, the result follows from (ii).

(vii) We argue by contradiction. Suppose that $a > b$. Using (iii) and (iv), $a_n - b_n \to a - b$ as $n \to \infty$, and so there exists n_0 such that $|(a_n - b_n) - (a - b)| < a - b$ for $n \geq n_0$. But this implies that $a_n - b_n > 0$, giving a contradiction. Thus $a \geq b$.

(viii) Given $\epsilon > 0$ there exists N such that $|a_n - a| < \epsilon$ for $n \geq N$, and there exists k_0 such that $n_k > N$ for $k \geq k_0$. Thus if $k \geq k_0$ then $|a_{n_k} - a| < \epsilon$. $\square$

As an example, if $0 < r < 1$, then the sequence $(r^n)_{n=1}^\infty$ is a decreasing sequence, bounded below by 0, and so it converges to a limit l, say, by Theorem 3.2.4. Then $r^{n+1} \to rl$ as $n \to \infty$, by (iv), and $r^{n+1} \to l$ as $n \to \infty$, by (viii). Thus $rl = l$, by Proposition 3.2.2, and so $l = 0$: $(r^n)_{n=0}^\infty$ is a null sequence. So also is $(r^n)_{n=0}^\infty$, for $-1 < r < 0$, by (ii).

Some care is needed using (vii). Suppose that $a_n \to a$ and $b_n \to b$ as $n \to \infty$ and that $a_n < b_n$ for all n. Then it does not follow that $a < b$. As an example, consider the sequences $(r^n)_{n=0}^\infty$ and $(-r^n)_{n=0}^\infty$, where $0 < r < 1$.

As another example, let us give another proof of Theorem 6.4.6: if y is a positive real number, if $k \in \mathbf{N}$ and if $k \geq 2$, then there exists a unique positive real number s such that $s^k = y$. We write $s = y^{1/k}$.

Let $a_0 = \max(1, y)$, so that $a_0^k \geq y$. We show that we can define the sequence $(a_n)_{n=0}^{\infty}$ recursively by setting

$$a_{n+1} = \frac{1}{k}\left((k-1)a_n - \frac{y}{a_n^{k-1}}\right) = a_n\left(1 - \frac{a_n^k - y}{ka_n^k}\right) \text{ for } n \in \mathbf{N},$$

and that $0 < a_{n+1} \leq a_n$ and $y \leq a_n^k$. Suppose that have defined a_n, and shown (if $n > 0$) that $0 < a_n \leq a_{n-1}$ and $y \leq a_n^k$. Since $a_n > 0$, a_{n+1} is properly defined. Since $ka_n^k > a_n^k - y \geq 0$, $0 < a_{n+1} \leq a_n$. In order to show that $a_{n+1}^k \geq y$, we use the following inequality, proved in Lemma 2.10.12:

$$\text{if } 0 < t < 1 \text{ then } (1-t)^k \geq 1 - kt.$$

Thus

$$a_{n+1}^k = a_n^k\left(1 - \frac{a_n^k - y}{ka_n^k}\right)^k \geq a_n^k\left(1 - \frac{a_n^k - y}{a_n^k}\right) = y.$$

Since the sequence $(a_n)_{n=0}^{\infty}$ is bounded below, it follows that it converges to a limit l as $n \to \infty$. Using Theorem 3.2.5, it follows that $l^k \geq y$, so that $l > 0$, and it then follows that

$$a_{n+1} = a_n\left(1 - \frac{a_n^k - y}{ka_n^k}\right) \to l\left(1 - \frac{l^k - y}{kl^k}\right) \text{ as } n \to \infty.$$

Since $a_n \to l$ as $n \to \infty$, it therefore follows from Proposition 3.2.2 that

$$l = l\left(1 - \frac{l^k - y}{kl^k}\right),$$

so that $l^k = y$. If $l^k = m^k$ then

$$0 = l^k - m^k = (l - m)(l^{k-1} + l^{k-2}m + \cdots + m^{k-1}),$$

so that $l = m$.

This may seem to be a proof that is too complicated to be interesting. But it is not. It is an example of the use of the Newton–Raphson method: the sequence $(a_n)_{n=0}^{\infty}$ converges very rapidly. Thus it not only proves the existence of $y^{1/k}$, but also enables good approximations to it to be calculated.

Here is an easy but useful result.

Proposition 3.2.6 (The sandwich principle) *Suppose that $a_n \leq b_n \leq c_n$ for all n, and that $a_n \to l$ and $c_n \to l$ as $n \to \infty$. Then $b_n \to l$ as $n \to \infty$.*

Proof Given $\epsilon > 0$ there exists n_0 such that $|a_n - l| < \epsilon$ and $|c_n - l| < \epsilon$ for $n \geq n_0$. Thus

$$l - \epsilon < a_n \leq b_n \leq c_n < l + \epsilon \text{ for } n \geq n_0,$$

so that $b_n \to l$ as $n \to \infty$. □

Corollary 3.2.7 *If $x \in \mathbf{R}$, there exist a strictly increasing sequence $(r_n)_{n=1}^\infty$ of rational numbers and a strictly decreasing sequence $(s_n)_{n=1}^\infty$ of rational numbers such that $r_n \to x$ and $s_n \to x$ as $n \to \infty$.*

Proof Using the notation following Theorem 3.1.1, let r_1 be the 'best' rational with $x - 1 < r_1 < x$ and let s_1 be the 'best' rational with $x < s_1 < x + 1$. Arguing recursively, let r_n be the 'best' rational with

$$\max\left(x - \frac{1}{n}, r_{n-1}\right) < r_n < x,$$

and let s_n be the 'best' rational with

$$x < s_n < \min\left(x + \frac{1}{n}, s_{n-1}\right).$$

Then $(r_n)_{n=1}^\infty$ is a strictly increasing sequence and $(s_n)_{n=1}^\infty$ is a strictly decreasing sequence. Since

$$x - \frac{1}{n} < r_n < s_n < x + \frac{1}{n},$$

$r_n \to x$ and $s_n \to x$ as $n \to \infty$, by the sandwich principle. □

The next result shows that to test for convergence we need only consider a sequence of values of ϵ.

Proposition 3.2.8 *Suppose that $(\epsilon_k)_{k=1}^\infty$ is a null sequence of positive numbers. Then a sequence $(a_n)_{n=0}^\infty$ converges to l if and only if for each k there exists n_k such that $|a_n - l| < \epsilon_k$ for $n \geq n_k$.*

Proof The condition is certainly necessary. If it is satisfied, and if $\epsilon > 0$, then there exists $k \in \mathbf{N}$ such that $0 < \epsilon_k < \epsilon$. If $n \geq n_k$, then $|a_n - l| < \epsilon_k < \epsilon$. □

It is often convenient to take $\epsilon_k = 1/k$, or to take $\epsilon_k = 1/2^k$.

Exercises

3.2.1 Suppose that $a > b > 0$. Find $\lim_{n\to\infty}(a^n - b^n)/(a^n + b^n)$.

3.2.2 Show that $\sqrt{n+1} - \sqrt{n} \to 0$ as $n \to \infty$.

3.2.3 Does $n^{1000000}/2^n$ converge, as $n \to \infty$?

3.2.4 Suppose that $x \in \mathbf{R}$ and that $x > 1$, and that $k \in \mathbf{N}$. Does x^n/n^k converge?

3.2.5 Suppose that $-1 < x < 1$ and that $\alpha \in \mathbf{R}$. Show that

$$\binom{\alpha}{n}x^n = \frac{\alpha(\alpha - 1)\ldots(\alpha - n + 1)}{n!}x^n \to 0 \text{ as } n \to \infty.$$

3.2.6 Suppose that $x \in \mathbf{R}$. Show that $x^n/n! \to 0$ as $n \to \infty$.

3.2.7 Let $a_n = \sqrt{n^2 + n} - n$, for $n \in \mathbf{N}$. Show that a_n converges as $n \to \infty$. What is the limit? Is the sequence $(a_n)_{n=1}^\infty$ monotonic?

3.2.8 Let $a_n = \sum_{j=n+1}^{2n}(1/j)$ and let $b_n = \sum_{j=n}^{2n}(1/j)$. Show that $(a_n)_{n=1}^\infty$ is an increasing sequence, that $(b_n)_{n=1}^\infty$ is a decreasing sequence, and that they tend to a common limit as $n \to \infty$.

3.2.9 Use a calculator, and the method described, to calculate $2^{1/3}$ to 5 decimal places.

3.2.10 Suppose that $a_1, \ldots, a_n$ are positive real numbers. Let $A = (a_1 + \cdots + a_n)/n$ be the *arithmetic mean* and let $G = (a_1 a_2 \ldots a_n)^{1/n}$ be the *geometric mean*. Suppose that $a_1, \ldots, a_n$ are not all equal. Show that there exist $1 \le i, j \le n$ such that $a_i > A$ and $a_j < A$. Show that $a_i a_j < A(a_i + a_j - A)$.
Let $a_i' = A$, $a_j' = a_i + a_j - A$ and let $a_k' = a_k$ for $k \ne i, j$. Let A' and G' be the corresponding means. Show that $A' = A$ and $G' > G$.
Show by induction on $|\{i : a_i \ne A\}|$ that $A \ge G$, with equality if and only if $a_1 = a_2 = \cdots = a_n$. (The *arithmetic mean-geometric mean* inequality.)

3.2.11 Use the arithmetic mean-geometric mean inequality to establish the following results.
(a) If $nt > -1$ then $(1 - t)^n \ge 1 - nt$.
(b) If $-x < n < m$ then $(1 + x/n)^n \le (1 + x/m)^m$.
(c) If $x > 0$ then $(1 - x/n)^n$ converges to a positive limit, as $n \to \infty$.
(d) If $x > 0$ then $(1 - x/n^2)^n \to 1$ as $n \to \infty$.
(e) If $x > 0$ then $(1 + x/n)^n$ converges to a finite limit, as $n \to \infty$.

3.2.12 Suppose that $-1 \le t \le 1$. Define t_n recursively by setting $t_0 = 0$ and $t_n = t_{n-1} + \frac{1}{2}(t - t_{n-1})^2$. Show that $0 \le t_{n-1} \le t_n \le |t|$, for all $n \in \mathbf{N}$. What is $\lim_{n\to\infty} t_n$?

3.2.13 Suppose that $0 < a_0 < b_0$. Define a_n and b_n recursively by setting $a_n = 2a_{n-1}b_{n-1}/(a_{n-1} + b_{n-1})$ and $b_n = (a_{n-1} + b_{n-1})/2$. Show that $a_{n-1} < a_n < b_n < b_{n-1}$. Determine $\lim_{n\to\infty} a_n$ and $\lim_{n\to\infty} b_n$.

3.2.14 Suppose that $0 < a_0 < b_0$. Define a_n and b_n recursively by setting $a_n = \sqrt{a_{n-1}b_{n-1}}$ and $b_n = (a_{n-1} + b_{n-1})/2$. Show that $a_{n-1} < a_n < b_n < b_{n-1}$, and show that the sequences $(a_n)_{n=0}^\infty$ and $(b_n)_{n=0}^\infty$ tend to a common limit as $n \to \infty$.

3.2.15 Let $R_n = F_{n+1}/F_n$, where F_n is the nth Fibonacci number, and $n \geq 1$. Show that (R_{2n-1}) is an increasing sequence and that (R_{2n}) is a decreasing sequence. Show that R_n tends to a limit as $n \to \infty$, and find the limit.

3.2.16 Give an example of a sequence $(x_n)_{n=1}^\infty$ such that x_{nk} converges as $n \to \infty$ for all $k \geq 2$, whereas x_n does not converge as $n \to \infty$.

3.2.17 Suppose that $(x_n)_{n=1}^\infty$ is a sequence such that each of the subsequences $(x_{2n})_{n=1}^\infty$, $(x_{2n+1})_{n=1}^\infty$ and $(x_{3n})_{n=1}^\infty$ converges as $n \to \infty$. Show that the sequence $(x_n)_{n=1}^\infty$ converges as $n \to \infty$.

3.3 The uniqueness of the real number system

In the Prologue, Dedekind cuts were used to construct the real number system **R**. As Exercise 3.3.2 shows, there are other ways of constructing the real numbers, and we need to show that the outcome is essentially the same.

First, let us introduce some terminology. Suppose that x is a positive real number. We set $\lfloor x \rfloor = \sup\{n \in \mathbf{N} : n \leq x\}$ and $\{x\} = x - \lfloor x \rfloor$, so that $x = \lfloor x \rfloor + \{x\}$, and $0 \leq \{x\} < 1$. $\lfloor x \rfloor$ is the *integral part* of x, and $\{x\}$ is the *fractional part* of x. The latter is not only bad notation (the context should however make it clear when $\{x\}$ is being used for the singleton set) but also bad terminology, since 'fractional' suggests incorrectly that $\{x\}$ must be a rational number.

Theorem 3.3.1 *Suppose that* **R**$'$ *is an ordered field with the supremum property. There exists a unique bijection* $j : \mathbf{R} \to \mathbf{R}'$ *such that if* $x, y \in \mathbf{R}$ *then*

(i) $j(x + y) = j(x) + j(y)$ *and* $j(xy) = j(x)j(y)$, *and*

(ii) *if* $x < y$ *then* $j(x) < j(y)$.

Proof We use the fact that each of **R** and **R**$'$ contains a copy of the rational numbers (Theorem 2.8.3). If r is a rational in **R**, let $j_\mathbf{Q}(r)$ be the corresponding rational in **R**$'$.

If x is a positive element of **R**, we set $x_n = \lfloor 2^n x \rfloor / 2^n$. Then $(x_n)_{n=0}^\infty$ is an increasing sequence of rationals in **R**, bounded above by $\lfloor x \rfloor + 1$. Further,

$0 \leq x - x_n \leq 1/2^n$, so that $x_n \to x$ as $n \to \infty$. The sequence $(j_{\mathbf{Q}}(x_n))_{n=0}^{\infty}$ is an increasing sequence of rationals in $\mathbf{R}'$, bounded above by $j_{\mathbf{Q}}(\lfloor x \rfloor + 1)$, and so it converges, by Theorem 3.2.4 (which can clearly be applied to $\mathbf{R}'$). We set $j(x) = \lim_{n \to \infty} j_{\mathbf{Q}}(x_n)$. Note that $j(x) > 0$, and that if $x \in \mathbf{Q}$ then $j_{\mathbf{Q}}(x_n) \to j_{\mathbf{Q}}(x)$, so that $j(x) = j_{\mathbf{Q}}(x)$. If $x < 0$, we set $j(x) = -j(-x)$.

If x and y are positive elements of $\mathbf{R}$, and $n \in \mathbf{Z}^+$ then

$$|j_{\mathbf{Q}}((x+y)_n) - j_{\mathbf{Q}}(x_n) - j_{\mathbf{Q}}(y_n)| = |(x+y)_n - x_n - y_n|$$
$$\leq (|x+y| - |(x+y)_n|) + |x - x_n| + |y - y_n| \leq 3/2^n,$$

so that

$$j(x+y) - j(x) - j(y) = \lim_{n \to \infty} (j_{\mathbf{Q}}((x+y)_n) - j_{\mathbf{Q}}(x_n) - j_{Q}Q(y_n)) = 0.$$

Thus $j(x+y) = j(x) + j(y)$. A similar argument shows that $j(xy) = j(x)j(y)$, and it then follows easily that (i) holds for all x and y in $\mathbf{R}$.

If $x < y$ then there exist rationals r and s such that $x < r < s < y$ then $j(x) \leq j(r) < j(s) \leq j(y)$, and so (ii) holds. It follows from this that j is injective.

Using the same procedure, we construct a mapping $k : \mathbf{R}' \to \mathbf{R}$ for which the results corresponding to (i) and (ii) hold. If $x \in \mathbf{R}$ and $x > 0$ then

$$k(j(x)) = \lim_{n \to \infty} k(j(x)_n) = \lim_{n \to \infty} x_n = x.$$

If $x < 0$ then $k(j(x)) = -k(-j(x)) = -k(j(-x)) = -(-x) = x$. A similar argument shows that $j(k(x')) = x'$ for all $x' \in \mathbf{R}'$. Thus j and k are bijections.

Finally, we show that j is unique. Suppose that $j_1 : \mathbf{R} \to \mathbf{R}'$ is another mapping satisfying (i) and (ii). Then $j_1(0) = 0$ and $j_1(1) = 1$, from which it follows that $j_1(r) = j(r)$ for r a rational in $\mathbf{R}$. If x is a positive element of $\mathbf{R}$, then $j_1(x_n) \to j_1(x)$ as $n \to \infty$. But $j_1(x_n) = j(x_n) \to j(x)$ as $n \to \infty$, and so $j_1(x) = j(x)$. If $x < 0$ then

$$j_1(x) = -j_1(-x) = -j(-x) = j(x).$$

Thus j is unique. $\square$

Exercises

3.3.1 Define the notion of a convergent sequence in an ordered field. Suppose that F is an ordered field in which each bounded increasing sequence converges. In this exercise we show that F has the supremum property, so that there exists a unique order-preserving field isomorphism of $\mathbf{R}$ onto F.

(a) Show that each bounded decreasing sequence converges.

(b) Show that if $\epsilon > 0$ then there exists n such that $1/n < \epsilon$.

(c) Suppose that A is a non-empty subset of F which is bounded above. Show that there exists $n \in \mathbf{N}$ which is an upper bound for A.

(d) Suppose that A is a non-empty subset of $F^+ = \{x \in F : x \geq 0\}$ which is bounded above. If $k \in \mathbf{N}$, let

$$b_k = \inf\{j \in \mathbf{N} : j \geq 2^k a \text{ for each } a \in A\},$$

and let $c_k = b_k/2^k$. Show that $(c_k)_{k=0}^\infty$ is a bounded decreasing sequence.

(e) Let $c = \lim_{k \to \infty} c_k$. Show that $c = \sup A$.

(f) Show that F has the supremum property.

3.3.2 This extended question provides an alternative construction of the real numbers.

(a) A *rational Cauchy sequence* is a sequence $\mathbf{r} = (r_n)_{n=1}^\infty$ in $\mathbf{Q}$ such that for each $j \in \mathbf{N}$ there exists n_j such that $|r_m - r_n| < 1/j$ for $m, n \geq n_j$. If $\mathbf{r}$ and $\mathbf{s}$ are rational Cauchy sequences, set $\mathbf{r} \leq \mathbf{s}$ if $r_n \leq s_n$ for all $n \in \mathbf{N}$. Show that $\leq$ is a partial order on the set C of rational Cauchy sequences.

(b) Define a relation $\mathbf{r} \sim \mathbf{s}$ on the set C by setting $\mathbf{r} \sim \mathbf{s}$ if the sequence $(r_1, s_1, r_2, s_2, \dots)$ is a rational Cauchy sequence. Show that $\sim$ is an equivalence relation on C.

(c) Let $D = C/\sim$ be the set of equivalence classes in C. Define a relation $\leq$ on D by setting $a \leq b$ if there exist $\mathbf{r} \in a$ and $\mathbf{s} \in b$ such that $\mathbf{r} \leq \mathbf{s}$. Show that this defines a *total* order on D. Show that D does not have a greatest or least element.

(d) If $r \in \mathbf{Q}$, let $\mathbf{r} = (r_n)_{n=1}^\infty$, where $r_n = r$ for all $n \in \mathbf{N}$, and let $j(r) = [\mathbf{r}]$ be its equivalence class in D. Show that j is an injective order-preserving mapping of $\mathbf{Q}$ into D.

(e) Define addition and multiplication of elements of D, and show that D is an ordered field.

(f) Show that a bounded increasing sequence in D converges. (This is the hardest part. Choose representatives, and use a diagonal argument to find the limit.)

3.4 The Bolzano–Weierstrass theorem

Before reading this section, it is advisable to read Section 2.4, possibly excluding Ramsey's theorem (Theorem 2.4.4). You need to understand the diagonal procedure (Theorem 2.4.2) and Theorem 2.4.3.

Sequences can behave in many different ways: consider for example a sequence $(q_n)_{n=1}^\infty$ which maps $\mathbf{N}$ onto the set of rational numbers between 0 and 1. The next theorem is therefore remarkable and is of great theoretical importance.

Theorem 3.4.1 (The Bolzano–Weierstrass theorem) *Suppose that* $(a_n)_{n=0}^\infty$ *is a bounded sequence of real numbers. Then there is a subsequence* $(a_{n_k})_{k=1}^\infty$ *which converges.*

Proof   We shall give two proofs here, and a third proof in the next section. (It is always worth giving more than one proof of important results; each proof can throw a different light on the result, and the ideas from a proof can often be used to prove other results.) Each of the proofs uses Theorem 3.2.4.

The first proof is very short. $(a_n)_{n=0}^\infty$ has a monotone subsequence (Corollary 2.4.3). This subsequence is bounded, and so it converges (Theorem 3.2.4).

The second proof, which is essentially the proof that Weierstrass gave, uses repeated subdivision, and a diagonal argument. Let us introduce some notation. If $b, c \in \mathbf{R}$ and $b \leq c$, then the *closed interval* $[b, c]$ is the set $\{x \in \mathbf{R} : b \leq x \leq c\}$. It has *length* $c - b$; it is *closed* because it contains its *endpoints* b and c. We shall discuss these notions further in Section 4.1.

Since $(a_n)_{n=0}^\infty$ is bounded, there exist b_0, c_0 with $b_0 \leq c_0$ such that $a_n \in [b_0, c_0]$, for all n. Let $d_0 = (b_0 + c_0)/2$ be the midpoint of the closed interval. Then there are two possibilities. First, there are infinitely many n for which $a_n \in [b_0, d_0]$; in this case we set $b_1 = b_0$ and $c_1 = d_0$, so that $[b_1, c_1] = [b_0, d_0]$. Secondly, there are only finitely many n for which $a_n \in [b_0, d_0]$; in this case, $a_n \in [d_0, c_0]$ for infinitely many n, and we set $b_1 = d_0$ and $c_1 = c_0$, so that $[b_1, c_1] = [d_0, c_0]$. Thus in either case $A_1 = \{n \in \mathbf{N} : a_n \in [b_1, c_1]\}$ is an infinite subset of $\mathbf{N}$. We have an infinite set of terms in an closed interval of half the length of the original closed interval.

We now iterate this procedure recursively. At the jth step, we obtain a closed interval $[b_j, c_j]$ such that $[b_j, c_j] \subseteq [b_{j-1}, c_{j-1}]$ and $c_j - b_j = (c_0 - b_0)/2^j$, and such that $A_j = \{n \in \mathbf{N} : a_n \in [b_j, c_j]\}$ is an infinite subset of A_{j-1}. For each $j \in \mathbf{N}$ let $(n_k^{(j)})_{k=1}^\infty$ be the standard enumeration of A_j.

Let $m_j = n_j^{(j)}$, so that $b_j \leq a_{m_j} \leq c_j$, for $j \in \mathbf{N}$. By the diagonal procedure (Theorem 2.4.2) the sequence $(m_j)_{j=1}^{\infty}$ is a subsequence of $\mathbf{N}$. The sequence $(b_j)_{j=0}^{\infty}$ is increasing, and is bounded above by c_0, and so it converges as $j \to \infty$, to b, say. Since $c_j = b_j + (c_0 - b_0)/2^j$, c_j converges to b as $j \to \infty$, as well. Since $b_j \leq a_{m_j} \leq c_j$, $a_{m_j} \to b$ as $j \to \infty$, by the sandwich principle. $\qquad\square$

Exercises

3.4.1 Let $(q_n)_{n=1}^{\infty}$ be a sequence which maps $\mathbf{N}$ onto the set of rational numbers between 0 and 1. Show that if $l \in [0, 1]$ then there exists a subsequence $(q_{n_j})_{j=1}^{\infty}$ which converges to l.

3.4.2 Suppose that $(a_n)_{n=0}^{\infty}$ is a bounded sequence with the property that there exists l such that if $(a_{n_j})_{j=1}^{\infty}$ is any convergent subsequence of $(a_n)_{n=0}^{\infty}$ then its limit is l. Show that $a_n \to l$ as $n \to \infty$.

3.4.3 Suppose that $(a_n)_{n=0}^{\infty}$ is a bounded sequence of real numbers which does not converge. Show that $(a_n)_{n=0}^{\infty}$ has two convergent subsequences which converge to different limits.

3.5 Upper and lower limits

Suppose that $(a_n)_{n=0}^{\infty}$ is a bounded sequence, and that $(a_{n_k})_{k=0}^{\infty}$ is a convergent subsequence, convergent to l, say. What can we say about l? First, we can say that $l \in [m_0, M_0]$ where $m_0 = \inf\{a_n : n \in \mathbf{Z}^+\}$ and $M_0 = \sup\{a_n : n \in \mathbf{Z}^+\}$. But it may happen, for example, that a_0 is much larger than all the other terms in the sequence. Then $a_0 = M_0$, and M_0 does not give us much information about l. Indeed, the value of l is not constrained in any way by any finite set of values of a_n.

Let us therefore set $M_j = \sup\{a_n : n \in \mathbf{Z}^+, n \geq j\}$. Then the sequence $(M_j)_{j=1}^{\infty}$ is decreasing (we take suprema over smaller and smaller sets), and is bounded below by m_0. It therefore converges to a limit as $j \to \infty$. This limit is called the *upper limit* or *limes superior* of the sequence $(a_n)_{n=0}^{\infty}$ and is denoted by $\limsup_{n\to\infty}(a_n)$. In exactly the same way, we define $m_j = \inf\{a_n : n \in \mathbf{N}, n \geq j\}$; $(m_j)_{j=1}^{\infty}$ is increasing and bounded above by M_0 and converges to the *lower limit* or *limes inferior* $\liminf_{n\to\infty}(a_n)$ of the sequence $(a_n)_{n=1}^{\infty}$. Since $m_j \leq M_j$ for all j, $\liminf_{n\to\infty}(a_n) \leq \limsup_{n\to\infty}(a_n)$.

Upper and lower limits are a little complicated, being defined in two stages; first we consider a sequence of suprema or infima, and secondly we take the limit of these sequences. Such a procedure will recur else-where! We can characterize upper and lower limits by their fundamental properties.

Theorem 3.5.1 *Suppose that $(a_n)_{n=0}^\infty$ is a bounded sequence, and that $S = \limsup_{n\to\infty}(a_n)$. Then S is the unique real number with the two following properties:*

(i) if $t > S$ then there exists n_0 such that $a_n < t$ for all $n > n_0$;

(ii) if $r < S$ and $n \in \mathbf{N}$ then there exists $m \in \mathbf{N}$ with $m \geq n$ such that $a_m > r$.

There is a similar characterization of $\liminf_{n\to\infty}(a_n)$.

We can express (i) by saying that a_n is *eventually* less than t, and (ii) by saying that a_n is *frequently* greater than r.

Proof First, we show that S satisfies (i) and (ii).

(i) Since $S = \inf\{M_j : j \in \mathbf{N}\}$ and $t > S$, t is not a lower bound for $\{M_j : j \in \mathbf{N}\}$. Thus there exists n_0 such that $M_{n_0} < t$: then $a_n \leq M_{n_0} < t$ for $n \geq n_0$.

(ii) Since $r < S \leq M_n$ and $M_n = \sup\{a_m : m \geq n\}$, r is not an upper bound for $\{a_m : m \geq n\}$. Thus there exists $m \geq n$ with $a_M > r$.

We now turn to uniqueness. Suppose that $T > S$. Let $U = (S + T)/2$, so that $T > U > S$. By (i), there exists n_0 such that $a_n < U$ for all $n \geq n_0$, and so (ii) does not hold for T.

Suppose that $R < S$. Let $Q = (S + R)/2$, so that $R < Q < S$. By (ii), if $n \in \mathbf{N}$ there exists $m \geq n$ with $a_m > Q$, and so (i) does not hold for R. $\square$

We can now answer the question that was raised at the beginning of the section, and give a third proof of the Bolzano–Weierstrass theorem.

Theorem 3.5.2 *Suppose that $(a_{n_j})_{j=0}^\infty$ is a convergent subsequence of a bounded sequence $(a_n)_{n=0}^\infty$. Then*

$$\liminf_{n\to\infty}(a_n) \leq \lim_{j\to\infty} a_{n_j} \leq \limsup_{n\to\infty}(a_n).$$

Further, there exist subsequences $(a_{l_j})_{j=0}^\infty$ and $(a_{m_j})_{j=0}^\infty$ such that

$$a_{l_j} \to \liminf_{n\to\infty}(a_n) \ \text{and} \ a_{m_j} \to \limsup_{n\to\infty}(a_n) \ \text{as} \ j \to \infty.$$

Proof Since $m_{n_j} \leq a_{n_j} \leq M_{n_j}$, and since $m_{n_j} \to \liminf_{n\to\infty}(a_n)$ and $M_{n_j} \to \limsup_{n\to\infty}(a_n)$, the first result follows from Theorem 3.2.5 (vii).

Let $S = \limsup_{n\to\infty}(a_n)$. By Theorem 3.5.1 (i), there exists a least n_0 such that $a_n < S + 1$ for all $n \geq n_0$, and by (ii) there exists a least $p_0 > n_0$

such that $a_{p_0} > S - 1$. Continuing recursively, there exists a least $n_j > p_{j-1}$ such that $a_n < S + 1/j$ for $n \geq n_j$, and there exists a least $p_j > n_j$ such that $a_{p_j} > S - 1/j$. Then $S - 1/j < a_{p_j} < S + 1/j$, so that $a_{p_j} \to S$ as $j \to \infty$, by the sandwich principle. An exactly similar proof works for the lower limit. $\qquad \square$

What happens when the upper and lower limits are equal?

Theorem 3.5.3 *Suppose that $(a_n)_{n=0}^{\infty}$ is a bounded sequence and that $l \in \mathbf{R}$. Then $a_n \to l$ as $n \to \infty$ if and only if $\limsup_{n\to\infty}(a_n) = \liminf_{n\to\infty}(a_n) = l$.*

Proof If $\limsup_{n\to\infty}(a_n) = \liminf_{n\to\infty}(a_n) = l$, then $m_j \to l$ and $M_j \to l$ as $j \to \infty$. Since $m_j \leq a_j \leq M_j$, $a_j \to l$, by the sandwich principle.

Conversely, suppose that $a_n \to l$ as $n \to \infty$. Then if $\epsilon > 0$ there exists n_0 such that $l - \epsilon/2 < a_n < l + \epsilon/2$ for $n \geq n_0$. Thus $l - \epsilon < M_n < l + \epsilon$ for $n \geq n_0$, so that $M_n \to l$ as $n \to \infty$. Thus $\limsup_{n\to\infty}(a_n) = l$. Similarly $\liminf_{n\to\infty}(a_n) = l$. $\qquad \square$

What do we do when $(a_n)_{n=0}^{\infty}$ is not bounded? If $(a_n)_{n=0}^{\infty}$ is not bounded above, then $M_j = \infty$ for all $j \in \mathbf{Z}^+$; we therefore define $\limsup_{n\to\infty} a_n$ to be $+\infty$. If $(a_n)_{n=0}^{\infty}$ is bounded above, but is not bounded below, then $(M_j)_{n=0}^{\infty}$ is a decreasing sequence. If this is bounded below, we define $\limsup_{n\to\infty} a_n = \lim_{n\to\infty} M_n$; if not, we set $\limsup_{n\to\infty} a_n$ to be $-\infty$. We treat $\liminf_{n\to\infty} a_n$ in a similar way.

Exercises

3.5.1 Consider the second proof of the Bolzano–Weierstrass theorem in the preceding section. What is $\lim_{j\to\infty} a_{m_j}$?

3.5.2 Suppose that $(a_n)_{n=0}^{\infty}$ is a bounded sequence. Let $s_n = a_0 + \cdots + a_n$. Show that

$$\liminf_{n\to\infty} a_n \leq \liminf_{n\to\infty} \frac{s_n}{n+1} \leq \limsup_{n\to\infty} \frac{s_n}{n+1} \leq \limsup_{n\to\infty} a_n.$$

Deduce that if $a_n \to l$ as $n \to \infty$ then $s_n/(n+1) \to l$ as $n \to \infty$. Give an example to show that the converse does not hold.

3.5.3 Suppose that $(a_n)_{n=0}^{\infty}$ is a bounded sequence. Let

$$U = \{x \in \mathbf{R} : \{n \in \mathbf{Z}^+ : a_n > x\} \text{ is finite}\}.$$

Show that $\inf U = \limsup_{n\to\infty} a_n$.

3.5.4 Suppose that $(a_n)_{n=0}^{\infty}$ and $(b_n)_{n=0}^{\infty}$ are bounded sequences. Show that

$$\lim_{n\to\infty} \inf a_n + \lim_{n\to\infty} \inf b_n \le \lim_{n\to\infty} \inf (a_n + b_n) \le \lim_{n\to\infty} \inf a_n + \lim_{n\to\infty} \sup b_n$$

$$\le \lim_{n\to\infty} \sup (a_n + b_n) \le \lim_{n\to\infty} \sup a_n + \lim_{n\to\infty} \sup b_n.$$

Show that equality holds in the last inequality if and only if there exists a strictly increasing sequence $(n_j)_{j=0}^{\infty} \in \mathbf{N}$ such that

$$a_{n_j} \to \lim_{n\to\infty} \sup a_n \text{ and } b_{n_j} \to \lim_{n\to\infty} \sup b_n \text{ as } j \to \infty.$$

Give an example where all the inequalities are strict.

3.5.5 Suppose that $(a_n)_{n=0}^{\infty}$ and $(b_n)_{n=0}^{\infty}$ are sequences of positive numbers, and that $a_n \to a$ as $n \to \infty$. Show that if $a > 0$ then $\lim\inf_{n\to\infty} a_n b_n = a \lim\inf_{n\to\infty} b_n$. Show that equality need not hold if $a = 0$.

3.5.6 Suppose that $(s_n)_{n=0}^{\infty}$ is a sequence of real numbers and that $(t_n)_{n=0}^{\infty}$ is a strictly increasing unbounded sequence of positive numbers. Show that $\lim\sup_{n\to\infty}(s_n/t_n) \le \lim\sup_{n\to\infty}((s_{n+1} - s_n)/(t_{n+1} - t_n))$.

3.5.7 Suppose that $(a_n)_{n=0}^{\infty}$ is a sequence of positive numbers. Show that $\lim\sup_{n\to\infty} a_n^{1/n} \le \lim\sup(a_{n+1}/a_n)$.

3.6 The general principle of convergence

Suppose that $(a_n)_{n=0}^{\infty}$ is a sequence of real numbers. How can we tell whether it converges or not? If we suspect that its limit is l, we can consider the behaviour of $|a_n - l|$ as n becomes large. But what if we do not know what l should be? We have seen that we can answer this question when $(a_n)_{n=0}^{\infty}$ is monotonic (Theorem 3.2.4), but this only happens in special circumstances. Here we provide a more general answer.

A sequence $(a_n)_{n=0}^{\infty}$ is a *Cauchy sequence* if whenever $\epsilon > 0$ there exists n_0 (usually dependent on ϵ) such that $|a_m - a_n| < \epsilon$ for $m, n \ge n_0$. The terms of the sequence become close as m and n become large.

Proposition 3.6.1 *A Cauchy sequence $(a_n)_{n=0}^{\infty}$ is bounded.*

Proof The proof is just like the proof of Proposition 3.2.3. There exists n_0 such that $|a_m - a_n| < 1$ for $m, n \ge n_0$. Let

$$M = \max\{|a_0|, |a_1|, \ldots, |a_{n_0}|, |a_{n_0}| + 1\}.$$

If $n > n_0$ then $|a_n| \le |a_n - a_{n_0}| + |a_{n_0}| \le |a_{n_0}| + 1$, so that $|a_n| \le M$ for all n. $\qquad\square$

Theorem 3.6.2 (The general principle of convergence) *A sequence* $(a_n)_{n=0}^{\infty}$ *of real numbers is convergent if and only if it is a Cauchy sequence.*

Proof First, suppose that $a_n \to l$ as $n \to \infty$. Given $\epsilon > 0$ there exists n_0 such that $|a_n - l| < \epsilon/2$ for $n \geq n_0$. If $m, n \geq n_0$ then

$$|a_m - a_n| \leq |a_m - l| + |a_n - l| < \epsilon/2 + \epsilon/2 = \epsilon,$$

so that $(a_n)_{n=0}^{\infty}$ is a Cauchy sequence.

Conversely, suppose that $(a_n)_{n=0}^{\infty}$ is a Cauchy sequence. Then it is bounded, and so, by the Bolzano–Weierstrass theorem, it has a convergent subsequence $(a_{n_k})_{k=0}^{\infty}$, convergent to l say. We shall show that $a_n \to l$ as $n \to \infty$. Suppose that $\epsilon > 0$. Then there exist k_0 such that $|a_{n_k} - l| < \epsilon/2$ for $k \geq k_0$ and N such that $|a_m - a_n| < \epsilon/2$ for $m, n \geq N$. There exists $k_1 \geq k_0$ such that $n_{k_1} > N$. If $n \geq N$ then

$$|a_n - l| \leq |a_n - a_{n_{k_1}}| + |a_{n_{k_1}} - l| < \epsilon/2 + \epsilon/2 = \epsilon. \qquad \square$$

A Cauchy sequence is a sequence that looks as if it should converge. A Cauchy sequence of rational numbers need not converge to a rational number (consider a sequence of rational numbers converging to $\sqrt{2}$), but does converge to a real number. This indicates again that the real numbers provide a good extension of the rational numbers.

Exercises

3.6.1 Show from the definitions that the upper and lower limits of a Cauchy sequence are equal. Use this to give another proof of the general principle of convergence.

3.6.2 Let $a_n = \sqrt{n}$. Show that if $\epsilon > 0$ then there exists n_0 such that $|a_{n+1} - a_n| < \epsilon$ for $n \geq n_0$. Is $(a_n)_{n=1}^{\infty}$ a Cauchy sequence?

3.7 Complex numbers

This volume is principally concerned with real analysis: the study of real-valued functions of a real variable, and sequences of real numbers. There are however topics, such as the theory of power series, where it is natural to consider complex-valued functions of a complex variable. This topic will be considered in much more detail in Volume III, but here, and in the next section, we introduce some of the basic properties of complex numbers.

Why do we need to consider complex numbers? Although the construction of the real numbers allows us to find roots of the polynomial $x^2 - 2$, there are plenty of polynomials with no real roots. For example, if $a \in \mathbf{R}$ then $a^2 \geq 0$, so that $a^2 + 1 \geq 1$, and so the polynomial $x^2 + 1$ has no real roots. We overcome this by enlarging the real field $\mathbf{R}$ to obtain the complex field $\mathbf{C}$.

This is a problem of algebra, rather than analysis. We want to adjoin an element i to $\mathbf{R}$ with the property that $i^2 = -1$. We shall describe a simple way of doing this; Exercises 3.7.1 and 3.7.2 provide other constructions. In each case, we are concerned with vector spaces. Suppose that K is a field. A *vector space* E over K is an abelian additive group $(E, +)$, with zero element 0, together with a mapping (*scalar multiplication*) $(\lambda, x) \to \lambda x$ of $K \times E$ into E which satisfies

- $1.x = x$,
- $(\lambda + \mu)x = \lambda x + \mu x$,
- $\lambda(\mu x) = (\lambda \mu)x$,
- $\lambda(x + y) = \lambda x + \lambda y$,

for $\lambda, \mu \in K$ and $x, y \in E$. The elements of E are called *vectors* and the elements of K are called *scalars*. A vector space over $\mathbf{R}$ is called a *real* vector space.

It then follows that $0.x = 0$ and $\lambda.0 = 0$ for $x \in E$ and $\lambda \in K$. [Note that we use the same symbol 0 for the additive identity element in E (the *zero vector*) and the zero element (the *zero scalar* in K).] We denote $E \setminus \{0\}$ by E^*.

Besides the element i, we want to consider elements bi, where $b \in \mathbf{R}$, and elements $a + bi$, where $a, b \in \mathbf{R}$. We therefore take $\mathbf{R}^2 = \{(x, y) : x, y \in \mathbf{R}\}$ as our underlying set. $\mathbf{R}^2$ is a real vector space:

$$(x_1, y_1) + (x_2, y_2) = (x_1 + x_2, y_1 + y_2) \text{ and } a(x, y) = (ax, ay) \text{ for } a \in \mathbf{R}.$$

We set $\mathbf{1} = (1, 0)$ and $\boldsymbol{i} = (0, 1)$, so that any element $(x, y) \in \mathbf{R}^2$ can be written as $(x, y) = x\mathbf{1} + y\boldsymbol{i}$. We want to define an associative and distributive multiplication in such a way that

$$\mathbf{1}^2 = \mathbf{1}, \ \mathbf{1}.\boldsymbol{i} = \boldsymbol{i}.\mathbf{1} = \boldsymbol{i} \quad \text{and} \quad \boldsymbol{i}^2 = -\mathbf{1}.$$

Thus we require that

$$(x_1\mathbf{1} + y_1\boldsymbol{i})(x_2\mathbf{1} + y_2\boldsymbol{i}) = x_1x_2\mathbf{1}.\mathbf{1} + x_1y_2\mathbf{1}.\boldsymbol{i} + y_1x_2\boldsymbol{i}.\mathbf{1} + y_1y_2\boldsymbol{i}.\boldsymbol{i}$$
$$= (x_1x_2 - y_1y_2)\mathbf{1} + (x_1y_2 + y_1x_2)\boldsymbol{i},$$

and so we define multiplication by setting

$$(x_1 1 + y_1 i)(x_2 1 + y_2 i) = (x_1 x_2 - y_1 y_2)1 + (x_1 y_2 + y_1 x_2)i.$$

We denote $\mathbf{R}^2$, with this multiplication, by $\mathbf{C}$. Note that if $a, b \in \mathbf{R}$ then $a1 + b1 = (a + b)1$ and $(a1)(b1) = ab1$, so that if we identify $\mathbf{R}$ with $\mathbf{R}.1 = \{(a, 0) : a \in \mathbf{R}\}$, then the addition and multiplication on $\mathbf{C}$ extends the addition and multiplication on $\mathbf{R}$.

We need to verify that this multiplication is commutative (this is clear from the definition) and associative. We verify associativity directly:

$$[(x_1 1 + y_1 i)(x_2 1 + y_2 i)](x_3 1 + y_3 i)$$
$$= [(x_1 x_2 - y_1 y_2)1 + (x_1 y_2 + y_1 x_2)i](x_3 1 + y_3 i)$$
$$= ((x_1 x_2 - y_1 y_2)x_3 - (x_1 y_2 + y_1 x_2)y_3)1$$
$$+ ((x_1 x_2 - y_1 y_2)y_3 + (x_1 y_2 + y_1 x_2)x_3)i$$
$$= (x_1(x_2 x_3 - y_2 y_3) - y_1(x_2 y_3 + y_2 x_3))1$$
$$+ (x_1(x_2 y_3 + y_2 x_3) + y_1(x_2 x_3 - y_2 y_3))i$$
$$= (x_1 1 + y_1 i)[(x_2 x_3 - y_2 y_3)1 + (x_2 y_3 + y_2 x_3)i]$$
$$= (x_1 1 + y_1 i)[(x_2 1 + y_2 i)(x_3 1 + y_3 i)].$$

It is equally straightforward to verify the distributive law:

$$(x_1 1 + y_1 i)[(x_2 1 + y_2 i) + (x_3 1 + y_3 i)]$$
$$= [(x_1 1 + y_1 i)(x_2 1 + y_2 i)] + [(x_1 1 + y_1 i)(x_3 1 + y_3 i)].$$

If $z = x1 + yi \neq 0$ then $x^2 + y^2 \neq 0$. We define

$$z^{-1} = \frac{x}{x^2 + y^2}1 - \frac{y}{x^2 + y^2}i,$$

and then $zz^{-1} = z^{-1}z = 1$, so that z^{-1} is the multiplicative inverse of z; it is also written as $1/z$. Thus $\mathbf{C}$ is a field, the *complex number field*, which has a subfield $\mathbf{R1}$ isomorphic to $\mathbf{R}$.

If $z = x1 + yi$, we define its *(complex) conjugate* $\bar{z}$ to be $\bar{z} = x1 - yi$.

Theorem 3.7.1 *If $z \in \mathbf{C}$ then $\bar{\bar{z}} = z$. The mapping $z \to \bar{z}$ is a field isomorphism of $\mathbf{C}$ onto itself: $\bar{1} = 1$, and if $z, w \in \mathbf{C}$ then $\overline{z + w} = \bar{z} + \bar{w}$, $\overline{zw} = \bar{z}.\bar{w}$ and $\overline{1/z} = 1/\bar{z}$. If $z = x1 + yi$ then $z\bar{z} = (x^2 + y^2)1$.*

Proof Easy direct verification. □

We now write 1 for $\mathbf{1}$, and i for $\mathbf{i}$, so that $(x, y) = x\mathbf{1} + y\mathbf{i} = x + iy$. x is the *real part* of z, denoted by $\Re z$, and y is the *imaginary part*, denoted by $\Im z$. If $y = 0$ then z is *real*, and if $x = 0$ then z is *pure imaginary*.

We have therefore embedded the real number field $\mathbf{R}$ in a larger field $\mathbf{C}$, in which the polynomial $x^2 + 1$ has two roots, i and $-i$, and we can factorize the polynomial $x^2 + 1$ as $(x - i)(x + i)$. The construction is straightforward, but the step is enormous. As we shall see, the real numbers, and real analysis, are fascinating. By comparison, the complex numbers, and complex analysis, are magical. Let us state one result to illustrate this. If $p(x) = a_n x^n + \cdots + a_0$ is a complex polynomial, with $n > 0$ and $a_n \neq 0$, then p has a root in $\mathbf{C}$, and we can express p as a product of linear factors: $p(x) = a_n(x - c_1) \ldots (x - c_n)$. We have extended the field to deal with one very simple quadratic polynomial, and the resulting extension is powerful enough to handle *all* polynomials.

We set $|z| = (x^2 + y^2)^{1/2}$, so that $|z|^2 = z\bar{z}$. The quantity $|z|$ is the *modulus*, or *absolute value*, of z; it measures the size of z. Note that $|\bar{z}| = |z|$. If x is real, its modulus as a real number is the same as its modulus as a complex number. If $z \neq 0$, then $|z| > 0$, and $z^{-1} = \bar{z}/|z|^2$.

Note that $z + \bar{z} = 2x$, and $z - \bar{z} = 2y$, so that $|z + \bar{z}| \leq 2|z|$, with equality if and only if z is real, and $|z - \bar{z}| \leq 2|z|$, with equality if and only if z is pure imaginary. Note also that

$$|zw|^2 = (zw)(\overline{zw}) = z\bar{z}w\bar{w} = |z|^2|w|^2, \quad \text{so that } |zw| = |z||w|.$$

Proposition 3.7.2 *If $z_1, z_2 \in \mathbf{C}$, set $d(z_1, z_2) = |z_1 - z_2|$. Then*

$$d(z_1, z_2) = d(z_2, z_1);$$
$$d(z_1, z_2) = 0 \text{ if and only if } z_1 = z_2;$$
$$d(z_1, z_3) \leq d(z_1, z_2) + d(z_2, z_3) \text{ (the triangle inequality).}$$

Proof The first two statements are obvious. For the third, let $v = z_1 - z_2$ and $w = z_2 - z_3$. Then we must show that $|v + w| \leq |v| + |w|$. Let $t = v\bar{w}$, so that $\bar{t} = \bar{v}w$ and $|t| = |v|.|w|$. Then

$$|v + w|^2 = (v + w)\overline{(v + w)} = v\bar{v} + t + \bar{t} + w\bar{w}$$
$$\leq |v|^2 + |t + \bar{t}| + |w|^2 \leq |v|^2 + 2|t| + |w|^2$$
$$= |v|^2 + 2|v||w| + |w|^2 = (|z| + |w|)^2. \qquad \square$$

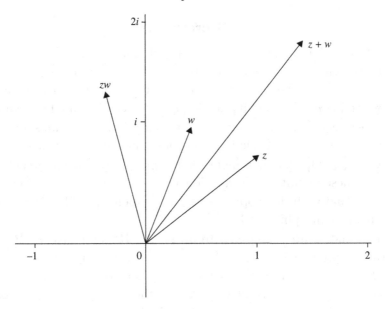

Figure 3.7. The Argand diagram.

Again, d is a metric on $\mathbf{C}$, which extends the metric on $\mathbf{R}$. We can consider a point $(x, y) \in \mathbf{R}^2$ as a point in the plane, with coordinates x and y. When we identify $\mathbf{C}$ with $\mathbf{R}^2$, the plane is called the *complex plane* or *Argand diagram*.

If $w = u + iv \in \mathbf{C}$, the mapping $z \to z + w$ is a represented by a shift, sending (x, y) to $(x+u, y+v)$. The mapping $z \to \bar{z}$ is represented by reflection in the real axis $\{(x, y) \in \mathbf{R}^2 : y = 0\}$. We shall consider the geometry of multiplication later, when we have established further properties of complex numbers.

In Figure 3.7, we take $z = 1 + (3/4)i$ and $w = 5/12 + i$.

We end this section by listing some of the subsets of $\mathbf{C}$ of particular importance.

- $\mathbf{C}^* = \{z \in \mathbf{C} : z \neq 0\} = \mathbf{C} \setminus \{0\}$ is the *punctured plane*.
- $\mathbf{D} = \{z \in \mathbf{C} : |z| < 1\}$ is the *open unit disc*.
- $\overline{\mathbf{D}} = \{z \in \mathbf{C} : |z| \leq 1\}$ is the *closed unit disc*.
- $\mathbf{T} = \{z \in \mathbf{C} : |z| = 1\}$ is the *unit circle*.
- $H_+ = \{z = x + iy : y > 0\}$ is the *upper half-plane*.
- $H_- = \{z = x + iy : y < 0\}$ is the *lower half-plane*.
- $H_r = \{z = x + iy : x > 0\}$ is the *right-hand half-plane*.
- $H_l = \{z = x + iy : x < 0\}$ is the *left-hand half-plane*.

Exercises

3.7.1 Let $\mathbf{R}[x]$ denote the set of all real polynomials. Let N be the set of all elements of $\mathbf{R}[x]$ which are divisible by $x^2 + 1$: $p(x) \in N$ if we can write $p(x) = (x^2 + 1)q(x)$, with $q(x) \in \mathbf{R}[x]$. Define a relation on $\mathbf{R}[x]$ by setting $p(x) \sim r(x)$ if $p(x) - r(x) \in N$. Verify that this is an equivalence relation. Show that each equivalence class contains an element of degree at most 1. Define operations on the equivalence classes by setting $[p(x)] + [r(x)] = [p(x) + q(x)]$, $[p(x)].[r(x)] = [p(x).r(x)]$. Show that these definitions do not depend on the choice of representatives. Show that with these operations, the quotient space $\mathbf{R}[x]/\sim$ becomes a field, isomorphic to $\mathbf{C}$.

3.7.2 If f and g are mappings from $\mathbf{R}^2$ to $\mathbf{R}^2$ and $a, b \in \mathbf{R}$, define the mapping $af + bg : \mathbf{R}^2 \to \mathbf{R}^2$ by setting $(af + bg)(z) = af(z) + bg(z)$ and define fg as $f \circ g$. Let $I((x, y)) = (x, y)$ and let $J((x, y)) = (-y, x)$. Show that with these laws of composition, the set of mappings $\{aI + bJ : (a, b) \in \mathbf{R}^2\}$ becomes a field, isomorphic to $\mathbf{C}$.

3.7.3 Suppose that $\theta : \mathbf{C} \to \mathbf{C}$ is a field isomorphism of $\mathbf{C}$ onto itself for which $\theta(x) = x$ for x real. Show that either $\theta(z) = z$ for all $z \in \mathbf{C}$ or $\theta(z) = \overline{z}$ for all $z \in \mathbf{C}$.

3.7.4 Show that if x is a non-zero element of an ordered field then $x^2 > 0$. Show that there is no total ordering of $\mathbf{C}$ which makes it an ordered field.

3.7.5 Suppose that $z = x + iy$, with $y > 0$. Show that there are positive real numbers u and v with $2u^2 = |z| + x$ and $2v^2 = |z| - x$. Calculate $(u + iv)^2$. Show that z has exactly two complex square roots. Show that the same holds when $y < 0$.

3.7.6 Suppose that $z_1 z_2 \in \mathbf{C}$. Show that $|z_1 + z_2|^2 + |z_1 - z_2|^2 = 2(|z_1|^2 + |z_2|^2)$ (the parallelogram law). Use induction to find a corresponding result for a finite set $\{z_1, \ldots, z_n\}$ of complex numbers.

3.7.7 Sketch the region $\{z \in \mathbf{C} : |z - 1| < 1\}$ in the Argand diagram.

3.7.8 Sketch the region $\{z \in \mathbf{C} : |z| < 2|z - 3|\}$ in the Argand diagram.

3.7.9 Sketch the sets $\Re(z^2) = c$ and $\Im(z^2) = d$, where c and d are real constants.

3.7.10 A triple (a, b, c) of integers is called a *Pythagorean triple* if $a^2 + b^2 = c^2$. Suppose that (a, b, c) and (m, n, p) are Pythagorean triples. Verify that $(am - bn, an + bm, cp)$ is a Pythagorean triple, and interpret this in terms of complex multiplication.

3.8 The convergence of complex sequences

In this volume, we concentrate almost exclusively on real analysis. In the next chapter, however, we consider infinite series, and, in particular, power series. Here it is appropriate to consider series with complex terms. In this section we consider the convergence of complex-valued sequences.

The definitions are very straightforward generalizations of the definitions in the real case. Suppose that $(z_n)_{n=1}^\infty$ is a sequence of of complex numbers. It *converges* to a complex number z if whenever $\epsilon > 0$ there exists $n_0 \in \mathbf{N}$ such that $|z_n - z| < \epsilon$ for all $n \geq n_0$, and it is a *Cauchy sequence* if whenever $\epsilon > 0$ there exists $n_0 \in \mathbf{N}$ such that $|z_n - z_m| < \epsilon$ for all $m, n \geq n_0$. We write $z_n \to z$ as $n \to \infty$ if z_n converges to z as $n \to \infty$. If z_n converges to 0 as $n \to \infty$, we say that $(z_n)_{n=1}^\infty$ is a *null sequence*.

These definitions can be expressed in terms of real sequences. If z or z_n is a complex number and we write $z = x + iy$ or $z_n = x_n + iy_n$, then x and x_n are always the real parts and y and y_n the imaginary parts of z and z_n, respectively.

Proposition 3.8.1 *Suppose that $(z_n)_{n=1}^\infty = (x_n + iy_n)_{n=1}^\infty$ is a sequence in $\mathbf{C}$ and that $z = x + iy \in \mathbf{C}$. Then $z_n \to z$ as $n \to \infty$ if and only if $x_n \to x$ and $y_n \to y$ as $n \to \infty$. The sequence $(z_n)_{n=1}^\infty$ is a Cauchy sequence if and only if each of the real sequences $(x_n)_{n=1}^\infty$ and $(y_n)_{n=1}^\infty$ is a Cauchy sequence.*

Proof First suppose that $z_n \to z$ as $n \to \infty$. Since $|x_n - x| \leq |z_n - z|$ and $|y_n - y| \leq |z_n - z|$, $x_n \to x$ and $y_n \to y$ as $n \to \infty$. Conversely, since $|z_n - z| \leq |x_n - x| + |y_n - y|$, it follows that if $x_n \to x$ and $y_n \to y$ as $n \to \infty$, then $z_n \to z$ as $n \to \infty$. The proof of the result concerning Cauchy sequences is essentially the same. □

These elementary results enable us the deduce the following results from the results of Section 3.1. A subset B of $\mathbf{C}$ is *bounded* if $\{|z| : z \in B\}$ is bounded in $\mathbf{R}$. A sequence $(z_n)_{n=0}^\infty$ is *bounded* if the set of values $\{z_n : n \in \mathbf{Z}^+\}$ is bounded.

Theorem 3.8.2 *Suppose that $(z_n)_{n=1}^\infty$ and $(w_n)_{n=1}^\infty$ are sequences in $\mathbf{C}$.*

(i) *If $z_n \to z$ as $n \to \infty$ and $z_n \to w$ as $n \to \infty$, then $z = w$.*

(ii) *If z_n is a convergent sequence in $\mathbf{C}$, then it is bounded.*

(iii) *If $z_n = z$ for all n, then $z_n \to z$ as $n \to \infty$.*

(iv) *If $(z_n)_{n=0}^\infty$ is a null sequence, and $(w_n)_{n=0}^\infty$ is bounded, then $(z_n w_n)_{n=0}^\infty$ is a null sequence.*

(v) *If $z_n \to z$ and $w_n \to w$ as $n \to \infty$ then $z_n + w_n \to z + w$ as $n \to \infty$.*

(v) *If $z_n \to z$ and $w_n \to w$ as $n \to \infty$ then $z_n w_n \to zw$ as $n \to \infty$.*

(vi) If $z_n \neq 0$ and $z \neq 0$ and $z_n \to z$ as $n \to \infty$ then $1/z_n \to 1/z$ as $n \to \infty$.

(viii) If $z_n \to z$ as $n \to \infty$ and if $(z_{n_k})_{k=0}^\infty$ is a subsequence, then $z_{n_k} \to z$ as $k \to \infty$.

Proof The reader should verify that these results follow from the results of Section 3.2 and Proposition 3.8.1.   □

Similarly, we have the following results.

Theorem 3.8.3 (The complex Bolzano–Weierstrass theorem) *Suppose that $(z_n)_{n=0}^\infty$ is a bounded sequence of complex numbers. Then there is a subsequence $(z_{n_k})_{k=1}^\infty$ which converges.*

Proof By the real Bolzano–Weierstrass theorem, there exists a subsequence $(z_{m_l})_{l=1}^\infty$ such that the real subsequence $(x_{m_l})_{l=1} \to \infty$ converges, and there exists a subsequence $(z_{n_k})_{k=1}^\infty$ of that for which $(y_{n_k})_{k=1}^\infty$ converges. Then $(z_{n_k})_{k=1}^\infty$ converges, by Proposition 3.8.1.   □

Theorem 3.8.4 (The complex general principle of convergence) *A sequence $(z_n)_{n=0}^\infty$ of complex numbers is convergent if and only if it is a Cauchy sequence.*

Proof This follows easily from Proposition 3.8.1.   □

Example 3.8.5 Suppose that $z \in \mathbf{C}$. Let $z_n = z^n$. Then $z_n \to 0$ as $n \to \infty$ if $|z| < 1$, $z_n \to 1$ if $z = 1$. Otherwise, the sequence $(z_n)_{n=1}^\infty$ does not converge.

For $|z_n - 0| = |z|^n$. If $|z| < 1$ then $|z|^n \to 0$ as $n \to \infty$, from which it follows that $z_n \to 0$ as $n \to \infty$. If $z = 1$ then $z_n = 1$ for all n, so that $z_n \to 1$ as $n \to \infty$. If $|z| \geq 1$ and $z \neq 1$ then $|z_{n+1} - z_n| = |z^n||z - 1| \geq |z - 1|$ for all $n \in \mathbf{N}$, so that $(z_n)_{n=1}^\infty$ is not a Cauchy sequence, and therefore does not converge.

Exercise

3.8.1 Suppose that $(z_n)_{n=1}^\infty$ is a sequence in $\mathbf{C}$ which converges to z. Show that $\bar{z}_n \to \bar{z}$ and $|z_n| \to |z|$ as $n \to \infty$.

4

Infinite series

4.1 Infinite series

The notion of convergence of a sequence allows us to consider infinite sums, or series. Once again, we take either $\mathbf{N}$ or $\mathbf{Z}^+$ as index set. We shall generally consider the case where the terms of the series are complex-valued; since $\mathbf{R} \subseteq \mathbf{C}$, the results will also apply to the case where all the terms are real-valued. Suppose that $(z_j)_{j=0}^{\infty}$ is a sequence of complex numbers. We set

$$s_n = \sum_{j=0}^{n} z_j = z_0 + \cdots + z_n,$$

where s_n is the nth *partial sum*. If $s_n \to s$ as $n \to \infty$, we say that the *infinite sum*, or *infinite series*, $\sum_{j=0}^{\infty} z_j$ *converges* to s. If s_n does not converge, then we say that $\sum_{j=0}^{\infty} z_j$ *diverges*.

Here are two easy examples: as we shall see, the first one is particularly useful. Suppose that $|z| < 1$. Let $z_j = z^j$ for $j \in \mathbf{Z}^+$. Then

$$(1 - z)s_n = (1 + z + \cdots + z^n) - (z + z^2 + \cdots + z^{n+1}) = 1 - z^{n+1},$$

so that

$$s_n = \frac{1 - z^{n+1}}{1 - z} = \frac{1}{1 - z} - \frac{z^{n+1}}{1 - z} \quad \text{and} \quad s_n \to \frac{1}{1 - z} \quad \text{as } n \to \infty.$$

Thus $\sum_{j=0}^{\infty} z^j = 1/(1 - z)$.

Secondly, let

$$a_j = \frac{1}{j(j+1)} = \frac{1}{j} - \frac{1}{j+1} \quad \text{for } j \in \mathbf{N}.$$

Then

$$s_n = \sum_{j=1}^{n} a_j = \left(1 - \frac{1}{2}\right) + \left(\frac{1}{2} - \frac{1}{3}\right) + \cdots + \left(\frac{1}{n} - \frac{1}{n+1}\right) = 1 - \frac{1}{n+1},$$

so that $s_n \to 1$: $\sum_{j=1}^{\infty} 1/j(j+1) = 1$.

We can apply the results that we have obtained about convergent sequences to infinite series. For example, a complex series is convergent if and only if the sum of the real parts and the sum of the imaginary parts of the terms both converge.

Proposition 4.1.1 *Suppose that $z_j = x_j + iy_j$ and that $s = \sigma + i\tau$. Then $\sum_{j=0}^{\infty} w_j$ converges to s if and only if $\sum_{j=0}^{\infty} x_j$ converges to σ and $\sum_{j=0}^{\infty} y_j$ converges to τ.*

The following result follows immediately from Theorems 3.8.2 and 3.2.5.

Theorem 4.1.2 *Suppose that $\sum_{j=0}^{\infty} z_j$ converges to s and that $\sum_{j=0}^{\infty} w_j$ converges to t.*

(i) *When it exists, the sum is unique: if $\sum_{j=0}^{\infty} z_j = s'$, then $s = s'$.*
(ii) *$\sum_{j=0}^{\infty} (z_j + w_j)$ converges to $s + t$.*
(iii) *If $c \in \mathbf{C}$ then $\sum_{j=0}^{\infty} c z_j$ converges to cs.*
(iv) *If z_j and w_j are real, and $z_j \leq w_j$ for all j, then $s \leq t$.*

Suppose that $(j_k)_{k=0}^{\infty}$ is a strictly increasing sequence in $\mathbf{Z}^+$. Set $b_0 = \sum_{j=0}^{j_0} z_j$, and set $b_k = \sum_{j=j_{k-1}+1}^{j_k} z_j$ for $k > 0$. Then the sequence $(b_k)_{k=0}^{\infty}$ is called a *block sequence*, or *bracketed sequence*, derived from $(a_j)_{j=0}^{\infty}$. The following result then follows immediately from Theorem 3.2.5 (viii).

Proposition 4.1.3 *If $\sum_{j=0}^{\infty} z_j$ converges to s and $(b_k)_{k=0}^{\infty}$ is a block sequence derived from it, then $\sum_{k=0}^{\infty} b_k$ converges to s.*

The converse is false in general (but see Corollary 4.2.3 below). Let $z_j = (-1)^j$, for $j \in \mathbf{N}^+$. Then $s_{2n} = 0$ and $s_{2n+1} = 1$ for all $n \in \mathbf{Z}^+$, so that $\sum_{j=0}^{\infty} z_j$ diverges. If we set $j_k = 2k + 1$, then $b_k = z_{2k} + z_{2k+1} = 0$ for $k \in \mathbf{N}^+$, so that $\sum_{k=1}^{\infty} b_k$ converges to 0.

We also have the following simple result.

Proposition 4.1.4 *If $\sum_{j=0}^{\infty} z_j$ converges, then $z_j \to 0$ as $j \to \infty$.*

Proof Suppose that the sum is s. Then $s_j \to s$ and $s_{j-1} \to s$ as $j \to \infty$, so that $z_j = s_j - s_{j-1} \to 0$ as $j \to \infty$.   □

The general principle of convergence takes the following form.

Theorem 4.1.5 (The general principle of convergence) *Suppose that* $(z_j)_{j=0}^\infty$ *is a sequence of complex numbers. Then* $\sum_{j=0}^\infty z_j$ *converges if and only if given* $\epsilon > 0$ *there exists* n_0 *such that* $|s_n - s_m| = |\sum_{j=m+1}^n z_j| < \epsilon$ *for* $n > m \geq n_0$.

Proof This follows immediately from the corresponding result for sequences.

$\square$

Exercises

4.1.1 Show that if $|z| < 1$ then the series $\sum_{j=0}^\infty (j+1)z^j$ converges, and find its sum.

4.1.2 Simplify $1/(1-z) - z/(1-z^2)$. Hence or otherwise show that if $z^2 \neq 1$ then $\sum_{n=0}^\infty z^{2^n}/(1-z^{2^{n+1}})$ converges, and find its sum.

4.1.3 Simplify $z/(1-z) - z/(1+z)$. Hence show that if $|z| < 1$ then

$$\frac{z}{1+z} + \frac{2z^2}{1+z^2} + \frac{4z^4}{1+z^4} + \frac{8z^8}{1+z^8} + \cdots$$

converges, and find its sum.

4.2 Series with non-negative terms

Series with real non-negative terms behave particularly well. Theorem 3.2.4 has the following immediate consequence.

Theorem 4.2.1 *Suppose that* $(a_j)_{j=0}^\infty$ *is a sequence of non-negative real numbers, and that* $s_n = \sum_{j=1}^n a_j$. *Then* $(s_n)_{n=0}^\infty$ *is an increasing sequence. Either* $(s_n)_{n=0}^\infty$ *is bounded, in which case* $\sum_{j=0}^\infty a_j$ *converges to* $\sup_n s_n$, *or* $s_n \to \infty$, *in which case we say that* $\sum_{j=0}^\infty a_j$ *diverges to* $+\infty$, *and write* $\sum_{j=0}^\infty a_j = +\infty$.

This theorem indicates that summing a series of non-negative terms is reasonably straightforward. Here are some of its consequences; the first is one of many tests for convergence.

Corollary 4.2.2 (The comparison test) *If* $0 \leq c_j \leq a_j$ *for all* $j \geq j_0$ *and* $\sum_{j=0}^\infty a_j$ *converges then* $\sum_{j=0}^\infty c_j$ *converges, and* $\sum_{j=0}^\infty c_j \leq \sum_{j=0}^\infty a_j$.

For example, $\sum_{j=1}^\infty 1/j^2$ converges, since $1/j^2 \leq 2/j(j+1)$. Note that this corollary does not tell us what the sum is, although we can deduce from Theorem 4.1.2 that it is at most 2. (In fact the sum is $\pi^2/6$; we shall prove this much later!)

Corollary 4.2.3 *If $(a_j)_{j=0}^{\infty}$ is a sequence of non-negative numbers and $(b_k)_{k=0}^{\infty}$ is a block sequence derived from it, then $\sum_{j=1}^{\infty} a_j$ converges to s if and only if $\sum_{k=1}^{\infty} b_k$ converges to s.*

Proof $s_n \to s$ as $n \to \infty$ if and only if $s_{j_l} \to s$ as $l \to \infty$, and $s_{j_l} = \sum_{k=0}^{l} b_k$. □

We can say more when $(a_j)_{j=0}^{\infty}$ is a decreasing sequence of non-negative numbers.

Corollary 4.2.4 (The compression principle) *If $(a_j)_{j=1}^{\infty}$ is a decreasing sequence of non-negative real numbers, then $\sum_{j=1}^{\infty} a_j$ converges if and only if $\sum_{k=0}^{\infty} 2^k a_{2^k}$ converges. If so then*

$$\frac{1}{2} \sum_{k=0}^{\infty} 2^k a_{2^k} \leq \sum_{j=0}^{\infty} a_j \leq \sum_{k=0}^{\infty} 2^k a_{2^k}.$$

Proof Let $(b_k)_{k=0}^{\infty}$ be the block sequence obtained by taking $j_k = 2^k$. Then

$$\frac{1}{2} 2^k a_{2^k} = 2^{k-1} a_{2^k} \leq b_k = a_{2^{k-1}+1} + \cdots + a_{2^k} \leq 2^{k-1} a_{2^{k-1}},$$

since $(a_j)_{j=0}^{\infty}$ is decreasing, and there are 2^{k-1} summands. Thus the convergence result follows from two applications of the comparison test and the inequalities from Theorem 4.1.2 (iv). □

Corollary 4.2.5 *The harmonic series $\sum_{j=1}^{\infty} 1/j$ diverges to $+\infty$.*

Proof For if $a_j = 1/j$ then $2^k a_{2^k} = 1$, so that the result follows from the preceding corollary. □

Corollary 4.2.6 (Cauchy's test) *Suppose that $(a_j)_{j=0}^{\infty}$ is a bounded sequence of non-negative real numbers. If $\limsup_{j \to \infty} a_j^{1/j} < 1$ then $\sum_{j=1}^{\infty} a_j$ converges, and if $\limsup_{j \to \infty} a_j^{1/j} > 1$ then $\sum_{j=1}^{\infty} a_j = +\infty$.*

Proof In the first case, choose r such that $\limsup_{j \to \infty} a_j^{1/j} < r < 1$. Then there exists j_0 such that $a_j^{1/j} < r$ for $j \geq j_0$. Thus $a_j \leq r^j$ for $j \geq j_0$ and so, using the comparison test, $\sum_{j=1}^{\infty} a_j$ converges.

In the second case, for each $j \in \mathbf{Z}^+$ there exists $k \geq j$ such that $a_k^{1/k} > 1$, so that $a_k > 1$. Thus $(a_j)_{j=0}^{\infty}$ is not a null sequence and so $\sum_{j=1}^{\infty} a_j$ diverges to $+\infty$. □

Corollary 4.2.7 (D'Alembert's ratio test) *Suppose that $(a_j)_{j=0}^{\infty}$ is a sequence of positive real numbers. If $\limsup_{j\to\infty} a_{j+1}/a_j < 1$ then $\sum_{j=1}^{\infty} a_j$ converges. If $\liminf_{j\to\infty} a_{j+1}/a_j > 1$ then $\sum_{j=1}^{\infty} a_j$ diverges to $+\infty$.*

Proof In the first case, choose r such that $\limsup_{j\to\infty} a_{j+1}/a_j < r < 1$. Then there exists j_0 such that $a_{j+1}/a_j < r$ for $j \geq j_0$. Thus if $j > j_0$ then

$$a_j = \left(\frac{a_j}{a_{j-1}}\right)\left(\frac{a_{j-1}}{a_{j-2}}\right) \cdots \left(\frac{a_{j_0+1}}{a_{j_0}}\right) a_{j_0} \leq r^{j-j_0} a_{j_0} = (a_{j_0} r^{j_0}) r^j,$$

and so, taking the terms $a_1, \ldots, a_{j_0}$ into account, there exists M such that $a_j \leq Mr^j$ for all j. By the comparison test, $\sum_{j=1}^{\infty} a_j$ converges.

In the second case, there exists j_1 such that $a_{j+1} > a_j$ for $j \geq j_1$, so that $a_j \geq a_{j_1}$ for $j \geq j_1$. Thus $(a_j)_{j=0}^{\infty}$ is not a null sequence, so that by Proposition 4.1.4, $\sum_{j=1}^{\infty} a_j$ diverges to $+\infty$. □

It is important to note that neither corollary gives any information when $\limsup_{j\to\infty} a_j^{1/j} = 1$ or when $\limsup_{j\to\infty} a_j/a_{j+1} = 1$. When $a_j = 1/j$, the sum diverges, and when $a_j = 1/j^2$, the sum converges. In either case, $a_j^{1/j} \to 1$ and $a_{j+1}/a_j \to 1$ as $j \to \infty$.

We use D'Alembert's ratio test to introduce the *exponential function*, one of the most important functions in analysis. Suppose that $x \geq 0$. Let $a_j = x^j/j!$. Then $a_{j+1}/a_j = x/(j+1)$ and $x/(j+1) \to 0$ as $j \to \infty$, so that $\sum_{j=0}^{\infty} x^j/j!$ converges, to $\exp(x)$, say. The mapping $x \to \exp(x)$ is the exponential function. We set $e = \exp(1) = \sum_{j=0}^{\infty} 1/j!$. Note that since $1/n! \leq 1/2^{n-1}$, it follows that

$$2 \leq e \leq 1 + \sum_{j=1}^{\infty} \frac{1}{2^{j-1}} = 3.$$

In fact, $e = 2.718281828\ldots$. We shall extend this definition for negative x in the next section and for complex x in Section 4.7.

Let us give an example, relating to the argument of Theorem 3.3.1. Suppose that x is a positive real number. As in Theorem 3.3.1, we set $x_n = \lfloor 2^n x \rfloor / 2^n$. Let $a_0 = x_0 = \lfloor x \rfloor$ and let $a_n = 2^n(x_n - x_{n-1})$ for $n \in \mathbf{N}$. Then $a_n = 0$ or 1, and

$$x_n = a_0 + \left(\frac{a_1}{2} + \frac{a_2}{2^2} + \cdots + \frac{a_n}{2^n}\right).$$

Thus $x = \sum_{n=0}^{\infty}(a_n/2^n)$. We can write this as

$$x = a_0 \cdot a_1 a_2 \ldots.$$

This is the *binary expansion* of x. Note that, with this procedure, recurrent 1s are avoided.

We can of course also consider expansions with bases other than 2. We can for example write $u = u_0 + \sum_{j=1}^{\infty} u_j/10^j$ where $0 \le u_j \le 9$, to obtain the familiar *decimal expansion* of u, and we can write $v = v_0 + \sum_{n=1}^{\infty} v_j/3^j$, where $v_j = 0, 1$ or 2; this is the *ternary expansion* of v. There are other possibilities: for example, we can write $w = w_0 + \sum_{j=2}^{\infty} w_j/j!$, where $0 \le w_j < j$.

We can use these ideas to show that **R** is uncountable.

Theorem 4.2.8 *The set* **R** *of real numbers is uncountable.*

Proof We give two proofs. The first was given by Cantor in 1891. It is enough to show that $[0, 1) = \{x \in \mathbf{R} : 0 \le x < 1\}$ is uncountable. Suppose that $(x_n)_{n=1}^{\infty}$ is a sequence in $[0, 1)$. We show that there exists $y \in [0, 1)$ which does not occur in the sequence, so that there can be no surjective mapping of **N** onto $[0, 1)$. Let $x_n = 0.x_{n1}x_{n2}\ldots$ be the decimal expansion of x_n. We set $y_n = 0$ if $x_{nn} \ne 0$, and $y_n = 2$ if $x_{nn} = 0$. The sum $\sum_{n=1}^{\infty} y_n/10^n$ converges, to y, say. From the construction, $|x_n - y| \ge 1/10^n$, and so $y \ne x_n$, for any n.

For the second proof, we define an injective map c from $P(\mathbf{N})$ into $[0, 1]$; since $P(\mathbf{N})$ is uncountable, so is $[0, 1]$. This time, let us use ternary expansions. Suppose that $A \subseteq \mathbf{N}$. Let $a_j = 2$ if $j \in A$, and let $a_j = 0$ if $j \notin A$. Then $\sum_{j=1}^{\infty} a_j/3^j$ converges, to $c(A)$, say. Suppose that $A \ne B$, and that k is the least integer in exactly one of A and B. Then $|c(A) - c(B)| \ge 2/3^k - \sum_{j=k+1}^{\infty} 2/3^j = 1/3^k$, and so $c(A) \ne c(B)$. Thus the mapping $C : A \to c(A) : P(\mathbf{N}) \to [0, 1]$ is injective. We shall meet this function again later. $\square$

Cantor's result, first proved by him in 1873, was very controversial. We know that the rationals are countable, and so there are 'many more' irrationals than rationals. We can say more. A real number x is *algebraic* if there exists a non-zero polynomial p with rational coefficients such that x is a root of p; otherwise it is *transcendental*. For example, radicals (numbers of the form $k^{1/n}$) are algebraic. So are the three real roots of the quintic $x^5 - 4x + 2$, although, following the results of Ruffini and Abel, these roots cannot be expressed in term of radicals. It can be hard to decide whether a particular number is algebraic or transcendental, and it was only in 1844 that Liouville first showed that any transcendental number existed. In 1851 he gave the first explicit example, showing that the number $\sum_{n=1}^{\infty} 1/10^{n!}$ is trancendental. It is easy to see that $e = \sum_{n=0}^{\infty} 1/n!$ is not rational. If $e = p/q$, then $q!e$ must be an integer; but

$$q!e = (q! + q! + q!/2! + \cdots + 1) + \left(\frac{1}{q} + \frac{1}{q(q+1)} + \cdots \right).$$

The first term is an integer, and the second is less than 1, giving a contradiction. It is much harder to determine whether e is algebraic or transcendental, and it was only in 1873 (the same year as the first proof of Cantor's theorem) that Hermite showed that e is transcendental, whereas the transcendence of π was only established by Lindemann nine years later, in 1882. But the set of algebraic numbers is countable (Exercise 4.2.15), and so there are 'many more' transcendental numbers than algebraic ones! One valid objection to this argument is that it is non-constructive; it does not give a method for producing transcendental numbers. It is however the case that many important results of analysis have this non-constructive property.

Exercises

4.2.1 Which of the following series converge, and which diverge?

$$\sum_{n=1}^{\infty} \frac{1}{1+n^2}; \quad \sum_{n=1}^{\infty} \frac{n!}{n^n}; \quad \sum_{n=1}^{\infty} \frac{1}{(n^2+n)^{1/2}}.$$

4.2.2 Suppose that $0 < a_n < 1$ for $n \in \mathbf{N}$. Show that if $\sum_{n=1}^{\infty} a_n$ converges, then so do $\sum_{n=1}^{\infty} a_n^2$ and $\sum_{n=1}^{\infty} a_n/(1 - a_n)$. Are the converse statements true?

4.2.3 Suppose that $0 < a < b$. Show that

$$1 + \frac{1+a}{1+b} + \frac{(1+a)(1+2a)}{(1+b)(1+2b)} + \cdots$$

converges.

4.2.4 Suppose that $(a_j)_{j=0}^{\infty}$ is a sequence of non-negative real numbers for which $\sum_{j=0}^{\infty} a_j$ converges. Show that there is a sequence $(m_j)_{j=0}^{\infty}$ of positive numbers such that $m_j \to \infty$ as $j \to \infty$ and $\sum_{j=0}^{\infty} m_j a_j$ converges.

4.2.5 Suppose that $(a_j)_{j=0}^{\infty}$ is a sequence of non-negative real numbers for which $\sum_{j=0}^{\infty} a_j$ diverges to $+\infty$. Show that there is a null sequence $(m_j)_{j=0}^{\infty}$ of positive numbers such that $\sum_{j=0}^{\infty} m_j a_j$ diverges to $+\infty$.

4.2.6 Suppose that $x > 0$. Use the binomial theorem to show that

$$e_n(x) = \sum_{j=0}^{n} \frac{x^j}{j!} \geq (1 + x/n)^n.$$

Recall (Exercise 3.1.9) that $(1 + x/m)^m$ is an increasing bounded sequence, which tends to a limit. Show that

$$\lim_{m \to \infty} (1 + x/m)^m \geq e_n(x).$$

Show that $(1 + x/m)^m \to \exp(x)$ as $m \to \infty$.

4.2.7 Suppose that $(a_j)_{j=1}^\infty$ is a decreasing sequence of non-negative real numbers. Show that $\sum_{j=1}^\infty a_j$ converges if and only if $\sum_{k=0}^\infty 3^k a_{3^k}$ converges, and if and only if $\sum_{k=0}^\infty k a_{k^2}$ converges.

4.2.8 Suppose that $(a_j)_{j=1}^\infty$ is a decreasing sequence of non-negative real numbers for which $\sum_{j=1}^\infty a_j$ converges. Show that $n a_n \to 0$ as $n \to \infty$.

4.2.9 Simplify $1 - a/(1 + a)$. Suppose that $a_j \geq 0$ for $j \in \mathbf{N}$. Show that $\sum_{j=1}^\infty a_j/(1 + a_1)(1 + a_2) \ldots (1 + a_j)$ converges, to s say, where $0 < s \leq 1$. Determine s when $\sum_{j=1}^\infty a_j = +\infty$.

4.2.10 Suppose that $(a_j)_{j=0}^\infty$ and $(b_j)_{j=0}^\infty$ are sequences of positive real numbers, and that there exists j_0 such that $a_{j+1}/a_j \leq b_{j+1}/b_j$ for $j \geq j_0$. Show that if $\sum_{j=0}^\infty b_j$ converges, then so does $\sum_{j=0}^\infty a_j$.

4.2.11 Suppose that $(a_j)_{j=1}^\infty$ is a sequence of non-negative real numbers for which $\sum_{j=1}^\infty a_j/j$ converges. Show that $(\sum_{j=1}^n a_j)/n \to 0$ as $n \to \infty$ (Kronecker's Lemma).

4.2.12 The following tests, due to Kummer and Dini, extend D'Alembert's ratio test. Suppose that $(a_j)_{j=0}^\infty$ and $(c_j)_{j=0}^\infty$ are sequences of positive real numbers.
Show that if

$$\limsup_{j \to \infty} \left(\frac{c_{j+1} a_{j+1}}{a_j} - c_j \right) < 0$$

then $\sum_{j=1}^\infty a_j$ converges.
Show that if $\sum_{j=0}^\infty (1/c_j)$ diverges to $+\infty$ and

$$\liminf_{j \to \infty} \left(\frac{c_{j+1} a_{j+1}}{a_j} - c_j \right) > 0$$

then $\sum_{j=1}^\infty a_j$ diverges to $+\infty$.

4.2.13 As a special case of the tests of the previous exercise, suppose that $(a_j)_{j=0}^\infty$ is a sequences of positive numbers for which $a_{j+1}/a_j \to 1$, so that D'Alembert's test gives no information.
Show that if

$$\limsup_{j \to \infty} \frac{j(a_{j+1} - a_j)}{a_j} < 0$$

then $\sum_{j=1}^\infty a_j$ converges.

Show that if

$$\liminf_{j\to\infty} \frac{j(a_{j+1} - a_j)}{a_j} > 0$$

then $\sum_{j=1}^{\infty} a_j$ diverges to $+\infty$.

4.2.14 Suppose that $0 < a < b$. Show that

$$1 + \frac{1+a}{1+b} + \frac{(1+a)(2+a)}{(1+b)(2+b)} + \cdots$$

converges if $b > a + 1$ and diverges if $b < a + 1$. What happens if $b = a + 1$?

4.2.15 Show that the set of polynomials of degree d with rational coefficients is countable. Show that the set of all polynomials with rational coefficients is countable. Show that the set of algebraic numbers is countable.

4.3 Absolute and conditional convergence

A series $\sum_{j=0}^{\infty} z_j$ is said to *converge absolutely* if $\sum_{j=0}^{\infty} |z_j|$ converges.

Proposition 4.3.1 *If $\sum_{j=0}^{\infty} z_j$ converges absolutely then it converges, and $|\sum_{j=0}^{\infty} z_j| \leq \sum_{j=0}^{\infty} |a_j|$.*

Proof If $z_j = x_j + iy_j$ then $|x_j| \leq |z_j|$ and $|y_j| \leq |z_j|$, so that $\sum_{j=0}^{\infty} z_j$ converges absolutely if and only if $\sum_{j=0}^{\infty} x_j$ and $\sum_{j=0}^{\infty} y_j$ do; it is enough to consider series with real terms. Suppose that $\sum_{j=0}^{\infty} a_j$ is an absolutely convergent real series. Let

$$a_j^+ = a_j \text{ if } a_j \geq 0 \text{ and } a_j^+ = 0 \text{ if } a_j < 0,$$

$$a_j^- = 0 \text{ if } a_j \geq 0 \text{ and } a_j^- = -a_j = |a_j| \text{ if } a_j < 0.$$

Since $\sum_{j=0}^{\infty} |a_j|$ converges, each of the series $\sum_{j=0}^{\infty} a_j^+$ and $\sum_{j=0}^{\infty} a_j^-$ converges. Since $a_j = a_j^+ - a_j^-$, $\sum_{j=0}^{\infty} a_j$ converges. Since $|\sum_{j=0}^{n} a_j| \leq \sum_{j=0}^{n} |a_j|$, for all n, $|\sum_{j=0}^{\infty} a_j| \leq \sum_{j=0}^{\infty} |a_j|$. $\square$

Absolutely convergent series are generally as well behaved as series with non-negative terms. The comparison test, D'Alembert's ratio test and Cauchy's test can clearly be used to test for absolute convergence. For example, if $z \in \mathbf{C}$ then $\sum_{j=0}^{\infty} z^j/j!$ converges absolutely; we again denote the sum by $\exp(z)$. Thus we have defined the exponential function for all complex z.

A series $\sum_{j=0}^{\infty} a_j$ is said to be *conditionally convergent* if it converges, but does not converge absolutely.

Proposition 4.3.2 *If $\sum_{j=0}^{\infty} a_j$ is a conditionally convergent real series then $\sum_{j=0}^{\infty} a_j^+ = +\infty$ and $\sum_{j=0}^{\infty} a_j^- = +\infty$.*

Proof At least one of the sums must diverge. Suppose that $\sum_{j=0}^{\infty} a_j^+ = +\infty$ and that $\sum_{j=0}^{\infty} a_j^-$ converges to s_-. Suppose that $M > 0$. There exists n_0 such that $\sum_{j=0}^{n} a_j^+ > M + s_-$ for $n \geq n_0$, so that

$$s_n = \sum_{j=0}^{n} a_j^+ - \sum_{j=0}^{n} a_j^- \geq \sum_{j=0}^{n} a_j^+ - s_- > M \quad \text{for } n \geq n_0.$$

Thus $(s_n)_{n=0}^{\infty}$ is unbounded, giving a contradiction. A similar argument applies if $\sum_{j=0}^{\infty} a_j^- = +\infty$ and $\sum_{j=0}^{\infty} a_j^+$ converges. □

Thus conditional convergence depends on cancellation of positive and negative quantities, and arguments are generally more delicate. Fortunately there are some useful tests for convergence; the conditions that are imposed are all-important.

Theorem 4.3.3 (The alternating series test) *Suppose that $(a_j)_{j=0}^{\infty}$ is a decreasing null sequence of positive real numbers. Then $\sum_{j=0}^{\infty} (-1)^j a_j$ converges, to s, say. Further, the sequence $(s_{2n+1})_{n=0}^{\infty}$ increases to s, and the sequence $(s_{2n})_{n=0}^{\infty}$ decreases to s.*

Proof Since

$$s_{2n+1} = s_{2n-1} + (a_{2n} - a_{2n+1}) \geq s_{2n-1} \text{ and}$$

$$s_{2n+2} = s_{2n} - (a_{2n+1} - a_{2n}) \leq s_{2n},$$

the sequence $(s_{2n+1})_{n=0}^{\infty}$ is increasing and the sequence $(s_{2n})_{n=0}^{\infty}$ is decreasing. Since

$$s_{2n+1} = s_{2n} - a_{2n+1} \leq s_{2n} \leq s_0$$

$$\text{and } s_{2n+2} = s_{2n+1} + a_{2n+1} \geq s_{2n+1} \geq s_1,$$

the sequence $(s_{2n+1})_{n=0}^{\infty}$ is bounded above and the sequence $(s_{2n})_{n=0}^{\infty}$ is bounded below. Consequently, they both converge, as $n \to \infty$. Since $s_{2n} - s_{2n+1} = a_{2n+1} \to 0$ as $n \to \infty$, the limits are the same, and s_n converges to the common limit. □

This result has the benefit that if we calculate s_{2n} and s_{2n+1} then we know that $s_{2n+1} \leq s \leq s_{2n}$, and so we have an estimate of the error. But in practice this estimate is usually too crude to be useful.

The next three tests extend this result, and also apply to complex series.

Theorem 4.3.4 (Hardy's test)   *Suppose that $(a_j)_{j=0}^\infty$ is a null sequence of complex numbers for which $\sum_{j=1}^\infty |a_j - a_{j-1}| < \infty$, and that $(z_j)_{j=0}^\infty$ is a sequence of complex numbers for which the sequence of partial sums $(\sum_{j=0}^n z_j)_{n=0}^\infty$ is bounded. Then $\sum_{j=0}^\infty a_j z_j$ converges.*

Proof   This is a result whose proof is almost forced upon us. Since we do not know what the sum should be, we use the general principle of convergence. Thus we consider a sum of the form

$$s_n - s_m = a_{m+1} z_{m+1} + \cdots + a_n z_n.$$

Let $t_n = \sum_{j=0}^n z_j$ and let $s_n = \sum_{j=0}^n a_j z_j$ for $n \in \mathbf{Z}^+$. We do not have information about the terms z_j, but we do know that there exists M such that $|t_n| \leq M$ for all $n \in \mathbf{Z}^+$. Now $z_j = t_j - t_{j-1}$. We therefore substitute, and rearrange:

$$s_n - s_m =$$
$$= a_{m+1}(t_{m+1} - t_m) + \cdots + a_n(t_n - t_{n-1})$$
$$= -a_{m+1} t_m + (a_{m+1} - a_{m+2}) t_{m+1} + \cdots + (a_{n-1} - a_n) t_{n-1} + a_n t_n.$$

This equation (and others of a similar form) is known as *Abel's formula*.

   Suppose that $\epsilon > 0$. There exists n_0 such that $\sum_{j=n_0+1}^\infty |a_j - a_{j-1}| < \epsilon/3(M+1)$ and $|a_n| < \epsilon/3(M+1)$, for $n \geq n_0$. If $n > m \geq n_0$ then

$$|s_n - s_m| \leq |a_{m+1}|.|t_m| + \left(\sum_{j=m+1}^{n-1} |(a_j - a_{j+1}|.|t_j|) \right) + |a_n|.|t_n|$$

$$\leq \left(|a_{m+1}| + \left(\sum_{j=m+1}^{n-1} |(a_j - a_{j+1}| \right) + |a_n| \right) M < \epsilon.$$

Convergence therefore follows from the general principle of convergence.   □

Theorem 4.3.5 (Dirichlet's test)   *Suppose that $(a_j)_{j=0}^\infty$ is a decreasing null sequence of positive real numbers and that $(z_j)_{j=0}^\infty$ is a sequence of complex numbers for which the sequence of partial sums $(\sum_{j=0}^n z_j)_{n=0}^\infty$ is bounded. Then $\sum_{j=0}^\infty a_j z_j$ converges, to s say. Further, if $s_m = \sum_{j=0}^m a_j z_j$ and $M = \sup_n |t_n|$ then $|s - s_m| \leq 2 a_{m+1} M$.*

Proof   Since $\sum_{j=1}^\infty |a_j - a_{j-1}| = \sum_{j=1}^\infty (a_{j-1} - a_j) = a_0$, the first statement follows from Hardy's test. Let $t_n = \sum_{j=0}^n z_j$. Using Abel's formula, we see that

$$|s_n - s_m| \leq$$
$$\leq |a_{m+1}t_m| + |(a_{m+1} - a_{m+2})t_{m+1}| + \cdots + |(a_{n-1} - a_n)t_{n-1}| + |a_n t_n|$$
$$\leq (a_{m+1} + (a_{m+1} - a_{m+2}) + \cdots + (a_{n-1} - a_n) + a_n)M = 2a_{m+1}M.$$

Thus $|s - s_m| = \lim_{n \to \infty} |s_n - s_m| \leq 2a_{m+1}M.$ □

Theorem 4.3.6 (Abel's test) *Suppose that $(a_j)_{j=0}^\infty$ is a decreasing sequence of positive numbers and that $\sum_{j=0}^\infty z_j$ converges. Then $\sum_{j=0}^\infty a_j z_j$ converges.*

Proof We deduce this from Dirichlet's test. The sequence $(a_j)_{j=0}^\infty$ converges: let a be its limit. Since the sequence of partial sums $(\sum_{j=0}^n z_j)_{n=0}^\infty$ is bounded, it follows from Dirichlet's test that $\sum_{j=0}^\infty (a_j - a)z_j$ converges. But $\sum_{j=0}^\infty a z_j$ converges. Adding, we obtain the result. □

Exercises

4.3.1 Do the following series converge?

$$\sum_{n=1}^\infty \frac{(-1)^n}{n - (-1)^n}; \quad \sum_{n=1}^\infty \frac{(-1)^n}{\sqrt{n} - (-1)^n}.$$

4.3.2 Prove Abel's test directly, without appealing to Dirichlet's test.

4.3.3 Suppose that $(a_j)_{j=0}^\infty$ and $(z_j)_{j=0}^\infty$ satisfy the conditions of Abel's test, and that $\sum_{j=0}^\infty a_j z_j = t$. Find an upper bound for $|\sum_{j=0}^n a_j z_j - t|$.

4.3.4 Suppose that $\sum_{j=0}^\infty z_j^2$ converges absolutely. Show that $\sum_{j=0}^\infty z_j/(j+1)$ converges absolutely.

4.4 Iterated limits and iterated sums

A real-valued function f on $\mathbf{N} \times \mathbf{N}$ or on $\mathbf{Z}^+ \times \mathbf{Z}^+$ is called a *double sequence*; we frequently write $(f_{m,n})_{m=1,n=1}^{\infty,\infty}$ for f, where $f_{m,n} = f(m, n)$. Suppose that $(f_{m,n})_{m=1,n=1}^{\infty,\infty}$ is a double sequence. Suppose that $f_{m,n} \to g_n$ as $m \to \infty$, for each $n \in \mathbf{N}$ and that $g_n \to g$ as $n \to \infty$. Suppose also that $f_{m,n} \to h_m$ as $n \to \infty$, for each $m \in \mathbf{N}$ and that $h_m \to h$ as $m \to \infty$. Does it follow that $g = h$?

Simple examples show that the answer is 'no'. For example, let $f_{m,n} = 1$ if $m \leq n$ and let $f_{m,n} = 0$ if $m > n$. Then

$$\lim_{m\to\infty} \left(\lim_{n\to\infty} f_{m,n} \right) = \lim_{m\to\infty} 1 = 1,$$

$$\lim_{n\to\infty} \left(\lim_{m\to\infty} f_{m,n} \right) = \lim_{n\to\infty} 0 = 0.$$

Thus even when the iterated limits exist, the value can depend on the order in which the limits are taken.

The same phenomenon occurs with sums. Let $f_{m,n} = 1$ if $m = n$, let $f_{m,n} = -1/2^{m-n}$ if $m > n$ and let $f_{m,n} = 0$ if $m < n$. Then

$$\sum_{m=1}^{\infty} \left(\sum_{n=1}^{\infty} f_{m,n} \right) = \sum_{m=1}^{\infty} 2^{1-m} = 2$$

$$\sum_{n=1}^{\infty} \left(\sum_{m=1}^{\infty} f_{m,n} \right) = \sum_{n=1}^{\infty} 0 = 0.$$

These examples show that we cannot always interchange the order in which we take limits. On the other hand, if certain conditions are satisfied, then the same value is obtained, independent of the order in which the limits are taken. In the exercises, examples of this are given.

Exercises

4.4.1 Suppose that $\{a_{jk} : (j,k) \in \mathbf{Z}^+ \times \mathbf{Z}^+\}$ is a double sequence of non-negative numbers. Show that the following are equivalent.

(a) $\sum_{k=0}^{\infty} a_{jk}$ converges for each $j \in \mathbf{Z}^+$, and $\sum_{j=0}^{\infty}(\sum_{k=0}^{\infty} a_{jk})$ converges.

(b) $\sum_{j=0}^{\infty} a_{jk}$ converges for each $k \in \mathbf{Z}^+$, and $\sum_{k=0}^{\infty}(\sum_{j=0}^{\infty} a_{jk})$ converges.

(c) The set $\{\sum_{j=0}^{n}(\sum_{k=0}^{n} a_{jk}) : n \in \mathbf{Z}^+\}$ is bounded.

Show that if these conditions are satisfied then

$$\sum_{j=0}^{\infty}(\sum_{k=0}^{\infty} a_{jk}) = \sum_{k=0}^{\infty}(\sum_{j=0}^{\infty} a_{jk}).$$

4.4.2 Suppose that $\{a_{jk} : (j,k) \in \mathbf{Z}^+ \times \mathbf{Z}^+\}$ is a double sequence of non-negative numbers. Find a sufficient condition, corresponding to the condition in the previous example, for the two limits

$$\lim_{m\to\infty} \left(\lim_{n\to\infty} a_{m,n} \right) \text{ and } \lim_{n\to\infty} \left(\lim_{m\to\infty} a_{m,n} \right)$$

to exist and be equal.

4.4.3 Suppose that $\{z_{jk} : (j,k) \in \mathbf{Z}^+ \times \mathbf{Z}^+\}$ is a double sequence of complex numbers. Show that the following are equivalent.

(a) $\sum_{k=0}^{\infty} |z_{jk}|$ converges for each $j \in \mathbf{Z}^+$, and $\sum_{j=0}^{\infty}(\sum_{k=0}^{\infty} |z_{jk}|)$ converges.

(b) $\sum_{j=0}^{\infty} |z_{jk}|$ converges for each $k \in \mathbf{Z}^+$, and $\sum_{k=0}^{\infty}(\sum_{j=0}^{\infty} |z_{jk}|)$ converges.

(c) The set $\{\sum\{z_{jk} : (j,k) \in F\} : F$ a finite subset of $\mathbf{Z}^+ \times \mathbf{Z}^+\}$ is bounded.

Show that if these conditions are satisfied then

$$\sum_{j=0}^{\infty}(\sum_{k=0}^{\infty} a_{jk}) = \sum_{k=0}^{\infty}(\sum_{j=0}^{\infty} a_{jk}).$$

4.4.4 Let $a_{jk} = 1/(j^2 - k^2)$ for $(j,k) \in \mathbf{N} \times \mathbf{N}$ with $j \neq k$, and let $a_{jj} = 0$ for $j \in \mathbf{N}$. By writing

$$\frac{1}{j^2 - k^2} = \frac{1}{2j}\left(\frac{1}{j+k} + \frac{1}{j-k}\right) \text{ for } j \neq k,$$

show that $\sum_{k=1}^{\infty} a_{jk} = -3/4j^2$, and show that the series converges absolutely. Deduce that $\sum_{j=1}^{\infty}(\sum_{k=1}^{\infty} a_{jk})$ converges. Is

$$\sum_{j=1}^{\infty}(\sum_{k=1}^{\infty} a_{jk}) = \sum_{k=1}^{\infty}(\sum_{j=1}^{\infty} a_{jk})?$$

4.5 Rearranging series

What happens if we try to add the terms of an infinite series in a different order?

Theorem 4.5.1 *Suppose that $\sum_{j=0}^{\infty} z_j$ converges absolutely, and that $\sum_{j=0}^{\infty} z_j = s$. If σ is a permutation of $\mathbf{Z}^+$ then $\sum_{j=0}^{\infty} z_{\sigma(j)}$ converges to s.*

Proof By considering real and imaginary parts, it is enough to consider an absolutely convergent real series $\sum_{j=0}^{\infty} a_j$. First consider the case where all the terms are non-negative. If $n \in \mathbf{Z}^+$ and $k = \sup\{\sigma(j) : 1 \leq j \leq n\}$ then $\sum_{j=0}^{n} a_{\sigma(j)} \leq \sum_{i=0}^{k} a_i \leq s$. Thus $\sum_{j=0}^{\infty} a_{\sigma(j)}$ converges, and $\sum_{j=0}^{\infty} a_{\sigma(j)} \leq s$. By the same token,

$$s = \sum_{j=0}^{\infty} a_j = \sum_{j=0}^{\infty} a_{\sigma^{-1}\sigma(j)} \leq \sum_{j=0}^{\infty} a_{\sigma(j)}.$$

In the general case, write $a_j = a_j^+ - a_j^-$. Then $a_{\sigma(j)} = a_{\sigma(j)}^+ - a_{\sigma(j)}^-$, so that $\sum_{j=0}^\infty a_{\sigma(j)}$ converges to $\sum_{j=0}^\infty a_{\sigma(j)}^+ - \sum_{j=0}^\infty a_{\sigma(j)}^-$, and

$$\sum_{j=0}^\infty a_{\sigma(j)} = \sum_{j=0}^\infty a_{\sigma(j)}^+ - \sum_{j=0}^\infty a_{\sigma(j)}^- = \sum_{j=0}^\infty a_j^+ - \sum_{j=0}^\infty a_j^- = \sum_{j=0}^\infty a_j. \qquad \square$$

When $\sum_{j=0}^\infty a_j$ converges conditionally, the situation is completely different. Let us give an example. Let $a_j = (-1)^{j+1}/\sqrt{j}$, for $j \in \mathbf{N}$. Then

$$1 - \frac{1}{\sqrt{2}} + \frac{1}{\sqrt{3}} - \frac{1}{\sqrt{4}} + \cdots + \frac{1}{\sqrt{2j-1}} - \frac{1}{\sqrt{2j}} + \cdots$$

converges, by the alternating series test. Let us rearrange the terms, taking two positive terms and one negative one, and repeating, to give the series

$$(1 + \frac{1}{\sqrt{3}} - \frac{1}{\sqrt{2}}) + (\frac{1}{\sqrt{5}} + \frac{1}{\sqrt{7}} - \frac{1}{\sqrt{4}}) + \cdots + (\frac{1}{\sqrt{4j+1}} + \frac{1}{\sqrt{4j+3}} - \frac{1}{\sqrt{2j}}) + \cdots.$$

Now

$$\sqrt{j}\left(\frac{1}{\sqrt{4j+1}} + \frac{1}{\sqrt{4j+3}} - \frac{1}{\sqrt{2j}}\right) \to 1 - \frac{1}{\sqrt{2}} \text{ as } j \to \infty$$

so that there exists j_0 such that

$$\frac{1}{\sqrt{4j+1}} + \frac{1}{\sqrt{4j+3}} - \frac{1}{\sqrt{2j}} > \frac{1}{4\sqrt{j}} \text{ for } j \geq j_0.$$

Thus the sum of the rearranged terms diverges to $+\infty$.

This sort of phenomenon is quite general. Let us illustrate this by giving one result for real series, which also indicates that there are many other possibilities.

Theorem 4.5.2 *Suppose that $\sum_{j=1}^\infty a_j$ is a conditionally convergent real series, and that $m < M$. Then there exists a rearrangement $\sum_{j=1}^\infty a_{\sigma(j)}$ such that, setting $t_n = \sum_{j=1}^n a_{\sigma(j)}$, $\liminf_{n\to\infty} t_n = m$ and $\limsup_{n\to\infty} t_n = M$.*

Proof We shall describe the idea of the proof, but omit the technical details. Let

$$P = \{j_1 < j_2 < \cdots\} = \{j \in \mathbf{N} : a_j > 0\}, \ Q = \{k_1 < k_2 < \cdots\} = \mathbf{N} \setminus P.$$

Then $\sum_{i=1}^\infty a_{j_i} = \sum_{j=1}^\infty a_j^+ = +\infty$ and $\sum_{l=1}^\infty (-a_{k_l}) = \sum_{j=1}^\infty a_j^- = +\infty$.

Let us suppose that $M \geq 0$. Let i_1 be the least integer such that $\sum_{i=1}^{i_1} a_{j_i} > M$. We define $\sigma(i) = j_i$ for $1 \leq i \leq i_1$. Next, let l_1 be the least integer such that $\sum_{i=1}^{i_1} a_{j_i} + \sum_{l=1}^{l_1} a_{k_l} < m$. We define $\sigma(i_1 + j) = k_j$ for $1 \le j \le l_1$. We now iterate this procedure, so that the partial sums oscillate between values greater than M and values less than m. The procedure does not terminate, since the sums $\sum_{i=1}^{\infty} a_{j_i}$ and $\sum_{l=1}^{\infty} (-a_{k_l})$ are infinite. The resulting mapping σ from $\mathbf{N}$ to $\mathbf{N}$ is then clearly bijective. Finally, since the sum $\sum_{j=1}^{\infty} a_j$ is convergent, the sequence $(a_j)_{j=1}^{\infty}$ is a null sequence. Thus the size of the 'overshoots' tends to 0, so that $\liminf_{n \to \infty} t_n = m$ and $\limsup_{n \to \infty} t_n = M$.

If $M < 0$, we start by finding a sum less than m, and then proceed as above.   □

In particular, we can rearrange the series to converge to any limit whatever.

Corollary 4.5.3   *If $l \in \mathbf{R}$, there exists a rearrangement $\sum_{j=1}^{\infty} a_{\sigma(j)}$ which converges to l.*

Proof   Take $m = M = l$.   □

Exercises

4.5.1 Let
$$1 - \frac{1}{2} + \frac{1}{3} - \frac{1}{4} + \frac{1}{5} - \frac{1}{6} \cdots = s.$$

Show that
$$1 - \frac{1}{2} + \frac{1}{3} + \frac{1}{5} - \frac{1}{4} + \frac{1}{7} + \frac{1}{9} - \frac{1}{6} + \cdots = \frac{3s}{2}.$$

4.5.2 Show that
$$1 + \frac{1}{3^2} + \frac{1}{5^2} + \frac{1}{7^2} + \cdots = \frac{3}{4}\left(1 + \frac{1}{2^2} + \frac{1}{3^2} + \frac{1}{4^2} + \cdots\right).$$

4.5.3 Suppose that $\sum_{j=0}^{\infty} a_j$ is convergent to s, and that σ is a permutation of $\mathbf{N}$.

(a) Suppose that $|\sigma(j) - j| \le K$ for all j. Show that $\sum_{j=0}^{\infty} a_{\sigma(j)}$ is convergent to s.

(b) Let $m_j = \sup\{|a_k| : k > j\}$. Suppose that $m_j|\sigma(j) - j| \to 0$ as $j \to \infty$. Show that $\sum_{j=0}^{\infty} a_{\sigma(j)}$ is convergent to s.

4.6 Convolution, or Cauchy, products

The results in this section relate to power series, which we shall consider in more detail in the next section. Suppose that $(a_j)_{j=0}^{\infty}$ and $(b_j)_{j=0}^{\infty}$ are sequences of complex numbers. We consider two formal power series

$$a(x) = a_0 + a_1 x + a_2 x^2 + \cdots, \qquad b(x) = b_0 + b_1 x + b_2 x^2 + \cdots.$$

If we formally multiply them, and collect terms together, we obtain

$$a(x)b(x) = c(x) = c_0 + c_1 x + c_2 x^2 + \cdots, \text{ where } c_j = \sum_{i=0}^{j} a_i b_{j-i}.$$

The sequence $(c_j)_{j=0}^{\infty}$ is the *convolution product*, or *Cauchy product*, of the sequences $(a_j)_{j=0}^{\infty}$ and $(b_j)_{j=0}^{\infty}$.

Suppose that $\sum_{j=0}^{\infty} a_j$ converges to s and that $\sum_{j=0}^{\infty} b_j$ converges to t. What can we say about the convergence of $\sum_{j=0}^{\infty} c_j$? First, if both converge conditionally then $\sum_{j=0}^{\infty} c_j$ need not converge. For example, if $a_j = b_j = (-1)^j/\sqrt{j+1}$, then $\sum_{j=0}^{\infty} a_j$ and $\sum_{j=0}^{\infty} b_j$ converge, by the alternating series test. But

$$c_j = (-1)^j \sum_{k=1}^{j+1} \frac{1}{\sqrt{k}\sqrt{j+2-k}}.$$

Since $k(j+2-k) \le (j+2)^2/4$, it follows that $|c_j| \ge 2(j+1)/(j+2) \ge 1$, and the series $\sum_{j=0}^{\infty} c_j$ does not converge.

On the other hand, we have the following.

Proposition 4.6.1 *If $\sum_{j=0}^{\infty} a_j$ and $\sum_{j=0}^{\infty} b_j$ are absolutely convergent, to s and t respectively, and $c_j = \sum_{i=0}^{j} a_i b_{j-i}$ then $\sum_{j=0}^{\infty} c_j$ is absolutely convergent to st.*

Proof First suppose that $a_j \ge 0$ and $b_j \ge 0$ for all j. Consider the terms $a_i b_k$ arranged in a semi-infinite array:

$$
\begin{array}{cccc}
a_0 b_0 & a_0 b_1 & a_0 b_2 & \cdots \\
a_1 b_0 & a_1 b_1 & a_1 b_2 & \cdots \\
a_2 b_0 & a_2 b_1 & a_2 b_2 & \cdots \\
\vdots & \vdots & \vdots & \ddots
\end{array}
$$

Then c_j is the sum of the terms on the diagonal line $\{(i,k) : i+k = j\}$. Thus $u_n = \sum_{j=0}^{n} c_j$ is the sum of the terms in the triangle on and above the line $\{(i,k) : i+k = n\}$. Thus if $m = [n/2]$ is the integral part of $n/2$ then

$$s_m t_m = \sum_{i=0}^{m} a_i \sum_{k=0}^{m} b_k \leq u_n \leq \sum_{i=0}^{n} a_i \sum_{k=0}^{n} b_k = s_n t_n,$$

so that $u_n \to st$, by the sandwich principle.

The result now extends to the case where $\sum_{j=0}^{\infty} a_j$ and $\sum_{j=0}^{\infty} b_j$ are absolutely convergent, by considering real and imaginary parts, and splitting these into positive and negative parts. □

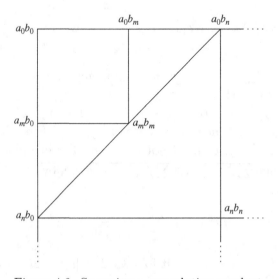

Figure 4.6. Summing a convolution product.

Let us apply this to the exponential function. Let $a_j = a^j/j!$ and $b_j = b^j/j!$. Then

$$c_j = \frac{b^j}{j!} + \frac{ab^{j-1}}{(j-1)!} + \cdots + \frac{a^i b^{j-i}}{i!(j-i)!} + \cdots + \frac{a_j}{j!} = (a+b)^j/j!,$$

by the binomial theorem. Thus $\exp(a)\exp(b) = \exp(a+b)$. Consequently $\exp(z)\exp(-z) = 1$, so that $e^z \neq 0$. In particular, if x is real and negative then $\exp(x) = 1/\exp(-x) > 0$. The mapping $z \to \exp(z)$ is a homomorphism of the additive group $(\mathbf{C}, +)$ into the multiplicative group $(\mathbf{C} \setminus \{0\}, \times)$ of non-zero complex numbers. For this reason, we frequently write e^z for $\exp(z)$.

What happens when one series is absolutely convergent and the other is conditionally convergent?

Theorem 4.6.2 *If $\sum_{j=0}^{\infty} a_j$ is absolutely convergent to s and $\sum_{j=0}^{\infty} b_j$ is conditionally convergent to t, and if $c_j = \sum_{i=0}^{j} a_i b_{j-i}$ then $\sum_{j=0}^{\infty} c_j$ is convergent to st.*

Proof Let s_n, t_n and u_n denote the nth partial sums of the three sequences. The sequence $(t_n)_{n=0}^{\infty}$ is bounded. Let $M = \sup_n |t_n|$, and let $L = \sum_{j=0}^{\infty} |a_j|$. Let $m = [n/2]$. Now

$$u_n = \sum_{j=0}^{n} c_j = \sum_{j=0}^{n} \left(\sum_{i=0}^{j} a_i b_{j-i} \right)$$

$$= \sum_{i=0}^{n} \left(\sum_{j=0}^{n-i} a_i b_{j-i} \right) = \sum_{i=0}^{n} a_i \left(\sum_{j=0}^{n-i} b_j \right)$$

$$= a_0 t_n + \cdots + a_n t_0.$$

Here we first add the rows of the triangle $\{(i, j) : i + j \le n\}$, and then add the resulting sums. Thus

$$u_n - s_n t = a_0(t_n - t) + \cdots + a_n(t_0 - t).$$

We split the sum into two parts: $u_n - s_n t = \lambda_1 + \lambda_2$, where

$$\lambda_1 = a_0(t_n - t) + \cdots + a_m(t_{n-m} - t),$$

$$\text{and } \lambda_2 = a_{m+1}(t_{n-m-1} - t) + \cdots + a_n(t_0 - t).$$

We consider the two sums separately. Given $\epsilon > 0$, there exists n_0 such that

$$\sum_{j=n_0}^{\infty} |a_j| < \frac{\epsilon}{3M + 1} \quad \text{and} \quad \sup_{n \ge n_0} |t_n - t| < \frac{\epsilon}{3L + 1} \quad \text{for } n \ge n_0.$$

If $n \ge 2n_0$, then $m \ge n_0$ and $n - j \ge n_0$ for $0 \le j \le m$, so that

$$|\lambda_1| \le \frac{\epsilon}{3L + 1} \left(\sum_{j=0}^{m} |a_j| \right) < \epsilon/3.$$

Further,

$$|\lambda_2| \le 2M \left(\sum_{j=m+1}^{\infty} |a_j| \right) < 2\epsilon/3$$

so that $u_n - s_n t \to 0$ as $n \to \infty$. Since $u_n - st = u_n - s_n t + (s_n - s)t$, it follows that $u_n \to st$ as $n \to \infty$. $\qquad \square$

The technique of this proof, where we divide a sum into two parts, and consider each part separately, is one that is used in many areas of analysis.

Exercises

4.6.1 Let $a_j = b_j = (-1)^j/(j+1)$ and let $c_j = \sum_{i=0}^j a_i b_{j-i}$. Show that

$$c_j = (-1)^j \frac{2}{j+2} \left(1 + \frac{1}{2} + \cdots + \frac{1}{j+1} \right).$$

Show that $(|c_j|)_{j=0}^\infty$ is a decreasing null sequence. Deduce that $\sum_{j=0}^\infty c_j$ converges.

4.6.2 Suppose that $a_n \to a$ as $n \to \infty$. Let $s_n = a_0 + \cdots + a_n$. Show that $s_n/(n+1) \to a$ as $n \to \infty$.

Suppose that $a_n \to a$ and that $b_n \to b$ as $n \to \infty$. Show that

$$\frac{1}{n+1}(a_0 b_n + \cdots + a_n b_0) \to ab \text{ as } n \to \infty.$$

Suppose that $(c_j)_{j=0}^\infty$ is the convolution product of the sequences $(a_j)_{j=0}^\infty$ and $(b_j)_{j=0}^\infty$, and that $\sum_{j=0}^\infty a_j$ is conditionally convergent to s and $\sum_{j=0}^\infty b_j$ is conditionally convergent to t. Let $u_n = c_0 + \cdots + c_n$.
(a) Show that $u_0 + \cdots + u_n = s_0 t_n + \cdots + s_n t_0$.
(b) Show that $(u_0 + \cdots + u_n)/(n+1) \to st$ as $n \to \infty$.
(c) Show that if $\sum_{j=0}^\infty c_j$ converges, then its sum must be st.

4.7 Power series

A *power series* is an expression of the form $\sum_{n=0}^\infty a_n(z - z_0)^n$, where $(a_n)_{n=0}^\infty$ is a sequence of complex numbers, z_0 is a complex number, and z is a complex number, which we also allow to vary. (In fact, in many circumstances we shall consider complex power series for which the coefficients a_n are real.) We are interested in the values of z for which the power series converges. For this it is clearly sufficient to consider the case where $z_0 = 0$.

We introduce some notation. If $0 < R < \infty$ we set $U_R = \{z \in \mathbf{C} : |z| < R\}$, the *open disc* of radius R with centre 0, and we set $U_\infty = \mathbf{C}$. Thus $U_1 = D$, the *open unit disc*.

Let us begin with a very simple example. Consider the power series $\sum_{n=0}^\infty z^n$. If $|z| \geq 1$ then $|z^n|$ does not tend to zero, and so the power series diverges. If $|z| < 1$, then

$$s_n = \sum_{j=0}^n z^j = \frac{1 - z^{n+1}}{1 - z}, \text{ so that } \left| s_n - \frac{1}{1-z} \right| = \frac{|z|^{n+1}}{|1-z|},$$

so that $\sum_{n=0}^\infty z^n$ converges to $1/(1-z)$. Thus $\sum_{n=0}^\infty z^n$ converges if and only if z is in the open unit disc $D = \{z : |z| < 1\}$.

We can however say more. If $|z| < 1$ then

$$\sum_{n=0}^{\infty} |z^n| = \sum_{n=0}^{\infty} |z|^n = 1/(1 - |z|),$$

so that the series $\sum_{n=0}^{\infty} z^n$ converges absolutely.

Suppose that $\sum_{n=0}^{\infty} a_n z^n$ is a complex power series. For what values of z does it converge? To answer this, it is convenient to consider the set

$$B = \{r \in [0, \infty) : (a_n r^n)_{n=0}^{\infty} \text{ is a bounded sequence}\}.$$

$0 \in B$, and if $r \in B$ then $[0, r] \subseteq B$. Thus B is an interval. If B is bounded we set $R = \sup B$. R may or may not belong to B. If $B = [0, \infty)$, we set $R = \infty$. R is called the *radius of convergence* of the power series $\sum_{n=0}^{\infty} a_n z^n$. The next theorem explains the reason for this name.

Theorem 4.7.1 *Suppose that $\sum_{n=0}^{\infty} a_n z^n$ is a complex power series with radius of convergence R. If $z \in U_R$ then $\sum_{n=0}^{\infty} a_n z^n$ converges absolutely. If $|z| > R$ then $\sum_{n=0}^{\infty} a_n z^n$ does not converge.*

Proof If $|z| > R$ then $(a_n z^n)_{n=0}^{\infty}$ is unbounded, and so the power series diverges. (In particular, if $R = 0$ then the series only converges when $z = 0$.) Suppose that $|z| < R$. There exists s such that $|z| < s < R$, and so $M_s = \sup_{n \in \mathbf{Z}^+} |a_n s^n| < \infty$. Let $r = |z|/s$, so that $0 \le r < 1$. Then

$$|a_n z^n| = |a_n s^n r^n| \le M_s r^n \text{ for } n \in \mathbf{N}.$$

By the comparison test, the series $\sum_{n=0}^{\infty} |a_n z^n|$ converges, and so $\sum_{n=0}^{\infty} a_n z^n$ converges absolutely. $\square$

Note that the proof depends only on the convergence of a geometric series. This simple idea is very powerful, and we shall use it, and the convergence of series such as $\sum_{n=0}^{\infty} n^k r^n$, where $0 \le r < 1$ and $k \in \mathbf{N}$, many times in the future.

We have the following formula for the radius of convergence.

Theorem 4.7.2 *Suppose that $\sum_{n=0}^{\infty} a_n z^n$ is a power series with radius of convergence R. Let $\Lambda = \limsup |a_n|^{1/n}$. If $\Lambda = 0$ then $R = \infty$. If $\Lambda = \infty$ then $R = 0$. Otherwise, $R = 1/\Lambda$.*

Proof This is just a matter of teasing out the definitions. Suppose that $\Lambda < \infty$ and that $S > \Lambda$. Then there exists n_0 such that $|a_n|^{1/n} < S$ for $n \ge n_0$. Thus $|a_n|/S^n < 1$ for $n \ge n_0$; the sequence $(a_n/S^n)_{n=0}^{\infty}$ is bounded, and so $1/S \le R$. Since this holds for all $S > \Lambda$, $R = \infty$ if $\Lambda = 0$, and

$R \geq 1/\Lambda$ otherwise. Suppose that $\Lambda > 0$ and $0 < s < \Lambda$. Let $s < t < \Lambda$. Then $|a_n|^{1/n} > t$ for infinitely many n. Thus $|a_n|/s^n \geq (t/s)^n$ for infinitely many n; the sequence $(a_n/s^n)_{n=0}^{\infty}$ is unbounded, and so $1/s \geq R$. Since this holds for all $s < \Lambda$, $R = 0$ if $\Lambda = \infty$, and $R \leq 1/\Lambda$ otherwise. □

The theorem says nothing about convergence on the *circle of convergence* $C_R = \{z \in \mathbf{C} : |z| = R\}$. There are many possibilities, as the following examples show.

1. $\sum_{n=0}^{\infty} n! z^n$. Since $(n! r^n)_{n=0}^{\infty}$ is unbounded for all $r > 0$, $R = 0$, and the series only converges when $z = 0$.
2. $\sum_{n=0}^{\infty} n z^n$. Here $B = [0, 1)$ and $R = 1$. The sequence $(n z^n)_{n=0}^{\infty}$ is unbounded for each $z \in C_1$.
3. $\sum_{n=0}^{\infty} z^n$. Here $B = [0, 1]$ and $R = 1$. $z^n \not\to 0$ as $n \to \infty$ for each $z \in C_1$.
4. $\sum_{n=0}^{\infty} z^n/n$. Here $B = [0, 1]$ and $R = 1$. $\sum z^n/n$ diverges when $z = 1$. If $z \in C_1$ and $z \neq 1$ then

$$\left| \sum_{j=0}^{n} z^j \right| = \left| \frac{1 - z^{n+1}}{1 - z} \right| \leq \left| \frac{2}{1 - z} \right|,$$

so that the sequence $(\sum_{j=0}^{n} z^j)_{n=1}^{\infty}$ is bounded. Consequently, the series $\sum_{n=0}^{\infty} z^n/n$ converges, by Dirichlet's test (Theorem 4.3.5).
5. $\sum_{n=0}^{\infty} z^n/n^2$. Here $B = [0, 1]$ and $R = 1$. The series converges uniformly on $\{z \in \mathbf{C} : |z| \leq 1\}$.
6. $\sum_{n=0}^{\infty} z^n/n!$. Here $B = [0, \infty)$ and $R = \infty$. The function $e^z = e(z) = \sum_{n=0}^{\infty} z^n/n!$ is the *exponential function*.

If $\sum_{n=0}^{\infty} a_n z^n$ and $\sum_{n=0}^{\infty} b_n z^n$ are power series, we can form the sum $\sum_{n=0}^{\infty} (a_n + b_n) z^n$.

Proposition 4.7.3 *Suppose that $\sum_{n=0}^{\infty} a_n z^n$ has radius of convergence R and $\sum_{n=0}^{\infty} b_n z^n$ has radius of convergence R'. If $R \neq R'$ then the radius of convergence of $\sum_{n=0}^{\infty} (a_n + b_n) z^n$ is $\min(R, R')$; if $R = R'$ the the radius of convergence is greater than or equal to R.*

Proof The proof is left as an exercise.   □

If $\sum_{n=0}^{\infty} a_n z^n$ and $\sum_{n=0}^{\infty} b_n z^n$ are power series, we can form the formal product $\sum_{n=0}^{\infty} c_n z^n$, where $c_n = \sum_{j=0}^{n} a_j b_{n-j}$, as in the previous section.

Theorem 4.7.4 *If $\sum_{n=0}^{\infty} a_n z^n$ has radius of convergence R and $\sum_{n=0}^{\infty} b_n z^n$ has radius of convergence R' then the formal product $\sum_{n=0}^{\infty} c_n z^n$ has radius* of convergence greater than or equal to $\min(R, R')$. If $|z| < \min(R, R')$ then

$$\left(\sum_{n=0}^{\infty} a_n z^n\right)\left(\sum_{n=0}^{\infty} b_n z^n\right) = \sum_{n=0}^{\infty} c_n z^n.$$

Proof Let R'' be the radius of convergence of $\sum_{n=0}^{\infty} c_n z^n$. If $|z| < \min(R, R')$ then all three series converge absolutely, by Proposition 4.6.1, and

$$\left(\sum_{n=0}^{\infty} a_n z^n\right)\left(\sum_{n=0}^{\infty} b_n z^n\right) = \sum_{n=0}^{\infty} c_n z^n.$$

Hence $R'' \geq \min(R, R')$ $\square$

We shall consider power series further in Section 6.6, and in Volume III.

Exercises

4.7.1 Prove Proposition 4.7.3.

4.7.2 Find the radii of convergence of the following power series:

$$\sum_{n=0}^{\infty}(2 + i^n)^n z^n; \quad \sum_{n=0}^{\infty}\frac{(2n)!}{(n!)^2}z^n; \quad \sum_{n=0}^{\infty}n^{\sqrt{n}}z^n; \quad \sum_{n=0}^{\infty}\frac{z^{3n}}{2^n(n+1)}.$$

4.7.3 What is the radius of convergence of the power series

$$\sum_{n=0}^{\infty}\frac{n^n}{n!}z^n?$$

At which points, if any, of the circle of convergence does it converge?

4.7.4 Suppose that $a_{n+1}/a_n \rightarrow \lambda$ as $n \rightarrow \infty$. What is the radius of convergence of $\sum_{n=0}^{\infty} a_n z^n$?

4.7.5 What are the radii of convergence of the power series

$$1 + z + 2z^2 + 4z^3 + 8z^4 + \cdots \text{ and } 1 - z - z^2 - z^3 - \cdots?$$

What is the radius of convergence of their product?

4.7.6 Suppose that the series $\sum_{n=0}^{\infty} a_n z^n$ has non-zero radius of convergence R. Let $f(z) = \sum_{n=0}^{\infty} a_n z^n$ for $|z| < R$.

(a) Show that if the coefficients a_n are real, then $f(\bar{z}) = \overline{f(z)}$ for $|z| < R$.

(b) Show that f is even – that is, $f(z) = -f(z)$ for $|z| < R$ – if and only if $a_n = 0$ for n odd.

(c) Show that f is odd – that is, $f(z) = -f(z)$ for $|z| < R$ – if and only if $a_n = 0$ for n even.

(d) Suppose that $f \neq 0$. Show that there exists $0 < r < R$ such that $f(z) \neq 0$ for $0 < |z| \leq r$.

4.7.7 Suppose that the power series $\sum_{n=0}^{\infty} a_n z^n$ has radius of convergence R. Let $s_n = \sum_{j=0}^{n} a_j$. Investigate the radius of convergence of the power series $\sum_{n=0}^{\infty} s_n z^n$.

4.7.8 Let

$$a_0 = 1, \ a_1 = -1, \ a_j = \frac{(-1)^n}{n2^n} \text{ for } 2^n \leq j < 2^{n+1} \text{ and } n \in \mathbf{N}.$$

Show that if $|z| = 1$ and $z \neq 1$ then

$$\left| \sum_{k=2^n}^{j} a_k z^k \right| \leq \frac{1}{n2^n |1 - z|} \text{ for } 2^n \leq j < 2^{n+1}.$$

Show that $\sum_{n=0}^{\infty} a_n z^n$ converges conditionally, for all z with $|z| = 1$.

5

The topology of **R**

In this chapter, we consider some particular sorts of subsets of **R**, and their relation to convergence. This involves many definitions; familiarity will only come with use. We study the ideas that arise here in a more general setting in Volume II.

5.1 Closed sets

We begin by considering intervals in **R**. A subset I of **R** is an *interval* if whenever two numbers belong to it, then so do all the intermediate points: that is, if $a < c < b$ and $a, b \in I$ then $c \in I$. **R** is an interval. The empty set and singleton sets are *degenerate* intervals. Other examples of intervals are the *semi-infinite intervals*

$$(-\infty, b) = \{x \in \mathbf{R} : x < b\}, \ (-\infty, b] = \{x \in \mathbf{R} : x \le b\},$$
$$(a, \infty) = \{x \in \mathbf{R} : a < x\}, \ [a, \infty) = \{x \in \mathbf{R} : a \le x\},$$

and the *bounded intervals*

$$(a, b) = (b, a) = \{x \in \mathbf{R} : a < x < b\}, (a, b] = [b, a) = \{x \in \mathbf{R} : a < x \le b\},$$
$$[a, b) = (b, a] = \{x \in \mathbf{R} : a \le x < b\}, [a, b] = [b, a] = \{x \in \mathbf{R} : a \le x \le b\},$$

where $a < b$. It is an easy exercise to show that every interval is of one of these forms. The *length* of a bounded interval is $b - a$; the length of **R** and of semi-infinite intervals is $+\infty$.

Note that if $\mathcal{I}$ is a set of intervals then $\cap_{I \in \mathcal{I}} I$ is an interval, and that if I_1 and I_2 are intervals with $I_1 \cap I_2 \neq \emptyset$ then $I_1 \cup I_2$ is an interval.

Next, we consider the closure of a subset of **R**. A real number b is called a *closure point* of a subset A of **R** if whenever $\epsilon > 0$ there exists $a \in A$ (which may depend upon ϵ) with $|b - a| < \epsilon$. Thus b is a closure point of A if there are points of A arbitrarily close to b. If $b \in A$, then b is a closure point of A, since we can take $a = b$ for any $\epsilon > 0$.

We can use convergent sequences to characterize closure points.

Proposition 5.1.1 *Suppose that A is a subset of* **R** *and that $b \in$* **R**. *b is a closure point of A if and only if there exists a sequence $(a_j)_{j=1}^{\infty}$ in A such that $a_j \to b$ as $j \to \infty$.*

Proof Suppose that there exists a sequence $(a_j)_{j=0}^{\infty}$ in A such that $a_j \to b$ as $j \to \infty$. Suppose that $\epsilon > 0$. There exists j_0 such that $|b - a_j| < \epsilon$ for $j \geq j_0$. Take $a = a_{j_0}$. Thus b is a closure point of A.

Conversely, if b is a closure point of A then for each $j \in$ **N** there exists $a_j \in A$ with $|b - a_j| < 1/j$. Then $a_j \to b$ as $j \to \infty$. □

The *closure $\overline{A}$* of A is the set of closure points of A. A is a subset of $\overline{A}$ since each point of A is a closure point of A. A subset A of **R** is said to be *closed* if $A = \overline{A}$:

Proposition 5.1.2 *A subset A of* **R** *is closed if and only if whenever $(a_n)_{n=1}^{\infty}$ is a sequence in A which converges to b, then $b \in A$.*

Proof This is an immediate consequence of Proposition 5.1.1. □

In other words, a subset A of **R** is closed if and only if it is *closed under taking limits*. For example, the interval $[a, b]$ is closed, since if $a \leq x_j \leq b$ and $x_j \to x$ as $j \to \infty$ then $a \leq x \leq b$, by Theorem 3.2.5. (This accords with our use of the term *closed interval* in Section 3.2.) If $a < b$, and $x_j = a + (b-a)/2^j$ for $j \in$ **Z**$^+$ then $x_j \in (a, b]$, and $x_j \to a$ as $j \to \infty$. Thus $(a, b]$ is not closed, since $a \notin (a, b]$. The set **Q** of rational numbers is not closed, since if x is any irrational number then by Corollary 3.2.7 there exists a sequence of rational numbers which converges to x, so that $\overline{\mathbf{Q}} = \mathbf{R}$. A subset A of a subset B of **R** is *dense in B* if $B \subseteq \overline{A}$. Thus **Q** is dense in **R**.

Proposition 5.1.3 *Suppose that A and B are subsets of* **R**.

(i) *If $A \subseteq B$ then $\overline{A} \subseteq \overline{B}$.*
(ii) *$\overline{A}$ is closed.*
(iii) *$\overline{A}$ is the smallest closed set containing A: if C is closed and $A \subseteq C$ then $\overline{A} \subseteq C$.*

Proof (i) follows trivially from the definition of closure.

(ii) Suppose that b is a closure point of $\overline{A}$ and suppose that $\epsilon > 0$. Then there exists $c \in \overline{A}$ such that $|b - c| < \epsilon/2$, and there exists $a \in A$ with $|c - a| < \epsilon/2$. Thus $|b - a| < \epsilon$, and so $b \in \overline{A}$.

(iii) By (i), $\overline{A} \subseteq \overline{C} = C$. □

Here are some fundamental properties of the collection of closed subsets of $\mathbf{R}$.

Proposition 5.1.4 *(i) The empty set $\emptyset$ and $\mathbf{R}$ are closed.*
(ii) If $\mathcal{A}$ is a set of closed subsets of $\mathbf{R}$ then $\cap_{A \in \mathcal{A}} A$ is closed.
(iii) If $\{A_1, \ldots, A_n\}$ is a finite set of closed subsets of $\mathbf{R}$ then $A = \cup_{j=1}^n A_j$
  is closed.

Proof (i) The empty set is closed, since it has no closure points, and $\mathbf{R}$ is trivially closed.

(ii) Suppose that b is a closure point of $\cap_{A \in \mathcal{A}} A$, and that $A \in \mathcal{A}$. If $\epsilon > 0$ then there exists $a \in \cap_{A \in \mathcal{A}} A$ with $|b - a| < \epsilon$. But then $a \in A$. Since this holds for all $\epsilon > 0$, $a \in \overline{A} = A$. Since this holds for all $A \in \mathcal{A}$, $b \in \cap_{A \in \mathcal{A}} A$.

(iii) Suppose that $b \notin A$. If $1 \leq j \leq n$ then $b \notin A_j = \overline{A_j}$, and so there exists $\epsilon_j > 0$ such that if $|b - c| < \epsilon_j$ then $c \notin A_j$. Let $\epsilon = \min\{\epsilon_j : 1 \leq j \leq n\}$. Then $\epsilon > 0$, and if $|b - c| < \epsilon$ then $c \notin \cup_{j=1}^n A_j = A$. Thus b is not a closure point of A; every closure point of A is in A, and so A is closed. $\square$

Corollary 5.1.5 *A finite subset of $\mathbf{R}$ is closed.*

Proof The empty set is closed, and a singleton set $\{a\}$ is closed, since if $b \neq a$ then, setting $\epsilon = |b - a|$, $\{a\} \cap \{x : |x - b| < \epsilon\} = \emptyset$. Now apply (iii). $\square$

Let us give another example.

Example 5.1.6 Suppose that $(a_j)_{j=0}^\infty$ is a sequence of real numbers convergent to a. Let $S = \{a_j : j \in \mathbf{Z}^+\}$. Then $\overline{S} = S \cup \{a\}$.

By Proposition 5.1.1, $a \in \overline{S}$. Suppose that $b \notin S \cup \{a\}$. We shall show that b is not a closure point of S. Let $\eta = |b - a|/2$: then $\eta > 0$. There exists j_0 such that $|a_j - a| < \eta$ for $j \geq j_0$. Then by the triangle inequality,

$$|b - a_j| \geq |b - a| - |a_j - a| \geq 2\eta - \eta = \eta, \text{ for } j \geq j_0.$$

Let $\epsilon = \min(\eta, \min\{|b - a_j| : 1 \leq j < j_0\})$. Then $\epsilon > 0$, and if $s \in S$ then $|b - s| \geq \epsilon$. Thus b is not a closure point of S.

Proposition 5.1.7 *If A is a non-empty subset of $\mathbf{R}$ which is bounded above then $\sup A \in \overline{A}$.*

Proof For each $j \in \mathbf{N}$ there exists $a_j \in A$ with $\sup A - 1/j < a_j \leq \sup A$. Then $a_j \to \sup A$, so that $\sup_A \in \overline{A}$. $\square$

We can also consider subsets of a subset X of $\mathbf{R}$. Suppose that A is subset of X. Then the *relative closure* of A in X is the set $\overline{A} \cap X$ of closure points of

A which are in X. The set A is *relatively closed* in X if it is equal to its relative closure. Relatively closed sets can be characterized in the following way.

Proposition 5.1.8 *Suppose that A is a subset of a subset X of $\mathbf{R}$. Then the following are equivalent:*

(i) *A is relatively closed in X;*
(ii) *there exists a closed subset F of $\mathbf{R}$ such that $A = F \cap X$;*
(iii) *if $(a_n)_{n=0}^{\infty}$ is a sequence in A which converges to a point b of X then $b \in A$.*

Proof This is a worthwhile exercise for the reader.   □

Exercises

5.1.1 Verify that every interval in $\mathbf{R}$ is of one of the forms described at the beginning of this section.

5.1.2 Show that a subset I of $\mathbf{R}$ is an interval if and only if whenever $a, b \in I$ and $0 \le t \le 1$ then $(1 - t)a + tb \in I$.

5.1.3 Suppose that $a \subseteq \mathbf{R}$. Show that the following are equivalent.
(a) A is closed.
(b) If $[a, b]$ is a closed interval for which $A \cap [a, b]$ is non-empty then $\sup(A \cap [a, b]) \in A$ and $\inf(A \cap [a, b]) \in A$.

5.1.4 Suppose that A is a non-empty closed subset of $\mathbf{R}$ and that $b \in \mathbf{R}$. Show that there exists $a \in A$ such that $|a - b| = \inf\{|x - a| : x \in A\}$. Is a unique?

5.1.5 If A and B are non-empty subsets of $\mathbf{R}$, we set $A + B = \{a + b : a \in A, b \in B\}$.
(a) Give an example of closed sets A and B for which $A + B$ is not closed.
(b) Show that if A is closed, and B is closed and bounded, then $A + B$ is closed. [*Hint*: Use the Bolzano–Weierstrass theorem.]

5.1.6 Suppose that $(A_j)_{j=0}^{\infty}$ is a sequence of subsets of $\mathbf{R}$. Show that

$$\overline{\bigcup_{j=0}^{n} A_j} = \bigcup_{j=0}^{n} \overline{A_j} \text{ and that } \overline{\bigcup_{j=0}^{\infty} A_j} \supseteq \bigcup_{j=0}^{\infty} \overline{A_j}.$$

Give an example to show that the inclusion can be strict. What about intersections?

5.1.7 Suppose that x is an irrational number. Let $a_n = \{nx\}$, the fractional part of nx. Use the pigeonhole principle to show that if $\epsilon > 0$ then there exist m, n such that $|a_m - a_n| < \epsilon$. Show that $\{a_n : n \in \mathbf{N}\}$ is dense in $[0, 1]$.

5.1.8 Using Proposition 5.1.2, give proofs of Propositions 5.1.3, 5.1.4 and 5.1.7 which use convergent sequences. Do the same for Example 5.1.6.

5.2 Open sets

Suppose that $a \in \mathbf{R}$ and $\epsilon > 0$. We define the *open ϵ-neighbourhood* of a to be the set of all numbers distant less than ϵ from a:

$$N_\epsilon(a) = \{x \in \mathbf{R} : |x - a| < \epsilon\}.$$

$N_\epsilon(a)$ is the interval $(a - \epsilon, a + \epsilon)$. We can express convergence in terms of ϵ-neighbourhoods; $a_j \to a$ as $j \to \infty$ if and only if for each $\epsilon > 0$ there exists j_0 such that $a_j \in N_\epsilon(a)$ for $j \geq j_0$. Similarly, the closure of a set is defined in terms of ϵ-neighbourhoods: $a \in \bar{A}$ if and only if $N_\epsilon(a) \cap A \neq \emptyset$, for all $\epsilon > 0$.

Suppose that A is a subset of $\mathbf{R}$. An element a of A is an *interior point* of A if there exists $\epsilon > 0$ such that $N_\epsilon(a) \subseteq A$. In other words, all the numbers sufficiently close to a are in A; we can move a little way from a without leaving A. The *interior* A° of A is the set of interior points of A. A subset U of $\mathbf{R}$ is *open* if $U = U^\circ$. The interval $(a, c) = \{b \in \mathbf{R} : a < b < c\}$ is open: if $b \in (a, c)$, we can take $\epsilon = \min(c - b, b - a)$. In particular, an open ϵ-neighbourhood $N_\epsilon(x)$ is open.

The collection of open sets of $\mathbf{R}$ is called the *topology* of $\mathbf{R}$. Properties that can be defined in terms of the open sets are called *topological properties*. The word 'topology' is also used to describe the study of topological properties.

The interval $(a, b]$ is not open: there is no suitable ϵ for b. Thus $(a, b]$ is an example of a set which is neither open nor closed. The set $\mathbf{Q}$ of rational numbers is also neither open nor closed: we have seen that it is not closed, and it is not open, since if $r \in \mathbf{Q}$ and $\epsilon > 0$ then the open ϵ-neighbourhood $N_\epsilon(r)$ contains irrational points (see Exercise 3.1.2).

'Interior' and 'closure', 'open' and 'closed', are closely related, as the next proposition shows. Recall that we denote the complement $\mathbf{R} \setminus S$ of a subset S of $\mathbf{R}$ by $C(S)$.

Proposition 5.2.1 *Suppose that A and B are subsets of $\mathbf{R}$, and that $a \in \mathbf{R}$.*

(i) *If $A \subseteq B$ then $A^\circ \subseteq B^\circ$.*
(ii) *$C(A^\circ) = \overline{C(A)}$.*
(iii) *A is open if and only if $C(A)$ is closed.*
(iii) *A° is open.*
(iv) *A° is the largest open set contained in A: if U is open and $U \subseteq A$ then $U \subseteq A^\circ$.*

Proof (i) This follows directly from the definition.

(ii) If $b \notin A^\circ$ then $N_\epsilon(b) \cap C(A) \neq \emptyset$ for all $\epsilon > 0$, and so $b \in \overline{C(A)}$. Conversely, if $b \in \overline{C(A)}$ then $N_\epsilon(b) \cap C(A) \neq \emptyset$ for all $\epsilon > 0$, and so $b \notin A^\circ$.

(iii) If A is open then $C(A) = C(A^\circ) = \overline{C(A)}$, by (ii), and so $C(A)$ is closed.
   If $C(A)$ is closed then $C(A^\circ) = \overline{C(A)} = C(A)$, so that $A^\circ = A$.
(iv) $C(A^\circ) = \overline{C(A)}$ is closed, so that A° is open, by (iii).
(v) By (i), $U = U^\circ \subseteq A^\circ$.   □

Corollary 5.2.2   *(i) The empty set* $\emptyset$ *and* **R** *are open.*
(ii) If $\mathcal{A}$ *is a set of open subsets of* **R** *then* $\cup_{A \in \mathcal{A}} A$ *is open.*
(iii) If $\{A_1, \dots, A_n\}$ *is a finite set of open subsets of* **R** *then* $\cap_{j=1}^{n} A_j$ *is open.*

Proof   Take complements.   □

   If A is a subset of **R** then the *frontier* or *boundary* ∂A of A is the set $\bar{A} \setminus A^\circ$. Since $\partial A = \bar{A} \cap \overline{C(A)}$, ∂A is closed. $x \in \partial A$ if and only if every open ϵ-neighbourhood of x contains an element of A and an element of $C(A)$.
   We can also consider subsets of a subset X of **R**. Suppose that A is subset of X. Then a point a of A is an *interior point of* A *relative to* X if there exists $\epsilon > 0$ such that $N_\epsilon(a) \cap X \subseteq A$. The set of interior points of A relative to X is then the *relative interior* of A in X, and A is *relatively open* in X if it is equal to its relative interior in X.
   Relatively open sets can be characterized in the following way.

Proposition 5.2.3   *Suppose that* A *is a subset of a subset* X *of* **R***. Then the following are equivalent:*

(i) A is relatively open in X;
(ii) there exists an open subset U of **R** *such that $A = U \cap X$;*
(iii) the set $X \setminus A$ is relatively closed in X.

Proof   Again, this is a worthwhile exercise for the reader.   □

Exercises

5.2.1 Suppose that A and B are subsets of **R** and that A is open. Show that $A + B$ is open.

5.2.2 Suppose that A is a subset of **R**. Show that by repeatedly taking closures and interiors, we can obtain at most six more different sets. Give an example to show that six more different sets can be obtained.

5.3 Connectedness

We now establish a fundamental characterization of non-empty intervals, in terms of open sets. This will allow us to say more about open sets. We need some more terminology. A non-empty subset A of **R** *splits* if there exist two disjoint open subsets U_1 and U_2 of $\mathbf{R}$ such that $A \subseteq U_1 \cup U_2$ and $A \cap U_1$ and $A \cap U_2$ are non-empty. If A does not split, it is *connected*.

Theorem 5.3.1   *A non-empty subset A of $\mathbf{R}$ is connected if and only if it is an interval.*

Proof Suppose first that A is not an interval. Then there exist $a < b < c$ such that $a, c \in A$ and $b \notin A$. Let $U_1 = (-\infty, b)$ and let $U_2 = (b, +\infty)$. Then U_1 and U_2 are disjoint open sets and $A \cap U_1$ and $A \cap U_2$ are non-empty. Thus A splits.

Conversely, suppose that A splits. Thus there exist disjoint open subsets U_1 and U_2 such that $A \subseteq U_1 \cup U_2$, and $A \cap U_1$ and $A \cap U_2$ are non-empty. Let $a_1 \in A \cap U_1$, and $a_2 \in A \cap U_2$. Without loss of generality, we can suppose that $a_1 < a_2$. Let $b = \sup(U_1 \cap [a_1, a_2])$. We shall show that $a_1 < b < a_2$ and that $b \notin A$, so that A is not an interval. First, there exists $0 < \theta \le a_2 - a_1$ such that $(a_1 - \theta, a_1 + \theta) \subseteq U_1$. Thus $b \ge a_1 + \theta > a_1$. Secondly, there exists $0 < \epsilon < a_2 - a_1$ such that $(a_2 - \epsilon, a_2 + \epsilon) \subseteq U_2$; thus $b < a_2$. Suppose if possible that $b \in U_1$. Then there exists $0 < \eta < a_2 - b$ such that $(b - \eta, b + \eta) \subseteq U_1$. Then $(b, b + \eta) \subseteq U_1 \cap [a_1, a_2]$, contradicting the definition of b. Thus $b \notin U_1$. Suppose if possible that $b \in U_2$. Then there exists $0 < \zeta < \theta$ such that $(b - \zeta, b + \zeta) \subseteq U_2$. Then $b - \zeta/2$ is an upper bound for $U_1 \cap [a_1, a_2]$, again contradicting the definition of b. Thus $b \notin U_1 \cup U_2$, and so $b \notin A$. $\square$

Corollary 5.3.2   *A subset A of $\mathbf{R}$ is both open and closed if and only if $A = \emptyset$ or $A = \mathbf{R}$.*

Proof We have seen that $\emptyset$ and $\mathbf{R}$ are both open and closed. If A is open and closed then $C(A)$ is open and closed. $\mathbf{R} = A \cup C(A)$; since $\mathbf{R}$ is connected, either $A = \emptyset$ or $C(A) = \emptyset$. $\square$

Open subsets of $\mathbf{R}$ can now be characterized as disjoint unions of open intervals.

Theorem 5.3.3   *Suppose that U is a non-empty subset of $\mathbf{R}$. U is open if and only if there is a countable set $\mathcal{I}$ of disjoint open intervals such that $U = \cup_{I \in \mathcal{I}} I$. The set $\mathcal{I}$ is uniquely determined.*

Proof A union of open intervals is open, by Corollary 5.2.2, and so the condition is sufficient. Suppose that U is open. We define an equivalence relation on U by setting $a \sim b$ if $[a, b] \subseteq U$ (here we allow the possibility that $a > b$, in which case $[a, b] = \{c \in \mathbf{R} : b \le c \le a\}$). If $a \sim b$ then $b \sim a$, since $[a, b] = [b, a]$; if $a \sim b$ and $b \sim c$ then $a \sim c$, since $[a, c] \subseteq [a, b] \cup [b, c] \subseteq U$. Thus $\sim$ is an equivalence relation on U. Let $\mathcal{I}$ be the set of equivalence classes.

If $I \in \mathcal{I}$, then I is an interval, and I and $U \setminus I$ are disjoint. If $a \in I$, then there exists $\epsilon > 0$ such that $N_\epsilon(a) \subseteq U$. If $b \in N_\epsilon(a)$ then $[a, b] \subseteq N_\epsilon(a)$ and so $b \in I$. Thus $N_\epsilon(a) \subseteq I$, and I is open. Thus U is a disjoint union of a set $\mathcal{I}$ of open intervals.

If $r \in U \cap \mathbf{Q}$, let I_r be the equivalence class to which r belongs. Since a non-empty open interval contains rational points (between any two real numbers, there is a rational number), the mapping $r \to I_r : U \cap \mathbf{Q} \to \mathcal{I}$ is surjective. Since $U \cap \mathbf{Q}$ is countable, so is $\mathcal{I}$.

The representation is unique. Suppose that $U = \cup_{J \in \mathcal{J}} J$, where $\mathcal{J}$ is a set of disjoint open intervals. Suppose that $J \in \mathcal{J}$, and that $x \in J$. Then $x \in I$, for some $I \in \mathcal{I}$. If $y \in J$ then $[x, y] \subseteq J \subseteq U$; hence $y \in I$ and $J \subseteq I$. Further, $I = J \cup ((U \setminus J) \cap I)$. Since J and $(U \setminus J)$ are disjoint open subsets of I, and I is connected, $I = J$. Hence $\mathcal{J} = \mathcal{I}$. $\square$

Exercises

5.3.1 Suppose that A is a non-empty subset of $\mathbf{R}$. Show that A is connected if and only if whenever F_1 and F_2 are closed subsets of $\mathbf{R}$ whose union is $\mathbf{R}$ then either $A \subseteq F_1$ or $A \subseteq F_2$.

5.3.2 Suppose that G is a proper closed subgroup of $(\mathbf{R}, +)$ and that $G \neq \{0\}$. Suppose that (a, b) is a connected component of $\mathbf{R} \setminus G$. Show that $G = \{n(b - a) : n \in \mathbf{Z}\}$.

5.3.3 Suppose that F and G are closed subsets of $\mathbf{R}$, that $[c_0, d_0] \subseteq F \cup G$ and that $c_0 \in F, d_0 \in G$. If $(c_0 + d_0)/2 \in F$, set $c_1 = (c_0 + d_0)/2$, $d_1 = d_0$; otherwise set $c_1 = c_0$, $d_1 = (c_0 + d_0)/2$. Repeat recursively. Show that there exists $b \in c_0, d_0$ such that $c_n \to b$ and $d_n \to b$ as $n \to \infty$. Show that $b \in F \cap G$. Use this to give another proof of Theorem 5.3.1.

5.3.4 Suppose that $\{O_\alpha\}_{\alpha \in A}$ is a family of disjoint non-empty open subsets of $\mathbf{R}$ (if $\alpha \neq \beta$ then $O_\alpha \cap O_\beta = \emptyset$).
(a) Show that A is countable.
(b) Suppose that, for each α, $O_\alpha = \cup_{I \in \mathcal{I}_\alpha} I$, where $\mathcal{I}_\alpha$ is a set of disjoint non-empty open intervals. Show that $\mathcal{J} = \cup_{\alpha \in A} \mathcal{I}_\alpha$ is a set of disjoint non-empty open intervals whose union is $\cup_{\alpha \in A} O_\alpha$.

5.4 Compact sets

We now use the Bolzano–Weierstrass theorem to obtain some important results about the bounded closed subsets of $\mathbf{R}$. Suppose that A is a subset of a set X and that $\mathcal{B}$ is a set of subsets of X. We say that $\mathcal{B}$ *covers* A, or that $\mathcal{B}$ is a *cover* of A, if $A \subseteq \cup_{B \in \mathcal{B}} B$. A subset $\mathcal{C}$ of $\mathcal{B}$ is a *subcover* if it covers A.

A cover $\mathcal{B}$ is *finite* if the set $\mathcal{B}$ is finite. If $X = \mathbf{R}$, a cover $\mathcal{B}$ is *open* if each $B \in \mathcal{B}$ is an open set.

Theorem 5.4.1 *Suppose that $\mathcal{U}$ is an open cover of the bounded closed interval $[a, b]$. Then there exists $\delta > 0$ such that if $x \in [a, b]$ then there exists $U \in \mathcal{U}$ such that $N_\delta(x) \subseteq U$.*

Proof Suppose not. Then for each $n \in \mathbf{N}$ there exists x_n such that $N_{1/n}(x_n)$ is not contained in any $U \in \mathcal{U}$. By the Bolzano–Weierstrass theorem, there exists a convergent subsequence $(x_{n_k})_{k=1}^\infty$, convergent to x, say. Since $[a, b]$ is closed, $x \in [a, b]$. Thus $x \in U$, for some $U \in \mathcal{U}$. Since U is open, there exists $\epsilon > 0$ such that $N_\epsilon(x) \subseteq U$. Since $x_{n_k} \to x$ as $k \to \infty$, there exists $K \in \mathbf{N}$, with $n_K > 2/\epsilon$, such that $|x_{n_k} - x| < \epsilon/2$ for $k \geq K$. If $y \in N_{1/n_K}(x_{n_K})$ then, by the triangle inequality

$$|y - x| \leq |y - x_{n_K}| + |x_{n_K} - x| < 1/n_K + \epsilon/2 < \epsilon,$$

so that $y \in U$. Thus $N_{1/n_K}(x_{n_K}) \subseteq U$, giving a contradiction. $\square$

A positive number δ which satisfies the conclusions of this theorem is called a *Lebesgue number* for the cover.

Theorem 5.4.2 (The Heine–Borel theorem for open sets) *Suppose that $\mathcal{U}$ is an open cover of the (non-degenerate) closed interval $[a, b]$. Then there is a finite subcover.*

Proof We give two proofs of this fundamental theorem. Another proof is given in Exercise 5.4.4. First, let δ be a Lebesgue number for the cover. We divide $[a, b]$ into finitely many intervals, each of length less than δ: choose $n \in \mathbf{N}$ such that $n > (b - a)/\delta$, and let $a_j = a + j(b - a)/n$, for $0 \leq j \leq n$. Then $a = a_0 < a_1 < \cdots < a_n = b$, and $a_j - a_{j-1} = (b - a)/n < \delta$. For each $0 \leq j \leq n$ there exists $U_j \in \mathcal{U}$ such that $N_\delta(a_j) \subseteq U_j$. Then $[a, b] \subseteq \cup_{j=0}^n U_j$.

For the second proof, let

$$C = \{x \in [a, b] : \text{there is a finite subcover of } [a, x]\}.$$

We must show that $b \in \mathbf{C}$. Since $a \in C$, C is not empty. Let $s = \sup C$. We take three steps.

First, $c > a$. For $a \in U$ for some $U \in \mathcal{U}$, and there exists $\epsilon > 0$ such that $N_\epsilon(a) \subseteq U$. Thus $N_\epsilon(a) \cap [a, b] \subseteq C$, so that if $c = \min(a + \epsilon/2, b)$ then $c \in C$. Hence $s \geq c > a$.

Secondly, $s \in C$. For $s \in V$ for some $V \in \mathcal{U}$, and there exists $\eta > 0$ such that $N_\eta(s) \subseteq V$. Then there exists $c \in C \cap (s - \eta, s]$. $[a, c]$ has a finite subcover $\mathcal{W}$, and $\mathcal{W} \cup \{V\}$ is a finite subcover of $[a, s]$.

Finally, $s = b$. For if not, and if $s < t < \min(s + \eta, b)$ then $\mathcal{W} \cup \{V\}$ is a finite subcover of $[a, t]$, so that $t \in C$. $\square$

A set B is said to be *compact* if every open cover of B has a finite subcover.

Proposition 5.4.3 *Suppose that $(U_n)_{n=1}^{\infty}$ is an increasing sequence of open sets in* **R** *which covers a compact subset K of* **R**. *Then there exists $n \in$* **N** *such that $K \subseteq U_n$.*

Proof There is a finite subcover $\{U_{n_1}, \ldots, U_{n_k}\}$. Then $K \subseteq U_N$, where $N = \max\{n_1, \ldots, n_k\}$. $\square$

Theorem 5.4.4 *A non-empty subset B of* **R** *is compact if and only if it is closed and bounded.*

Proof Suppose first that B is closed and bounded. There exists $[a, b]$ such that $B \subseteq [a, b]$. Then $\mathcal{U} \cup \{C(B)\}$ is an open cover of $[a, b]$, and so there is a finite subcover $\{U_1, \ldots, U_n, C(B)\}$ of $[a, b]$. Then $\{U_1, \ldots, U_n\}$ covers B.

Conversely, suppose that B is compact. Let $U_n = (-n, n)$. Then $(U_n)_{n=1}^{\infty}$ is an increasing sequence of open sets which covers B, and so, by Proposition 5.4.3, there exists $N \in$ **N** such that $B \subseteq U_N$; B is bounded.

Finally, we show that B is closed. Suppose that $a \notin B$. We shall show that $a \notin \overline{B}$. For each $n \in$ **N** let

$$V_n = \{x \in \mathbf{R} : |x - a| > 1/n\} = (-\infty, a - 1/n) \cup (a + 1/n, \infty).$$

Then $(V_n)_{n=1}^{\infty}$ is an increasing sequence of open sets which covers B, and so, by Proposition 5.4.3, there exists $N \in$ **N** such that $B \subseteq V_N$. Then $N_{1/N}(a) \cap B = \emptyset$, so that $a \notin \overline{B}$. Thus B is closed. $\square$

We can formulate the Heine–Borel theorem in terms of closed sets: this version is quite as useful as the 'open sets' version. We need more terminology. A set $\mathcal{F}$ of subsets of a set X has the *finite intersection property* if whenever $\{F_1, \ldots, F_n\}$ is a finite subset of $\mathcal{F}$ then $\cap_{j=1}^{n} F_j$ is non-empty.

Theorem 5.4.5 (The Heine–Borel theorem for closed sets) *Suppose that B is a bounded closed subset of* **R**, *and that $\mathcal{F}$ is a set of closed subsets of B with the finite intersection property. Then the total intersection $\cap_{F \in \mathcal{F}} F$ is non-empty.*

Proof This is just a matter of taking complements. Suppose that $\cap_{F \in \mathcal{F}} F = \emptyset$. Then $\{C(F) : F \in \mathcal{F}\}$ is an open cover of B, and so there is a finite subcover $\{C(F_1), \ldots, C(F_n)\}$. Thus

$$B \subseteq C(F_1) \cup \ldots \cup C(F_n) = C(F_1 \cap \ldots \cap F_n),$$

so that $F_1 \cap \ldots \cap F_n = \emptyset$, contradicting the finite intersection property. $\square$

Exercises

5.4.1 Let $(r_n)_{n=1}^{\infty}$ be an enumeration of the rational numbers in $(0, 1)$, and let $0 < \epsilon < 1$. For each n let I_n be an open interval in $(0, 1)$ containing r_n and of length at most $\epsilon/2^n$. Let $U = \cup_{n=1}^{\infty} I_n$. Show that $\overline{U} = [0, 1]$. Suppose that $U = \cup_{n=1}^{\infty} J_n$, where $(J_n)_{n=1}^{\infty}$ is a sequence of disjoint open intervals; let $l(J_n)$ be the length of J_n. Show that

$$\sum_{n=1}^{\infty} l(J_n) \le \epsilon.$$

5.4.2 The set $\mathbf{Q} \cap [0, 1]$ is not compact. Find an open cover of $\mathbf{Q} \cap [0, 1]$ with no finite subcover.

5.4.3 Suppose that $\mathcal{F}$ is a finite set of open intervals which covers the closed interval $[a, b]$, and that $\mathcal{F}$ is *minimal*; no proper subset of $\mathcal{F}$ covers $[a, b]$. Show that $\mathcal{F}$ can be listed as $I_1, \ldots, I_n$ in such a way that

$a \in I_1$, $\inf I_j < \sup I_{j-1} \le \inf I_{j+1} < \sup I_j$ for $1 < j < n$, and $b \in I_n$.

Deduce that $I_{j-1} \cap I_j \ne \emptyset$ for $2 \le j \le n$, and that no point of $[a, b]$ is in three members of $\mathcal{F}$.

5.4.4 Suppose that $\mathcal{U}$ is an open cover of the closed interval $[c_0, d_0]$, and suppose, if possible, that there is no finite subcover. If there is no finite subcover of $[c_0, (c_0 + d_0)/2]$ set $c_1 = c_0$, $d_1 = (c_0 + d_0)/2$; otherwise set $c_1 = (c_0 + d_0)/2$, $d_1 = d_0$. Show that $[c_1, d_1]$ has no finite subcover. Repeat recursively. Show that there exists $b \in [c_0, d_0]$ such that $c_n \to b$ and $d_n \to b$ as $n \to \infty$. Use this to give another proof of the Heine–Borel theorem.

5.4.5 Suppose that U is an open subset of $\mathbf{R}$ and that $x \in U$. Show that there exist rational numbers r and s such that $x \in N_r(s) \subseteq U$. Show that if $\mathcal{U}$ is an open cover of a subset A of $\mathbf{R}$ then there exists a countable subcover.

5.5 Perfect sets, and Cantor's ternary set

We now introduce another idea, similar enough to the notion of a closure point to be confusing. Suppose that A is a subset of $\mathbf{R}$ and that $a \in \mathbf{R}$.

A real number b is called a *limit point*, or *accumulation point*, of A if whenever $\epsilon > 0$ there exists $a \in A$ (which may depend upon ϵ) with $0 < |b - a| < \epsilon$. Thus b is a limit point of A if there are points of A, different from b, which are arbitrarily close to b.

If $a \in \mathbf{R}$ and $\epsilon > 0$ then the *punctured ϵ-neighbourhood* $N_\epsilon^*(a)$ of a is defined as

$$N_\epsilon^*(a) = \{x \in \mathbf{R} : 0 < |x - a| < \epsilon\} = (a - \epsilon, a) \cup (a, a + \epsilon) = N_\epsilon(a) \setminus \{a\}.$$

Thus b is a limit point of A if and only if $N_\epsilon^*(b) \cap A \neq \emptyset$, for each $\epsilon > 0$.

Proposition 5.5.1 *Suppose that A is a subset of **R** and that $b \in \mathbf{R}$. b is a limit point of A if and only if there exists a sequence $(a_j)_{j=1}^\infty$ in $A \setminus \{b\}$ such that $a_j \to b$ as $j \to \infty$.*

Proof The proof is just like the proof of Proposition 5.1.1, with obvious modifications.   □

The set of limit points of A is called the *derived set* of A, and is denoted by A'. It follows from the definitions that $A' \subseteq \overline{A}$. If $A = \{a\}$ then $A' = \emptyset$, so that A need not be a subset of A'. If $A = A'$, we say that A is *perfect*. For example, a non-degenerate closed interval is perfect, as is a finite union of non-degenerate closed intervals.

Suppose that A is a subset of **R** and that $a \in A$. a is an *isolated point* of A if there exists $\epsilon > 0$ such that $N_\epsilon^*(a) \cap A = \emptyset$.

Proposition 5.5.2 *Suppose that A' is the derived set of a subset A of **R**, and that $a \in \mathbf{R}$. Let $i(A)$ be the set of isolated points of A.*
  (i) $\overline{A}$ is the disjoint union of A' and $i(A)$.
  (ii) A' is closed.

Proof (i) Clearly A' and $i(A)$ are disjoint subsets of $\overline{A}$. Suppose that $a \in \overline{A} \setminus i(A)$. There are two possibilities. First, $a \in A$. Since a is not an isolated point of A, it must belong to A'. Secondly, $a \in \overline{A} \setminus A$. There is a sequence $(a_n)_{n=1}^\infty$ in A which tends to a as $n \to \infty$. Since $a_n \neq a$, for $n \in \mathbf{N}$, it follows that $a \in A'$.

(ii) Suppose, if possible, that $b \in \overline{A'} \setminus A'$. Then $b \in \overline{A} \setminus A'$, and so b is an isolated point of A, by (i). Thus there exists $\epsilon > 0$ such that $N_\epsilon^*(b) \cap A = \emptyset$. Since $b \in \overline{A'}$, there exists $c \in A'$ with $|b - c| < \epsilon/2$. Then $N_{\epsilon/2}^*(c) \subseteq N_\epsilon^*(b)$, so that $N_{\epsilon/2}^*(c) \cap A = \emptyset$, giving a contradiction.   □

As an example, let $A = \{1/j : j \in \mathbf{N}\}$. Then $\overline{A} = A \cup \{0\}$ and $A' = \{0\}$. Note that $(A')' = \emptyset \neq A'$. This example is taken further in Exercises 5.5.2–5.5.4.

We now give an example of a bounded non-empty perfect set which contains no non-degenerate intervals. This set is known as *Cantor's ternary set*, although it was first described by the Irish-born mathematician Henry Smith. We begin with $C_0 = [0, 1]$. First, we remove the middle third of C_0, to obtain

$$C_1 = [0, 1/3] \cup [2/3, 1] = I_L \cup I_R;$$

I_L is the *left interval* of C_1 and I_R is the *right interval*. We then remove the middle thirds of I_L and I_R to obtain

$$C_2 = ([0, 1/9] \cup [2/9, 1/3]) \cup ([2/3, 7/9] \cup [8/9, 1])$$
$$= (I_{LL} \cup I_{LR}) \cup (I_{RL} \cup I_{RR});$$

C_2 is the union of 2^2 disjoint closed intervals, each of length $(1/3)^2$; I_{LL} and I_{RL} are left intervals, and I_{LR} and I_{RR} are right intervals. We then repeat the process recursively, to obtain a decreasing sequence $(C_n)_{n=0}^{\infty}$ of closed sets; C_n is the union of 2^n disjoint closed intervals, each of length $(1/3)^n$, and each interval is either a left subinterval or a right subinterval of an interval of C_{n-1}. We then define Cantor's ternary set C to be $\cap_{n=0}^{\infty} C_n$.

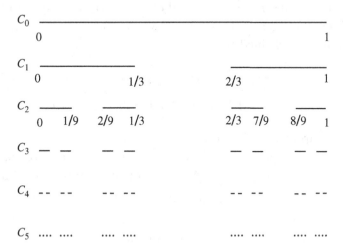

Figure 5.5. Construction of Cantor's ternary set.

Here are some of the properties of C.

Theorem 5.5.3 *Cantor's ternary set C is a perfect subset of $[0,1]$ with empty interior. There exists a bijection $c : P(\mathbf{N}) \to C$, and so C is uncountable.*

Proof C is closed, and is non-empty, by the Heine–Borel theorem for closed sets. But in fact, the end points of all the intervals that occur in the construction are in C. We use this to show that C is perfect. Suppose that $x \in C$ and that $\epsilon > 0$. Choose $j \in \mathbf{Z}^+$ such that $(1/3)^j < \epsilon/2$. There exists an interval I_j of C_j to which x belongs; both its end-points are in C, and at least one of them is different from x. Thus there exists $c \in N_\epsilon^*(x) \cap C$, and so C is perfect.

Further, the ϵ-neighbourhood $N_\epsilon(x)$ is not contained in an interval of C_j, and so $N_\epsilon(x)$ is not contained in C_j; thus $N_\epsilon(x)$ is not contained in C. Since this holds for all $x \in C$ and all $\epsilon > 0$, C has an empty interior.

If $A \in P(\mathbf{N})$, let $a_j = 2$ if $j \in A$ and let $a_j = 0$ otherwise. Let $c_n(A) = \sum_{j=1}^{n} a_j/3^j$, and let $c(A) = \sum_{j=1}^{\infty} a_j/3^j$. Then $c_n(A) \in C_n$, and $c_n(A) \to c(A)$ as $n \to \infty$, and so $c(A) \in C$, since C is closed. As in the proof of Cantor's theorem, if $A \subset B$ then $c(A) < c(B)$, and c is injective. Conversely, suppose that $x \in C$. Let

$$A = \{n \in \mathbf{N} : x \text{ is in a right-hand interval of } C_n\}.$$

Then $x = c(A)$. $\square$

Cantor's ternary set has a great deal of symmetry and self-similarity. For example, the mapping $x \to 3x$ is a bijective mapping of $C \cap [0, 1/3]$ onto C, and the mapping s_j defined by

$$s_j(x) = x + 2/3^j \text{ for } 0 \le x < 1 - 2/3^j,$$
$$= x + 2/3^j - 1 \text{ for } 1 - 2/3^j \le x \le 1,$$

is a bijective mapping of C onto itself.

There are many constructions similar to the construction of Cantor's ternary set. Suppose that $\epsilon = (\epsilon_n)_{n=1}^{\infty}$ is a sequence of positive numbers with $\sum_{n=0}^{\infty} \epsilon_n = s \le 1$. Let $s_n = \sum_{j=0}^{n-1} \epsilon_j$. We construct a sequence $(C_n^{(\epsilon)})_{n=0}^{\infty}$ of closed sets $C_n^{(\epsilon)}$ recursively; $C_0^{(\epsilon)} = [0,1]$, and if $n \in \mathbf{N}$ then $C_n^{(\epsilon)}$ consists of 2^n disjoint closed subintervals of $[0,1]$, each of length $(1 - s_n)/2^n$. We construct $C_{n+1}^{(\epsilon)}$ by removing an open interval of length $\epsilon_n/2^n$ from the middle of each of these intervals. Then $C_{n+1}^{(\epsilon)}$ consists of 2^{n+1} disjoint closed subintervals of $[0,1]$, each of length $(1 - s_{n+1})/2^{n+1}$. Finally we set $C^{(\epsilon)} = \cap_{n=0}^{\infty} C_n^{(\epsilon)}$. Then

$C^{(\epsilon)}$ is a perfect subset of $[0, 1]$, with empty interior. As we shall see, $C^{(\epsilon)}$ is of interest when $s < 1$. In such a case, we call $C^{(\epsilon)}$ a *fat Cantor set*.

Exercises

5.5.1 Suppose that U is an open subset of $\mathbf{R}$. Show that $\overline{U} = U'$.

5.5.2 Let $B = \{1/j + 1/k : j, k \in \mathbf{N}, k > j^2\}$. What is B'? What is $(B')'$? What is $((B')')'$?

5.5.3 Show that for each $k \in \mathbf{N}$ there exists a strictly increasing sequence $B_0 \subset B_1 \subset \cdots \subset B_k$ of subsets of $\mathbf{R}$ such that $B'_j = B_{j-1}$, for $1 \leq j \leq k$.

5.5.4 Construct a subset C of $\mathbf{R}$ such that, if $C_1 = C'$, and $C_j = C'_{j-1}$ for all $j \geq 2$ then $(C_j)_{j=0}^\infty$ is a strictly decreasing sequence of non-empty subsets of $\mathbf{R}$.

5.5.5 Suppose that $0 < \lambda < 9$. If $x \in [0, 1)$, let $a_n(x) = (x_1 + \cdots + x_n)/n$, where $x = 0.x_1 x_2 \ldots$ is the decimal expansion of x (without recurrent 9s), and let $a_n(1) = 0$. Let $E_n = \{x \in [0, 1] : a_n(x) \leq \lambda\}$. Show that E_n is closed. Show that $E = \cap_{n=1}^\infty E_n$ is a perfect subset of $[0, 1]$ with an empty interior.

5.5.6 Let $(C_n)_{n=0}^\infty$ be the sequence of closed sets that appears in the construction of Cantor's ternary set C. Suppose that $x \in [0, 2]$. Show that for each $n \in \mathbf{N}$ there exist $u_n, v_n \in C_n$ such that $x = u_n + v_n$. Use the Bolzano–Weierstrass theorem to show that there exist $u, v \in C$ such that $x = u + v$. Show that if $x \in [-1, 1]$ there exist $y, z \in C$ such that $x = y - z$.

5.5.7 Suppose that A is a non-empty bounded closed subset of $\mathbf{R}$. Let

$$C(A) = (-\infty, \inf A) \cup (\cup_{I \in \mathcal{J}} I) \cup (\sup A, +\infty),$$

where $\mathcal{J}$ is a set of disjoint open intervals contained in $(\inf A, \sup A)$. Order $\mathcal{J}$ by setting $I < J$ if $\sup I \leq \inf J$. $\mathcal{J}$ has the *intermediate property* if whenever I and J are in $\mathcal{J}$ and $I < J$ then there exists $K \in \mathcal{J}$ with $I < K < J$.

(a) Show that if $\mathcal{J}$ has the intermediate property, then A is perfect.

(b) Suppose that A is perfect and that $A^\circ = \emptyset$. Show that $\mathcal{J}$ has the intermediate property. Show that there is a bijection of $P(\mathbf{N})$ onto A.

5.5.8 Suppose that A is a non-empty perfect subset of $[0, 1]$ with empty interior. Show that there is a bijective mapping of $P(\mathbf{N})$ onto A, using a construction as in Cantor's theorem. (*Hint*: After n steps there are

2^n closed intervals whose union contains A. For the next step, remove a largest possible open interval from each.)

Deduce that there is an order-preserving bijection ϕ of Cantor's ternary set C onto A.

Deduce that a non-empty perfect subset of **R** is uncountable.

5.5.9 Suppose that C is a closed subset of **R**, with complement $\cup_j I_j$, where the I_j are disjoint open intervals. Show that C is perfect if and only if $\overline{I}_j \cap \overline{I}_k = \emptyset$ when $I_j \neq I_k$.

Is it possible to find a sequence of disjoint non-degenerate closed intervals whose union is $(0, 1)$?

5.5.10 If A is a subset of **R** then a point a of **R** is a *condensation point* of A if $N_\epsilon(a) \cap A$ is uncountable, for every $\epsilon > 0$. Show that if A is uncountable, then the set C of condensation points of A is closed, and $A \setminus C$ is countable. Show that C is the set of condensation points of itself.

5.5.11 Show that every point of Cantor's ternary set C is a condensation point of C.

6

Continuity

6.1 Limits and convergence of functions

So far we have considered the limits of sequences of real numbers. These sequences are real-valued functions defined on $\mathbf{Z}^+$ or $\mathbf{N}$. We now consider real-valued functions defined on a non-empty subset A of $\mathbf{R}$. It is useful to make definitions for a general set A, but the reader should have in mind examples such as an open interval, a closed interval, the set $\mathbf{Q}$ of rational numbers and the set $\{1/n : n \in \mathbf{N}\}$.

The notion of limit extends naturally to this setting. Suppose that $f : A \to \mathbf{R}$ is a function, that b is a limit point of A (which may or may not be an element of A) and that $l \in \mathbf{R}$. We say that $f(x)$ *converges to* l, or *tends to* l, as x *to* b if whenever $\epsilon > 0$ there exists $\delta > 0$ (which usually depends on ϵ) such that $|f(x) - l| < \epsilon$ for those $x \in A$ for which $0 < |x - b| < \delta$ (that is, for $x \in N_\delta^*(b) \cap A$). That is to say, as x gets close to b, $f(x)$ gets close to l. We say that l *is the limit of* f *as* x *tends to* b, write '$f(x) \to l$ as $x \to b$' and also write $l = \lim_{x \to b} f(x)$. Note that in the case where $b \in A$, we do not consider the value of $f(b)$, but only the values of f at points nearby.

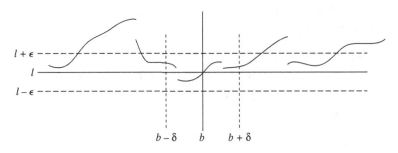

Figure 6.1a. Convergence of functions.

We now have the following elementary results, which correspond exactly to Propositions 3.2.2, 3.2.3 and 3.2.6, together with Theorem 3.2.5. We say that f is *bounded* on A if the image set $f(A)$ is a bounded set.

Theorem 6.1.1   *Suppose that f, g and h are real-valued functions on a subset A of $\mathbf{R}$ and that b is a limit point of A.*

   (i) If $f(x) \to l$ as $x \to b$ and $f(x) \to m$ as $x \to b$, then $l = m$.

   (ii) If $f(x) \to l$ as $x \to b$ then there exists $\delta > 0$ such that f is bounded on $N_\delta^*(b) \cap A$.

   (iii) If $f(x) = l$ for all $x \in A$, then $f(x) \to l$ as $x \to b$.

   (iv) If $f(x) \to 0$ as $x \to b$, and $g(x)$ is bounded on $N_\delta^*(b) \cap A$ for some $\delta > 0$, then $f(x)g(x) \to 0$ as $x \to b$.

   (v) If $f(x) \to l$ and $g(x) \to m$ as $x \to b$ then $f(x) + g(x) \to l + m$ as $x \to b$.

   (vi) If $f(x) \to l$ as $x \to b$ and $c \in \mathbf{R}$ then $cf(x) \to cl$ as $x \to b$.

   (vii) If $f(x) \to l$ and $g(x) \to m$ as $x \to b$ then $f(x)g(x) \to lm$ as $x \to b$.

   (viii) If $f(x) \neq 0$ for $x \in A$, $l \neq 0$ and $f(x) \to l$ as $x \to b$ then $1/f(x) \to 1/l$ as $x \to b$.

   (ix) If $f(x) \to l$ and $g(x) \to m$ as $x \to b$, and if $f(x) \leq g(x)$ for all $x \in N_\delta^*(b) \cap A$ for some $\delta > 0$, then $l \leq m$.

   (x) (The sandwich principle) Suppose that $f(x) \leq g(x) \leq h(x)$ for all $x \in N_\delta^*(b) \cap A$, for some $\delta > 0$, and that $f(x) \to l$ and $h(x) \to l$ as $x \to b$. Then $g(x) \to l$ as $x \to b$.

Proof   Since the definition of limit is so similar to the limit of a sequence, the proofs are simple modifications of the proofs of corresponding results for sequences, established in Section 3.1. The details are left as exercises for the reader.   □

Note that in several cases we have restricted attention to the behaviour of f in a set $N_\delta^*(b) \cap A$. This is clearly appropriate, since we are only concerned with the behaviour of f as x approaches b.

It is a useful fact that we can characterize convergence in terms of convergent sequences.

Proposition 6.1.2   *Suppose that f is a real-valued function on a subset A of $\mathbf{R}$, that b is a limit point of A and that $l \in \mathbf{R}$. Then $f(x) \to l$ as $x \to b$ if and only if whenever $(a_n)_{n=0}^\infty$ is a sequence in $A \setminus \{b\}$ which tends to b as $n \to \infty$ then $f(a_n) \to l$ as $n \to \infty$.*

Proof   Suppose that $f(x) \to l$ as $x \to b$ and that $(a_n)_{n=0}^\infty$ is a sequence in $A \setminus \{b\}$ which tends to b as $n \to \infty$. Given $\epsilon > 0$, there exists $\delta > 0$ such that if $x \in N_\delta^*(b) \cap A$ then $|f(x) - l| < \epsilon$. There then exists n_0 such that $|a_n - b| < \delta$ for $n \geq n_0$. Then $|f(a_n) - l| < \epsilon$ for $n \geq n_0$, so that $f(a_n) \to l$ as $n \to \infty$.

Suppose that $f(x)$ does not converge to l as $x \to b$. Then there exists $\epsilon > 0$ for which we can find no suitable $\delta > 0$. If $n \in \mathbf{N}$ then $1/n$ is not suitable, and so there exists $x_n \in N_{1/n}^*(b) \cap A$ with $|f(x_n) - l| \geq \epsilon$. Then $x_n \to x$ as $n \to \infty$ and $f(x_n)$ does not converge to l as $n \to \infty$. □

We have the following *general principle of convergence*.

Theorem 6.1.3 *Suppose that f is a real-valued function on a subset A of $\mathbf{R}$, that b is a limit point of A and that $l \in \mathbf{R}$. Then the following are equivalent.*

(i) *There exists l such that $f(x) \to l$ as $x \to b$.*
(ii) *Whenever $(a_n)_{n=0}^\infty$ is a sequence in $A \setminus \{b\}$ which tends to b as $n \to \infty$ then $(f(a_n))_{n=0}^\infty$ is a Cauchy sequence.*
(iii) *Given $\epsilon > 0$ there exists $\delta > 0$ such that if $x, y \in N_\delta^*(b)$ then $|f(x) - f(y)| < \epsilon$.*

Proof Suppose that (i) holds, and that $(a_n)_{n=0}^\infty$ is a sequence in $A \setminus \{b\}$ which tends to b as $n \to \infty$. By Proposition 6.1.2, $f(a_n) \to l$ as $n \to \infty$. Since a convergent sequence is a Cauchy sequence, (ii) holds.

Suppose that (iii) fails. Then there exists $\epsilon > 0$ for which for each $n \in \mathbf{N}$ there exist $a_n, a_n' \in N_{1/n}^*(b) \cap A$ with $|f(a_n) - f(a_n')| \geq \epsilon$. Let $c_{2n-1} = a_n$ and $c_{2n} = a_n'$, for $n \in \mathbf{N}$. Then $c_n \to b$ as $n \to \infty$, and $(f(c_n))_{n=0}^\infty$ is not a Cauchy sequence. Thus (ii) fails: (ii) implies (iii).

Finally suppose that (iii) holds, and that $\epsilon > 0$. There exists $\delta > 0$ such that if $x, y \in N_\delta^*(b) \cap A$ then $|f(x) - f(y)| < \epsilon/2$. Suppose that $(a_n)_{n=0}^\infty$ is a sequence in $A \setminus \{b\}$ which tends to b as $n \to \infty$. Then there exists n_0 such that $a_n \in N_\epsilon^*(b)$ for $n \geq n_0$. Thus if $m, n \geq n_0$ then $|f(a_n) - f(a_m)| < \epsilon/2$, and so $(f(a_n))_{n=0}^\infty$ is a Cauchy sequence. By the general principle of convergence, there exists l such that $f(a_n) \to l$ as $n \to \infty$, and if $n \geq n_0$ then $|f(a_n) - l| \leq \epsilon/2$. Thus if $x \in N_\delta^*(b) \cap A$ then $|f(x) - l| \leq |f(x) - f(a_{n_0})| + |f(a_{n_0}) - l| < \epsilon$; $f(x) \to l$ as $x \to b$. Thus (iii) implies (i). □

We now turn to a result which corresponds to Theorem 3.2.4. First we must introduce the idea of one-sided convergence. Suppose that f is a real-valued function on A and that $b \in \mathbf{R}$. Let $A_+ = A \cap (b, \infty)$ and let $A_- = A \cap (-\infty, b)$. Suppose that b is a limit point of A_+ – that is, $(b, b + \delta) \cap A$ is non-empty, for each $\delta > 0$. Then we say that $f(x)$ *tends to l as $x \to b$ from the right* if whenever $\epsilon > 0$ there exists $\delta > 0$ such that if $x \in A \cap (b, b + \delta)$ then

$|f(x)-l| < \epsilon$. We then write $f(x) \to l$ as $x \searrow b$, and denote l by $\lim_{x \searrow b} f(x)$, or, more briefly, by $f(b+)$. Similarly if $f(x)$ tends to l as $x \to b$ from the left, we denote the limit l by $\lim_{x \nearrow b} f(x)$, or $f(b-)$. Why do we use this terminology? If we consider the graph of f, drawn in the usual way, the variable x increases from left to right, and the values that the function f takes increase in an upwards direction. We therefore use 'left' and 'right' for the variable x, and reserve words such as 'upper' or 'lower' for the values of the function.

Theorem 6.1.4 *Suppose that f is a real-valued increasing function on A and that b is a limit point of $A_+ = A \cap (b, \infty)$. If f is bounded below on A_+ then $f(x) \to \inf\{f(y) : y \in A_+\}$ as $x \to b$ from the right.*

Similar results hold for 'convergence from the left', and for decreasing functions.

Proof Let $l = \inf\{f(y) : y \in A_+\}$. Suppose that $\epsilon > 0$. Then $l + \epsilon$ is not a lower bound for f on A_+, and so there exists $a \in A_+$ with $f(a) < l + \epsilon$. Let $\delta = a - b$. If $x \in A \cap (b, b + \delta) = A \cap (b, a)$ then $l \le f(x) \le f(a) < l + \epsilon$, so that $f(x) \to l$ as $x \to b$ from the right. □

This theorem is quite as important as Theorem 3.2.4.

Corollary 6.1.5 *If b is a limit point of A_+ and A_- then $f(b-) \le f(b+)$.*

Proof For $\sup\{f(x) : x \in A_-\} \le \inf\{f(x) : x \in A^+\}$. □

Suppose again that b is a limit point of a subset A of $\mathbf{R}$, and suppose that f is a real-valued function which is bounded on $N_\delta^*(b) \cap A$, for some $\delta > 0$. We can then define the upper and lower limits of f at b. For $0 < t < \delta$, let $M(t) = \sup\{f(x) : x \in N_t^*(b)\}$. Then $M(t)$ is an increasing function on $(0, \delta)$ which is bounded below. By Theorem 6.1.4 it follows that $M(t)$ converges to $M(0+) = \inf\{M(s) : 0 < s < \delta\}$ as $t \searrow 0$. $M(0+)$ is the *upper limit* or *limes superior* of f at b, and is denoted by $\limsup_{x \to b} f(x)$. The *lower limit*, or *limes inferior* $\liminf_{x \to b} f(x)$ is defined in a similar way.

The next theorem corresponds to Theorem 3.5.3.

Theorem 6.1.6 *Suppose that b is a limit point of a subset A of $\mathbf{R}$, and suppose that f is a real-valued function which is bounded on $N_\delta^*(b) \cap A$, for some $\delta > 0$. Then $f(x) \to l$ as $x \to b$ if and only if $\limsup_{x \to b} f(x) = \liminf_{x \to b} f(x) = l$.*

Proof Another exercise for the reader. □

As an example, suppose that $x \in (0,1]$. If $0 < 1/(n+1) < x \le 1/n$, set $f(x) = n(n+1)(x - 1/(n+1))$. Then $\limsup_{x \to 0} f(x) = 1$ and $\liminf_{x \to 0} f(x) = 0$. The function f does not tend to a limit as $x \to 0$, but oscillates between the values 0 and 1.

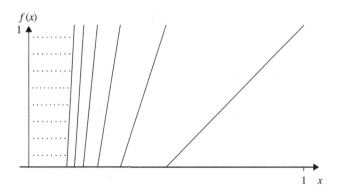

Figure 6.1b. $\limsup_{x \to 0} f(x) \ne \liminf_{x \to 0} f(x) = 0$.

We can also consider limits as $x \to +\infty$ or as $x \to -\infty$. Suppose that A is a subset of $\mathbf{R}$ which is not bounded above, that f is a real-valued function on A and that $l \in \mathbf{R}$. Then we say that $f(x) \to l$ as $x \to +\infty$ if whenever $\epsilon > 0$ there exists $x_0 \in \mathbf{R}$ such that if $x \in A$ and $x \ge x_0$ then $|f(x) - l| < \epsilon$. Similarly, if there exists x_0 such that f is bounded on $A \cap [x_0, \infty)$, and we define $M(x) = \sup\{f(a) : a \in A \cap [x, \infty)\}$ for $x \ge x_0$, then $M(x)$ is a decreasing function on $[x_0, \infty)$ which is bounded below; we define $\limsup_{x \to \infty} f(x)$ as $\inf\{M(x) : x \in [x_0, \infty)\}$. The lower limit is defined similarly. Limits as $x \to -\infty$ are defined in the same way. The reader should verify that all the results of this section, with appropriate modifications, extend without difficulty to these situations.

Exercises

6.1.1 Show that $\lim_{x \to 0} \sqrt{1 + x + x^2} = 1$.

6.1.2 Show that $(\sqrt{1 + x + x^2} - 1)/(\sqrt{1+x} - \sqrt{1-x})$ tends to a limit as $x \to 0$, and evaluate the limit.

6.2 Orders of magnitude

This section is a digression; it introduces some notation that is frequently used, though we shall use it sparingly.

Suppose that f is a function defined on a subset A of $\mathbf{R}$, and that b is a limit point of A. Frequently, the principal point of interest is the behaviour of

f near b, rather than its actual value. The O *(big O)* and o *(little o)* notation is used to describe the magnitude of f near b in terms of another, usually simpler, function g.

Suppose that g is another real-valued function on A. We write

$$f(x) = O(g(x)) \text{ as } x \to b$$

if there exists $\delta > 0$ and $M \in \mathbf{R}$ such that $|f(x)| \le M|g(x)|$ for $x \in N_\delta^*(b) \cap A$.

Suppose that there exists $\delta > 0$ such that $g(x) \ne 0$ for $x \in N_\delta^*(b) \cap A$. Then we write

$$f(x) = o(g(x)) \text{ as } x \to b$$

if $f(x)/g(x) \to 0$ as $x \to b$ and write

$$f(x) \sim g(x) \text{ as } x \to b$$

if $f(x)/g(x) \to 1$ as $x \to b$.

We use the same notation when $x \to \infty$; thus $f(x) = o(g(x))$ as $x \to \infty$ if $f(x)/g(x) \to 0$ as $x \to \infty$. As a particular example, if $(a_n)_{n=1}^\infty$ and $(b_n)_{n=1}^\infty$ are real-valued sequences then $a_n \sim b_n$ if $a_n/b_n \to 1$ as $n \to \infty$.

For example, suppose that $p(x) = a_0 + a_1 x + \cdots + a_n x^n$ is a polynomial function of degree n on $\mathbf{R}$ (with $a_n \ne 0$). Then

$$p(x) = O(x^n) \text{ as } x \to \infty,$$
$$p(x) = o(x^{n+1}) \text{ as } x \to \infty,$$
$$p(x) \sim a_n x^n \text{ as } x \to \infty,$$
$$p(x) = O(1) \text{ as } x \to 0.$$

This notation arose in analytic number theory, where a complicated expression f is approximated by a simpler function g, and the interest lies in estimating the magnitude of the difference. Thus it might be shown that $f(x) - g(x) = O(h(x))$ as $x \to \infty$; in this case we write $f(x) = g(x) + O(h(x))$. For example, if p is the polynomial above, then

$$p(x) = a_n x^n + O(x^{n-1}) = a_n x^n + o(x^n) \text{ as } x \to \infty,$$
$$\text{and } p(x) = a_0 + a_1 x + O(x^2) = a_0 + o(1) \text{ as } x \to 0.$$

Although this notation is expressive, its use requires care; in practice, it is frequently advisable to expand any statement involving it into a more standard form.

6.3 Continuity

We now introduce the fundamental concept of *continuity*. Suppose that f is a real-valued function defined on a subset A of $\mathbf{R}$, and that $a \in A$. f is *continuous* at a if whenever $\epsilon > 0$ there exists $\delta > 0$ (which usually depends on ϵ) such that $|f(x) - f(a)| < \epsilon$ for those $x \in A$ for which $|x - a| < \delta$ (that is for $x \in N_\delta(a) \cap A$). That is to say, as x gets close to a, $f(x)$ gets close to $f(a)$. If f is not continuous at a, we say that f has a *discontinuity* at a.

Compare this definition with the definition of convergence. First, a must be an element of A, so that $f(a)$ is defined. Secondly, a need not be a limit point of a. If it is, then f is continuous at a if and only if $f(x) \to f(a)$ as $x \to a$. If a is not a limit point, then it is an isolated point of A. In this case, there exists $\delta > 0$ such that $N_\delta(a) \cap A = \{a\}$, so that if $x \in N_\delta(a) \cap A$ then $f(x) = f(a)$, and f is continuous at a; functions are always continuous at isolated points.

We now have the following elementary results, which correspond exactly to Theorem 6.1.1.

Theorem 6.3.1 *Suppose that f, g and h are real-valued functions on a subset A of $\mathbf{R}$ and that $a \in A$.*

(i) *If f is continuous at a then there exists $\delta > 0$ such that f is bounded on $N_\delta(a) \cap A$.*

(ii) *If $f(x) = l$ for all $x \in A$, then f is continuous at a.*

(iii) *If $f(a) = 0$, f is continuous at a, and $g(x)$ is bounded on $N_\delta(a) \cap A$ for some $\delta > 0$, then fg is continuous at a.*

(iv) *If f and g are continuous at a then $f + g$ is continuous at a.*

(v) *If f and g are continuous at a then fg is continuous at a.*

(vi) *If $f(x) \neq 0$ for $x \in A$, and if f is continuous at a then $1/f$ is continuous at a.*

(vii) *(The sandwich principle) Suppose that $f(x) \leq g(x) \leq h(x)$ for all $x \in N_\delta(a) \cap A$, for some $\delta > 0$, that $f(a) = g(a) = h(a)$ and that f and h are continuous at a. Then g is continuous at a.*

Proof These results follow directly from Theorem 6.1.1, and the remarks above. $\square$

Similarly, we have the following consequence of Proposition 6.1.2.

Proposition 6.3.2 *Suppose that f is a real-valued function on a subset A of $\mathbf{R}$ and that $a \in A$. Then f is continuous at a if and only if whenever $a_n \to a$ as $n \to \infty$ then $f(a_n) \to f(a)$ as $n \to \infty$.*

Continuity behaves well under composition.

Theorem 6.3.3   *Suppose that f is a real-valued function on a subset A of $\mathbf{R}$ and that g is a real-valued function on a subset B of $\mathbf{R}$ which contains $f(A)$. If f is continuous at $a \in A$ and g is continuous at $f(a)$, then $g \circ f$ is continuous at a.*

Proof   Suppose that $\epsilon > 0$. Then there exists $\eta > 0$ such that if $b \in B$ and $|b - f(a)| < \eta$ then $|g(b) - g(f(a))| < \epsilon$. Similarly there exists $\delta > 0$ such that if $a' \in A$ and $|a' - a| < \delta$ then $|f(a') - f(a)| < \eta$. Thus if $a' \in A$ and $|a' - a| < \delta$ then $|g(f(a')) - g(f(a))| < \epsilon$.   □

The proof is trivial: the theoretical importance and practical usefulness are enormous.

Continuity is a local phenomenon. Nevertheless, there are many important cases where f is continuous at every point of A. In this case we say that f is *continuous on A*, or more simply, that f is *continuous*. Continuity on A can be characterized in terms of open sets, and in terms of closed sets.

Proposition 6.3.4   *Suppose that f is a real-valued function on a subset A of $\mathbf{R}$. The following are equivalent:*
  (i) *f is continuous on A;*
  (ii) *if U is an open subset of $\mathbf{R}$ then $f^{-1}(U)$ is a relatively open subset of A;*
  (iii) *for each $c \in \mathbf{R}$ the sets $U_c = \{x \in A : f(x) > c\}$ and $L_c = \{x \in A : f(x) < c\}$ are relatively open in A;*
  (iv) *if F is a closed subset of $\mathbf{R}$ then $f^{-1}(F)$ is a relatively closed subset of A;*
  (v) *for each $c \in \mathbf{R}$ the sets $F_c = \{x \in A : f(x) \geq c\}$ and $G_c = \{x \in A : f(x) \leq c\}$ are relatively closed in A.*

Proof   Suppose that f is continuous on A, that U is an open subset of $\mathbf{R}$ and that $x \in f^{-1}(U)$. Since U is open, there exists $\epsilon > 0$ such that $N_\epsilon(f(x)) \subseteq U$. Since f is continuous at x, there exists $\delta > 0$ such that if $y \in N_\delta(x) \cap A$ then $|f(y) - f(x)| < \epsilon$. Thus $N_\delta(x) \cap A \subseteq f^{-1}(U)$, and so $f^{-1}(U)$ is a relatively open subset of A. Thus (i) implies (ii). Since $U_c = f^{-1}((c, \infty))$ and $L_c = f^{-1}((-\infty, c))$, and (c, ∞) and $(-\infty, c)$ are open, (ii) implies (iii).

Suppose that (iii) holds. Suppose that $x \in A$ and that $\epsilon > 0$. Then the sets $U_{f(x)-\epsilon}$ and $L_{f(x)+\epsilon}$ are relatively open in A, and x is in each of them. Thus $U_{f(x)-\epsilon} \cap L_{f(x)+\epsilon}$ is relatively open, and there exists $\delta > 0$ such that $N_\delta(x) \cap A \subseteq U_{f(x)-\epsilon} \cap L_{f(x)+\epsilon}$; thus if $y \in N_\delta(x) \cap A$ then $f(y) > f(x) - \epsilon$

and $f(y) < f(x) + \epsilon$, so that $|f(y) - f(x)| < \epsilon$. Thus f is continuous at x. Since this holds for all $x \in A$, (iii) implies (i).

Finally, the equivalences of (ii) and (iv), and of (iii) and (v), follow by considering complements. □

It is important that these conditions involve *inverse images* of open and closed sets. Here is a simple example to show that similar results need not hold for direct images. Let $f(x) = 1/(1 + x^2)$. Then f is continuous on $\mathbf{R}$, $\mathbf{R}$ is both open and closed, and $f(\mathbf{R}) = (0, 1]$, which is neither open nor closed. The continuous image of an open set need not be open, and the continuous image of a closed set need not be closed.

We now consider some simple examples of continuous real-valued functions, and of discontinuities, which will enable us to introduce some more ideas.

1. Take $A = \mathbf{R}$, and set $i(x) = x$. Then i is continuous on $\mathbf{R}$; if $|x - a| < \epsilon$ then $|i(x) - i(a)| < \epsilon$, so that we can take $\delta = \epsilon$, for each $x \in \mathbf{R}$. Combining this with the results of Theorem 6.3.1, we see that all polynomial functions on $\mathbf{R}$ are continuous.

2. The exponential function is continuous on $\mathbf{R}$. First, note that if $|h| < 1$ then
$$|e^h - 1| = |h + \frac{h^2}{2!} + \frac{h^3}{3!} + \cdots| \le |h|(1 + \frac{1}{2} + \frac{1}{2^2} + \cdots) = 2|h|.$$
Suppose that $a \in \mathbf{R}$ and $\epsilon > 0$. Let $\delta = \epsilon/2e^a$. If $|x - a| < \delta$ then
$$|e^x - e^a| = |e^a e^{x-a} - e^a| = e^a|e^{x-a} - 1| < 2|x - a|e^a < 2\delta e^a = \epsilon.$$

3. Take $A = \mathbf{R}$, and set $f(x) = x$ if $x \neq 0$ and set $f(0) = 1$. Then f is continuous at every point of $\mathbf{R}$ except 0. The discontinuity at 0 is the simplest sort of discontinuity; if we change the value at 0 to 0, we remove the discontinuity. More generally, a real-valued function f on A has a *removable discontinuity* at a if $f(x) \to l$ as $x \to a$, and $l \neq f(a)$. If we redefine $f(a)$ as l, then the discontinuity disappears.

4. Suppose that f is a real-valued function on a subset A of $\mathbf{R}$, and that $a \in A$. We say that f is *continuous on the right* at a if whenever $\epsilon > 0$ there exists $\delta > 0$ (which usually depends on ϵ) such that $|f(x) - f(a)| < \epsilon$ for those $x \in A$ with $a \le x < a + \delta$. *Continuity on the left* is defined in a similar way. f is continuous on the right if and only if either $f(a+) = \lim_{x \searrow a} f(x)$ exists and is equal to $f(a)$, or there exists $\delta > 0$ such that $(a, a+\delta) \cap A = \emptyset$. We say that f has a *jump discontinuity* at a if one of the following cases holds:

(i) $f(a-)$ and $f(a+)$ both exist and are different, and $f(a) \in [f(a-), f(a+)]$ – in this case we have a jump of (positive or negative) size $f(a+) - f(a-)$;

(ii) $f(a-)$ exists and is different from $f(a)$, and f is continuous on the right at a – in this case we have a jump of (positive or negative) size $f(a) - f(a-)$;

(iii) $f(a+)$ exists and is different from $f(a)$, and f is continuous on the left at a – in this case we have a jump of (positive or negative) size $f(a+) - f(a)$.

[We give this cumbersome definition to allow for the possibility that $A \cap (a, a + \delta)$ or $A \cap (a - \delta, a)$ may be empty for some $\delta > 0$.]

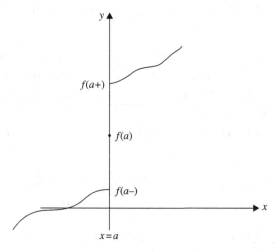

Figure 6.3. A jump discontinuity.

Theorem 6.3.5 *The only discontinuities of a monotonic function are jump discontinuities, and the set of discontinuities is countable.*

Proof   The first statement follows from Theorem 6.1.4. For the second, let D be the set of discontinuities of f. If $d \in D$, let $i(d) = (f(d-), f(d+))$ [in case (i) above] or $((f(d-), f(d))$ [in case (ii)], or $((f(d), f(d+))$ [in case (iii)]. Then the open intervals $\{i(d) : d \in D\}$ are disjoint, and their union is open, and so D is countable, by Theorem 5.3.3.   □

5. Suppose that A is a subset of **R**. Let I_A be the indicator function of A: $I_A(x) = 1$ if $x \in A$ and $I_A(x) = 0$ if $x \notin A$. If $x \in A°$ then there exists $\delta > 0$ such that $N_\delta(x) \subseteq A$, and then $I_A(y) = I_A(x) = 1$ for $y \in N_\delta(x)$. Thus I_A is continuous at each point of $A°$. Similarly, I_A is continuous at

each point of $(C(A))° = C(\overline{A})$. What happens if $x \in \partial A$? If $x \in \partial A$ and $\delta > 0$ then there exit $y \in N_\delta(x) \cap A$ and $z \in N_\delta(x) \cap C(A)$, so that $I_A(y) = 1$ and $I_A(z) = 0$. Thus I_A is not continuous at x.

For example, the indicator function of Cantor's ternary set is discontinuous at points of C, and continuous at points of the complement of C. The indicator function of the rationals has no points of continuity, since $\partial \mathbf{Q} = \mathbf{R}$.

6. Let f be the *saw-tooth function*

$$f(x) = \begin{cases} \{x\} & \text{for } 2k \le x < 2k+1, \\ 1 - \{x\} & \text{for } 2k+1 \le x < 2k+2, \end{cases}$$

for $k \in \mathbf{Z}$. Let $g(x) = f(1/x)$ for $x \neq 0$, and let $g(0) = 0$. Then g has a discontinuity at 0: $g(x)$ oscillates in value between 0 and 1 as $x \to 0$.

These examples by no means exhaust the ways in which a real-valued function can be discontinuous.

We have seen that the continuous image of a closed set need not be closed. The situation is different for bounded closed sets. We now use the Bolzano–Weierstrass theorem to obtain some results of fundamental importance.

Theorem 6.3.6 *Suppose that f is a continuous real-valued function on a non-empty bounded closed subset A of $\mathbf{R}$. The image $f(A) = \{f(x) : x \in A\}$ is a bounded and closed subset of $\mathbf{R}$. In particular, f attains its bounds: there exist $y, z \in A$ such that $f(y) = \sup\{f(x) : x \in A\}$ and $f(z) = \inf\{f(x) : x \in A\}$.*

Proof First, suppose, if possible, that f is not bounded. Then for each $n \in \mathbf{N}$ there exists $a_n \in A$ with $|f(a_n)| \ge n$. By the Bolzano–Weierstrass theorem there exists a subsequence $(a_{n_k})_{k=1}^\infty$ which converges to an element $a \in A$ as $k \to \infty$. Since f is continuous, $f(a_{n_k}) \to f(a)$ as $k \to \infty$ (Proposition 6.3.2), and so $(f(a_{n_k}))_{k=1}^\infty$ is bounded, giving a contradiction.

Secondly, suppose that $b \in \overline{f(A)}$. Then there exists a sequence $(a_n)_{n=1}^\infty$ in A such that $f(a_n) \to b$ as $n \to \infty$. By the Bolzano–Weierstrass theorem there exists a subsequence $(a_{n_k})_{k=1}^\infty$ which converges to an element $a \in A$ as $k \to \infty$. Since f is continuous, $f(a_{n_k}) \to f(a)$ as $k \to \infty$ (Proposition 6.3.2). But $f(a_{n_k}) \to b$ as $k \to \infty$, and so $b = f(a) \in f(A)$. Thus $\overline{f(A)} = f(A)$, and $f(A)$ is closed. $\square$

Suppose that f is a real-valued function defined on an interval I and that a is an interior point of I. f has a *local maximum* at a if there exists $\delta > 0$ such that $(a - \delta, a + \delta) \subseteq I$ and $f(x) \le f(a)$ for all $x \in (a - \delta, a + \delta)$. A *local minimum* is defined similarly.

Corollary 6.3.7 *Suppose that f is a continuous real valued function on an interval I which has no local maximum or local minimum. Then f is a monotonic function on I.*

Proof Suppose that f is not monotonic. Then there exist $a < d < b$ in I such that either $f(a) < f(d) > f(b)$ or $f(a) > f(d) < f(b)$. Consider the restriction of f to $[a, b]$. In the former case, f attains its supremum at a point c of $[a, b]$. Since $f(c) \geq f(d) > \max(f(a), f(b))$, c is an interior point of $[a, b]$, and c is a local maximum of f. In the second case, f has a local minimum in $[a, b]$; the proof is exactly similar.   □

Theorem 6.3.8 *Suppose that f is an injective continuous real-valued function on a non-empty bounded closed subset A of $\mathbf{R}$. Then the inverse mapping $f^{-1} : f(A) \to A$ is continuous.*

Proof Let $h = f^{-1}$. If F is a closed subset of $\mathbf{R}$, then $h^{-1}(F) = f(F \cap A)$, which is closed in $\mathbf{R}$, by Theorem 6.3.6, and is therefore closed in $f(A)$. Thus h is continuous, by Theorem 6.3.4.   □

If f is a continuous real-valued function on a set A, and $\epsilon > 0$, then for each $x \in A$ there exists a $\delta > 0$ such that $f(N_\delta(x) \cap A) \subseteq N_\epsilon(f(x))$. In general, the value of δ depends on x. To take a very easy example, consider the continuous real-valued function $f(x) = x^2$ on $\mathbf{R}$. Then if $\epsilon > 0$ and $x > 0$ then

$$(x + \epsilon/2x)^2 = x^2 + \epsilon + \epsilon^2/4x^2 > x^2 + \epsilon,$$

and so δ must be smaller than $\epsilon/2x$. Thus it is not possible to find a single $\delta > 0$ that will work for all x. There are however important cases where for each $\epsilon > 0$ a single δ will do. This merits a definition. Suppose that f is a real-valued function defined on a subset A of $\mathbf{R}$. f is *uniformly continuous* on A if whenever $\epsilon > 0$ there exists $\delta > 0$ (which usually depends on ϵ) such that if $x, y \in A$ and $|x - y| < \delta$ then $|f(x) - f(y)| < \epsilon$.

Theorem 6.3.9 *Suppose that f is a continuous real-valued function on a non-empty bounded closed subset A of $\mathbf{R}$. Then f is uniformly continuous on A.*

Proof Suppose not. Then there exists $\epsilon > 0$ for which we can find no suitable $\delta > 0$. Thus for each $n \in \mathbf{N}$ there exist elements a_n and b_n in A with

$|a_n - b_n| < 1/n$ and $|f(a_n) - f(b_n)| \geq \epsilon$. By the Bolzano–Weierstrass theorem there exists a subsequence $(a_{n_k})_{k=1}^\infty$ which converges to an element $a \in A$ as $k \to \infty$. Since $a_{n_k} - b_{n_k} \to 0$ as $k \to \infty$, $b_{n_k} \to a$, as well. Since f is continuous at a, $f(a_{n_k}) \to f(a)$ and $f(b_{n_k}) \to f(a)$ as $k \to \infty$, so that $f(a_{n_k}) - f(b_{n_k}) \to 0$ as $k \to \infty$. As $|f(a_{n_k}) - f(b_{n_k})| \geq \epsilon$ for all $k \in \mathbf{N}$, we have a contradiction. $\qquad\square$

We have seen that it is not always possible to exchange limiting procedures when we consider a double sequence. Similar phenomena occur when we consider a sequence of functions of a real variable, or a function of two real variables. For example, let $f(x, y) = e^{-x/y}$ for $x > 0, y > 0$. Then

$$\lim_{x \to \infty} \left(\lim_{y \to \infty} f(x, y) \right) = \lim_{x \to \infty} 1 = 1$$

$$\lim_{y \to \infty} \left(\lim_{x \to \infty} f(x, y) \right) = = \lim_{y \to \infty} 0 = 0.$$

Similarly, let $f_n(x) = x^n$, for $x \in [0, 1]$ and $n \in \mathbf{N}$. Then each function f_n is continuous on $[0, 1]$. Let $f(x) = 0$ in $0 \leq x < 1$ and let $f(1) = 1$. Then $f_n(x) \to f(x)$ for each $x \in [0, 1]$, and e is not continuous at 1.

There is however one easy and important case, where limits are taken of increasing functions or sequences, and sums are taken of positive elements. We shall prove just one case, which we shall need later.

Theorem 6.3.10 *Suppose that $f_n(x)$ is a sequence of non-negative increasing functions on an interval $[a, b]$, each of which is continuous on the left at b. Then*

$$\sum_{n=1}^\infty f_n(b) = \lim_{x \to b} \left(\sum_{n=1}^\infty f_n(x) \right).$$

(Here the sums and limit can be finite or infinite.)

Proof If $\sum_{n=1}^\infty f_n(c) = \infty$ for some $c \in [a, b)$, then $\sum_{n=1}^\infty f_n(x) = \infty$ for $x \in [c, b]$, and the result holds. Otherwise, the mapping $x \to \sum_{n=1}^\infty f_n(x)$ is increasing, and so

$$\sup_{a \leq x < b} \sum_{n=1}^\infty f_n(x) = \lim_{x \to b} \sum_{n=1}^\infty f_n(x).$$

Hence

$$\sum_{n=1}^{\infty} f_n(b) = \sup_{m\in\mathbf{N}} \sum_{n=1}^{m} f_n(b)$$

$$= \sup_{m\in\mathbf{N}} \left(\sup_{a\le x<b} \sum_{n=1}^{m} f_n(x) \right)$$

$$= \sup_{a\le x<b} \left(\sup_{m\in\mathbf{N}} \sum_{n=1}^{m} f_n(x) \right)$$

$$= \sup_{a\le x<b} \sum_{n=1}^{\infty} f_n(x) = \lim_{x\to b} \sum_{n=1}^{\infty} f_n(x).$$

$\square$

Exercises

6.3.1 A real-valued function on $\mathbf{R}$ is *periodic* if there exists a non-zero number t such that $f(x + t) = f(x)$ for all $x \in \mathbf{R}$. Suppose that f is periodic. Show that

$$\{t \in \mathbf{R} : f(x+t) = f(x) \text{ for all } x \in \mathbf{R}\}$$

is a subgroup of $\mathbf{R}$. Show that if f is continuous and not constant, then it is a proper closed subgroup of $\mathbf{R}$, and that there is a least positive t such that $f(x + t) = f(x)$ for all $x \in \mathbf{R}$; t is the *period* of f.

6.3.2 Show that the exponential function is strictly increasing on $\mathbf{R}$, that it is not uniformly continuous on $\mathbf{R}$, but that its restriction to any semi-infinite interval $(-\infty, A]$ is uniformly continuous.

6.3.3 Define a real-valued function f on $(0, 1)$ as follows. If r is rational and $r = p/q$ in lowest terms then $f(r) = 1/q$; if x is irrational, then $f(x) = 0$. Show that f is continuous at every irrational point of $(0, 1)$, and that f is discontinuous at every rational point of $(0, 1)$.

6.3.4 Suppose that f and g are continuous real-valued functions on A and that $h(x) = \max(f(x), g(x))$, for $x \in A$. Show that h is continuous. Give an example of a sequence $(f_n)_{n=0}^{\infty}$ of non-negative continuous real-valued functions on $[0, 1]$ for which $\inf_{n\in\mathbf{N}} f_n$ is not continuous.

6.3.5 Suppose that A is a non-empty subset of $\mathbf{R}$. Let $d_A(x) = \inf\{|x - a| : a \in A\}$, for $x \in \mathbf{R}$.
   (a) Show that $|d_A(x) - d_A(y)| \le |x - y|$.
   (b) Show that $\{x \in \mathbf{R} : d_A(x) = 0\} = \overline{A}$.

(c) Suppose that A is closed and that B is compact. Show that there exist $a \in A$ and $b \in B$ such that $|a - b| = \inf\{|x - y| : a \in A, y \in B\}$.

(d) Suppose further that A and B are disjoint. Show that there exist disjoint open sets U and V such that $A \subseteq U$ and $B \subseteq V$.

(e) Suppose that A and C are disjoint closed subsets of $\mathbf{R}$. Show that there exist disjoint open sets U and V such that $A \subseteq U$ and $C \subseteq V$.

6.3.6 Let K be a closed subset of $[0, 1]$ containing 0 and 1, and let f be a continuous real-valued function on K. Extend f to $[0, 1]$ by linear interpolation: if $x \in [0, 1] \setminus K$ let $l = \sup\{k \in K : k < x\}$, let $r = \inf\{k \in K : k > x\}$, and let

$$f(x) = \frac{r - x}{r - l} f(l) + \frac{x - l}{r - l} f(r).$$

Show that the extended real-valued function f is continuous on $[0, 1]$.

6.3.7 Give an example of a continuous injective real-valued function f on a closed subset A of $\mathbf{R}$ for which the inverse function is not continuous.

6.3.8 Show that if f is a uniformly continuous real-valued function on a subset A of $\mathbf{R}$ then f extends to a uniformly continuous function g on $\overline{A}$, and that the extension is unique. Give an example to show that the corresponding result for continuous real-valued functions is false.

6.3.9 At the jth stage in the construction of Cantor's ternary set C, we remove 2^{j-1} intervals, each of length $1/3^j$. List these intervals from left to right as $I_{1,j}, \ldots I_{2^{j-1},j}$ – that is, $\sup(I_{i,j}) < \inf(I_{i+1,j})$ for $1 \leq i < 2^{j-1}$. Define a function f on $[0, 1] \setminus C$ by setting $f(x) = (2i - 1)/2^j$ for $x \in I_{i,j}$. Verify that f is an increasing function on $[0, 1] \setminus C$. Set $f(1) = 1$, and if $x \in C$ and $x \neq 1$, set $f(x) = \inf(\{f(y) : y > x, y \in [0, 1] \setminus C\}$. Show that f is a continuous increasing function on $[0, 1]$. This is the *Cantor–Lebesgue function*.

6.3.10 Suppose that f is a real-valued function on $\mathbf{R}$ which satisfies $f(x+y) = f(x) + f(y)$ for all $x, y \in \mathbf{R}$, and which is continuous at 0. Show that there exists $\lambda \in \mathbf{R}$ such that $f(x) = \lambda x$ for all $x \in \mathbf{R}$.

6.3.11 Suppose that f is a continuous real-valued function on $[0, 1]$ with $f(0) = f(1) = 0$. Suppose that for each $0 < x < 1$ there exists $h > 0$, with $0 \leq x - h < x < x + h \leq 1$, such that

$$f(x) = (f(x - h) + f(x + h))/2.$$

Show that $f(x) = 0$ for all $x \in [0, 1]$.

6.3.12 If $\mathbf{Q} \in \mathbf{N}$, let Q_q be the set of rationals p/q in $[0,1]$, where p and q are coprime. Suppose that $n \in \mathbf{N}$ and that $x \in [0,1]$. If there exists $p/q \in Q_q$ such that $|x - p/q| \leq 1/2n^2$, let $f_n(x) = (1 - 2n^2|x - p/q|)/q$; otherwise let $f_n(x) = 0$. Show that f_n is a well-defined function on $[0,1]$ and that f is continuous. Show that for each $x \in [0,1]$, $f_n(x)$ converges to $f(x)$ as $n \to \infty$, where f is the function defined in Exercise 6.3.3. The point-wise limit of continuous functions can have a dense set of discontinuities.

6.4 The intermediate value theorem

We now consider a continuous real-valued function defined on an interval I. Suppose that f is continuous on I, that $a, b \in I$, and that $f(a)$ is negative and $f(b)$ is positive. Then intuition suggests that $f(c) = 0$ for some point c in the interval $[a, b]$. This is indeed so, but we must prove it. The result is a consequence of the connectedness of the interval $[a, b]$.

Theorem 6.4.1 *Suppose that f is a continuous function on an interval I, that a, b are points of I with $a < b$, and that $f(a) < v < f(b)$. Then there exists $a < c < b$ such that $f(c) = v$.*

Proof We give two proofs. The first uses the connectedness of I. Let $L = \{x \in [a, b] : f(x) < v\}$ and let $G = \{x \in [a, b] : f(x) > v\}$. Then L and G are disjoint non-empty relatively open subsets of I. Since $[a, b]$ is connected, it follows that $[a, b] \neq L \cup G$; if $c \in [a, b] \setminus (L \cup G)$ then $f(c) = v$.

For the second proof, we use repeated dissection, as in the second proof of the Bolzano–Weierstrass theorem. Set $a_0 = a$ and $b_0 = b$. Let $d_0 = (a_0 + b_0)/2$. If $f(d_0) \geq v$, we set $a_1 = a_0$ and $b_1 = d_0$. Otherwise, $f(d_0) < v$, and we set $a_1 = d_0$ and $b_1 = b_0$. Thus $b_1 - a_1 = (b_0 - a_0)/2$, and $f(a_1) \leq v \leq f(b_1)$. We now iterate this procedure recursively. At the jth step, we obtain a closed interval $[a_j, b_j]$ contained in $[a_{j-1}, b_{j-1}]$ with $b_j - a_j = (b_0 - a_0)/2^j$, and with $f(a_j) \leq v \leq f(b_j)$. Then the sequence $(a_j)_{j=0}^{\infty}$ is increasing, the sequence $(b_j)_{j=0}^{\infty}$ is decreasing, and both converge to a common limit c. Then $f(c) = \lim_{j \to \infty} f(a_j) \leq v$ and $f(c) = \lim_{j \to \infty} f(b_j) \geq v$, so that $f(c) = v$. □

Of course, a similar result holds if $f(a) > f(b)$.

Corollary 6.4.2 *If f is a continuous function on an interval I then $f(I)$ is an interval.*

Corollary 6.4.3 *If f is a continuous strictly monotonic function on an open interval I then $f(I)$ is an open interval.*

Proof If I is open and $x \in I$, there exist $a, b \in I$ with $a < x < b$, so that $f(x) \in (f(a), f(b)) \subseteq f(I)$; $f(I)$ is open. $\qquad \square$

Proposition 6.4.4 *If f is a continuous function on an interval I then f is injective if and only if f is strictly monotonic.*

Proof If f is strictly monotonic, then certainly f is injective. Suppose that f is not strictly monotonic, and suppose for example that $a < b < c$ while $f(a) < f(c) < f(b)$. Then there exists $d \in [a, b]$ such that $f(d) = f(c)$, contradicting the fact that f is injective. Other possibilities are dealt with in the same way. $\qquad \square$

Proposition 6.4.5 *If f is a strictly monotonic function on an interval I then $f^{-1} : f(I) \to I$ is continuous.*

Proof Suppose without loss of generality that f is strictly increasing. Suppose that $b \in f(I)$ and that $\epsilon > 0$. Suppose that $a = f^{-1}(b)$ is an interior point of I. There exist $c, d \in I$ with $a - \epsilon < c < a < d < a + \epsilon$. Then $f(c) < f(a) = b < f(d)$; let $\delta = \min(b - f(c), f(d) - b)$. If $|y - b| < \delta$, then $f(c) < y < f(d)$, so that $c < f^{-1}(y) < d$, and $|f^{-1}(y) - f^{-1}(b)| < \epsilon$. The case where a is an end-point of I is left to the reader. $\qquad \square$

Note that in this last proposition we do not require f to be continuous.

We can now establish the existence of nth roots of positive numbers without the need for any subsidiary calculations.

Corollary 6.4.6 *If $a > 0$ and $k \in \mathbf{N}$ then there exists a unique $y > 0$ such that $y^n = a$. Let $y = a^{1/n}$. The the mapping $a \to a^{1/n} : (0, \infty) \to (0, \infty)$ is continuous.*

Proof The function $f(x) = x^n$ is a strictly increasing continuous function on $(0, \infty)$, so that $f((0, \infty))$ is an interval. Since $f(x) \to 0$ as $x \to 0$ and $f(x) \to \infty$ as $x \to \infty$, $f((0, \infty)) = (0, \infty)$. Thus f^{-1} is a continuous bijection of $(0, \infty)$ onto $(0, \infty)$. If $a \in (0, \infty)$ then $y = f^{-1}(a)$ is the unique positive nth root of a. $\qquad \square$

We also have the following.

Proposition 6.4.7 *Suppose that p is a real polynomial of odd degree n. Then there exists $x \in \mathbf{R}$ with $p(x) = 0$.*

Proof Without loss of generality, we can suppose that p is monic, so that $p = x^n + a_{n-1}x^{n-1} + \cdots + a_0$. We shall show that p takes both positive and negative values. If $x \neq 0$ then $p(x) = x^n(1 + q(x))$, where $q(x) = a_{n-1}/x + \cdots + a_0/x^n$. Since $a_j/x^{n-j} \to 0$ as $x \to \infty$ and as $x \to -\infty$, for $0 \leq j \leq n-1$, there exists

$R > 0$ such that $|q(x)| < 1/2$ for $|x| \geq R$. Then $1 + q(x) > 1/2$ for $|x| \geq R$, so that $p(-R) \leq -R^n/2 < 0$ and $p(R) \geq R^n/2 > 0$. By the intermediate value theorem there exists $x \in [-R, R]$ for which $p(x) = 0$.   □

We have the following fixed-point theorem.

Theorem 6.4.8   *Suppose that $[a, b]$ is a closed bounded interval and that $f : [a, b] \to [a, b]$ is continuous. Then there exists $c \in [a, b]$ with $f(c) = c$.*

Proof If $f(a) = a$ or if $f(b) = b$, there is nothing to prove. Otherwise, let $g(x) = x - f(x)$. Then $g(a) = a - f(a) < 0$ and $g(b) = b - f(b) > 0$. By the intermediate value theorem, there exists $c \in [a, b]$ with $g(c) = c - f(c) = 0$.   □

Exercises

6.4.1 Suppose that $0 < a < b$. Find $\lim_{n \to \infty} (a^n + b^n)^{1/n}$.

6.4.2 Show that $n^{1/n} \to 1$ as $n \to \infty$.

6.4.3 Does $(n!)^{1/n}$ converge, as $n \to \infty$?

6.4.4 Give an example of a continuous bijective map of $(0, 1)$ onto itself with no fixed point.

6.4.5 Let $f(x) = x$ for x rational and $f(x) = 1 - x$ for f irrational. Show that f is a bijection of $[0, 1]$ onto itself, and that f has exactly one point of continuity. Can you find a bijection of $[0, 1]$ onto itself with no points of continuity?

6.4.6 Suppose that f is a continuous periodic function on $\mathbf{R}$, and that $t > 0$. Show that there exists $x \in \mathbf{R}$ with $f(x) = \frac{1}{2}(f(x + t) + f(x - t))$.

6.4.7 Suppose that $f(x)$ is a continuous function on $[0, 1]$ with $f(0) = f(1)$.

(a) Use the intermediate value theorem to show that there exists $0 \leq x \leq 1/2$ with $f(x) = f(x + 1/2)$.

(b) Suppose that $n \in \mathbf{N}$ and that $n > 1$. By considering the sequence $(f((j-1)/n) - f(j/n))_{j=1}^n$ show that there exists $0 \leq x \leq 1 - 1/n$ such that $f(x) = f(x + 1/n)$.

(c) Suppose that $0 < \lambda < 1$ and that $1/\lambda$ is not an integer. Let $h(x) = f(2/\lambda)x - f(2x/\lambda)$, where f is the saw-tooth function of Section 6.3. Show that there exists no $x \in [0, 1 - \lambda]$ with $h(x) = h(x + \lambda)$.

6.5 Point-wise convergence and uniform convergence

Suppose that $(f_n)_{n=1}^\infty$ is a sequence of real-valued functions on a set S, and that f is another such function. The sequence $(f_n)_{n=1}^\infty$ *converges point-wise* to f if $f_n(s) \to f(s)$ for each $s \in S$. More formally,

- the sequence $(f_n)_{n=1}^\infty$ converges *point-wise* to f if for each $s \in S$ and each $\epsilon > 0$ there exists $n_0 \in \mathbf{N}$ such that $|f_n(s) - f(s)| < \epsilon$ for all $n \geq n_0$.

Note that the choice of n_0 depends on both ϵ and s. This is a very natural idea to consider, but it turns out that point-wise convergence is too weak for many purposes, and is awkward to work with. A stronger, and more tractable notion is that of *uniform convergence*. Here the number n_0 depends only on ϵ: the same value works for all $s \in S$. Formally,

- the sequence $(f_n)_{n=1}^\infty$ converges *uniformly* to f on S if for each $\epsilon > 0$ there exists $n_0 \in \mathbf{N}$ such that $|f_n(s) - f(s)| < \epsilon$ for all $n \geq n_0$ and all $s \in S$.

Let

$$f_n(x) = \begin{cases} 2nx & \text{for } 0 \leq x \leq 1/2n, \\ 2 - 2nx & \text{for } 1/2n \leq x \leq 1/n, \\ 0 & \text{otherwise.} \end{cases}$$

Then $f_n(0) = 0$, and if $0 < x \leq 1$ then $f_n(x) = 0$ if $n > 1/x$, so that f_n converges point-wise to 0 on $[0,1]$. It does not converge uniformly, since $f_n(1/2n) = 1$, for $n \in \mathbf{N}$.

Uniform convergence is particularly useful when we consider continuity. Here is the fundamental result connecting continuity and uniform convergence: it is very easy, but very important.

Theorem 6.5.1 *Suppose that $(f_n)_{n=1}^\infty$ is a sequence of continuous real-valued functions defined on a subset A of $\mathbf{R}$ and that f_n converges uniformly to a function f, as $n \to \infty$. Then f is continuous on A.*

Proof Suppose that $z_0 \in A$ and that $\epsilon > 0$. Then there exists $n_0 \in \mathbf{N}$ such that $|f_n(z) - f(z)| < \epsilon/3$ for $n \geq n_0$ and for all $z \in A$. Since f_{n_0} is continuous at z_0, there exists $\delta > 0$ such that if $z \in A$ and $|z - z_0| < \delta$ then $|f_{n_0}(z) - f_{n_0}(z_0)| < \epsilon/3$. For such z,

$$|f(z) - f(z_0)| \leq |f(z) - f_{n_0}(z)| + |f_{n_0}(z) - f_{n_0}(z_0)| + |f_{n_0}(z_0) - f(z_0)|$$
$$< \epsilon/3 + \epsilon/3 + \epsilon/3 = \epsilon. \qquad \square$$

Let $f_n(x) = x^n$ for $x \in [0,1]$. Then $f_n(x) \to 0$ for $0 \leq x < 1$, and $f_n(1) = 1$, so that f_n converges point-wise on $[0,1]$ to a discontinuous function; the point-wise limit of continuous functions need not be continuous.

An infinite series $\sum_{n=0}^\infty f_n$ of real-valued functions on a set S converges point-wise, or uniformly, if the sequence of partial sums does. It is said to converge *absolutely uniformly* if $\sum_{n=0}^\infty |f_n|$ converges uniformly.

Proposition 6.5.2 *If an infinite series $\sum_{n=0}^{\infty} f_n$ of real-valued functions on a set S converges absolutely uniformly, then it converges uniformly.*

Proof For each $s \in S$, $\sum_{n=0}^{\infty} f_n(s)$ converges absolutely, and therefore converges to $t(s)$, say. Suppose that $\epsilon > 0$. Then there exists n_0 such that $\sum_{j=m+1}^{n} |f(s)| < \epsilon$, for $n_0 \leq m < n$ and for all $s \in S$. If $s \in S$ and $m > n_0$ then

$$|\sum_{j=0}^{m} f_j(s) - t(s)| = \lim_{n \to \infty} |\sum_{j=0}^{m} f_j(s) - \sum_{j=0}^{n} f_j(s)|$$

$$= \lim_{n \to \infty} |\sum_{j=m+1}^{n} f_j(s)| \leq \lim_{n \to \infty} \sum_{j=m+1}^{n} |f_j(s)| \leq \epsilon.$$

Since this holds for all $s \in S$, $\sum_{n=0}^{\infty} f_n$ converges uniformly to t. □

Here is a simple test for absolute uniform convergence.

Proposition 6.5.3 (Weierstrass' uniform M test) *Suppose that $\sum_{n=0}^{\infty} f_n$ is an infinite series of real-valued functions on a set S, and that $(M_n)_{n=0}^{\infty}$ is a sequence in $\mathbf{R}^+$ for which $|f_n(s)| \leq M_n$ for all $s \in S$ and all $n \in \mathbf{Z}^+$. If $\sum_{n=0}^{\infty} M_n < \infty$, then $\sum_{n=0}^{\infty} f_n$ converges absolutely uniformly.*

Proof An easy exercise. □

We shall consider uniform convergence in a more general setting in Volume II.

Exercises

6.5.1 Let $(r_n)_{n=0}^{\infty}$ be an enumeration of the rationals in $[0, 1]$, with $r_0 = 0$, $r_1 = 1$. If $x \in [0, 1]$, let $f_0(x) = x$, let $f_1(x) = 1 - x$ and let

$$f_k(x) = \begin{cases} x/r_k & \text{if } 0 \leq x \leq r_k \\ (1-x)/(1-r_k) & \text{if } r_k \leq x \leq 1 \end{cases}$$

for $k > 1$. Let $g_n(x) = \sum_{k=0}^{\infty} (f_k(x))^n / 2^k$, for $n \in \mathbf{N}$.
(a) Show that the sum converges uniformly on $[0, 1]$, so that g_n is a continuous function on $[0, 1]$.
(b) Show that $g_n(r_k) \to 2r_k$ as $n \to \infty$, for each $k \in \mathbf{Z}^+$, and that $g_n(y) \to 0$ as $n \to \infty$, for each irrational y in $[0, 1]$.
(c) Let $h(x) = \lim_{n \to \infty} g_n(x)$. Show that H is discontinuous at the rational points of $[0, 1]$ and is continuous at the irrational points.

6.5.2 Construct a sequence $(f_n)_{n=0}^\infty$ of continuous functions such that $\sum_{n=0}^\infty |f_n|$ converges point-wise and $\sum_{n=0}^\infty f_n$ converges uniformly, but not absolutely uniformly.

6.5.3 Prove Weierstrass' uniform M test.

6.5.4 *Dirichlet's test for uniform convergence.* Suppose that $(f_j)_{j=0}^\infty$ is a decreasing sequence of non-negative real-valued functions on a set S which converges uniformly to 0 and that $(z_j)_{j=0}^\infty$ is a sequence of real-valued functions on S for which the sequence of partial sums $(\sum_{j=0}^n z_j)_{n=0}^\infty$ is uniformly bounded: there exists M such that $|\sum_{j=0}^n z_j(s)| \le M$ for all $n \in \mathbf{Z}^+$ and all $s \in S$. Use Abel's formula to show that $\sum_{j=0}^\infty a_j z_j$ converges uniformly.

6.6 More on power series

We now consider the continuity of functions defined by power series. These are complex-valued functions, and we need to introduce the notion of the continuity of a complex-valued function of a complex variable. The definition is essentially the same as the definition of continuity of a real-valued function of a real variable. Suppose that f is a complex-valued function defined on a subset A of $\mathbf{C}$, and that $z_0 \in A$. Then f is continuous at z_0 if whenever $\epsilon > 0$ there exists $\delta > 0$ such that if $z \in A$ and $|z - z_0| < \delta$ then $|f(z) - f(z_0)| < \epsilon$. f is *continuous on* A if it is continuous at each point of A. (Of course, a real-valued function defined on a subset A of $\mathbf{R}$ can be considered as a complex-valued function, and A can be considered as a subset of $\mathbf{C}$: the two definitions of continuity are then trivially the same.) The reader should convince himself or herself that, except for the sandwich principle, which has no obvious analogue, the statements for complex-valued functions of a complex variable which correspond to the statements of Theorems 6.3.1 and 6.3.3 and Proposition 6.3.2 are also true. In particular, polynomial functions on $\mathbf{C}$ are continuous on $\mathbf{C}$.

There is also a complex version of Theorem 6.5.1. The proof is the same as in the real case.

Theorem 6.6.1 *Suppose that $(f_n)_{n=1}^\infty$ is a sequence of continuous real-valued functions defined on a subset A of $\mathbf{R}$ and that f_n converges uniformly to a function f, as $n \to \infty$. Then f is continuous on A.*

Complex versions of the Weierstrass M test and Dirichlet's test also hold (Exercises 6.5.3 and 6.5.4).

Suppose that $\sum_{n=0}^\infty a_n z^n$ is a complex power series with non-zero radius of convergence R. If $|z| < R$, let $f(z) = \sum_{n=0}^\infty a_n z^n$.

Theorem 6.6.2 *Suppose that $\sum_{n=0}^{\infty} a_n z^n$ is a complex power series with radius of convergence R. If $r < R$ then $\sum_{n=0}^{\infty} a_n z^n$ converges absolutely uniformly on $\{z : |z| \leq r\}$ and the function $f(z) = \sum_{n=0}^{\infty} a_n z^n$ on $z : |z| < R$ is continuous on $z : |z| < R$.*

Proof Choose $r < s < R$, and let $M_s = \sup_{n \in \mathbf{Z}^+} |a_n| s^n$. Then

$$\sum_{n=0}^{\infty} |a_n| r^n = \sum_{n=0}^{\infty} |a_n| \left(\frac{r}{s}\right)^n s^n \leq \sum_{n=0}^{\infty} M_s \left(\frac{r}{s}\right)^n = \frac{M_s s}{r - s} < \infty,$$

and if $|z| \leq r$ then $|a_n z^n| \leq |a_n| r^n$. Applying Weierstrass' uniform M test (Exercise 6.5.3) and Theorem 6.6.1, it follows that $\sum_{n=0}^{\infty} a_n z^n$ converges absolutely uniformly on the set $\{z : |z| \leq r\}$ to a function which is continuous on $\{z : |z| \leq r\}$. If $|z| < R$, choose r with $|z| < r < R$. Then f is continuous on the set $\{z : |z| \leq r\}$, and so, considered as a function on the set $\{z : |z| < R\}$, it is continuous at z. □

Note that the proof depends only on the convergence of a geometric series. This simple idea is very powerful, and we shall use it, and the convergence of series such as $\sum_{n=0}^{\infty} n^k r^n$, where $0 \leq r < 1$ and $k \in \mathbf{N}$, many times in the future.

Provided that their radii of convergence are positive, different power series define different functions.

Theorem 6.6.3 *Suppose that the power series $\sum_{n=0}^{\infty} a_n z^n$ and $\sum_{n=0}^{\infty} b_n z^n$ each have radius of convergence greater than or equal to $R > 0$. Let $f(z) = \sum_{n=0}^{\infty} a_n z^n$ and $g(z) = \sum_{n=0}^{\infty} b_n z^n$, for $|z| < R$. Suppose that $(z_k)_{k=1}^{\infty}$ is a null sequence of non-zero complex numbers in $\{z : |z| < R\}$ such that $f(z_k) = g(z_k)$ for all $k \in \mathbf{N}$. Then $a_n = b_n$ for all $n \in \mathbf{Z}^+$.*

Proof If not, let N be the least integer for which $a_N \neq b_N$, Let

$$f_N(z) = f(z) - \sum_{n=0}^{N-1} a_n z^n = \sum_{n=N}^{\infty} a_n z^n = z^N \left(\sum_{n=0}^{\infty} a_{n+N} z^n\right) = z^N F_N(z),$$

and let

$$g_N(z) = f(z) - \sum_{n=0}^{N-1} b_n z^n = \sum_{n=N}^{\infty} b_n z^n = z^N \left(\sum_{n=0}^{\infty} b_{n+N} z^n\right) = z^N G_N(z).$$

Then $f_N(z_k) = g_N(z_k)$ for all $k \in \mathbf{N}$, and so $F_N(z_k) = G_N(z_k)$ for all $k \in \mathbf{N}$. Since $F_N(z) = \sum_{n=0}^{\infty} a_{n+N} z^n$ for $|z| < R$, F_N is continuous at 0. So is G_N,

and so
$$a_N = f_N(0) = \lim_{k \to \infty} F_N(z_k) = \lim_{k \to \infty} G_N(z_k) = G_N(0) = b_N,$$

giving a contradiction. □

This means that if we obtain two power series for the same function, we can 'equate coefficients'.

Suppose that that the power series $\sum_{n=0}^{\infty} a_n z^n$ has radius of convergence 1. What can we say about $\sum_{n=0}^{\infty} a_n$? We begin with an easy result.

Proposition 6.6.4 *Suppose that the power series $\sum_{n=0}^{\infty} a_n z^n$ has radius of convergence 1. The following are equivalent.*

(i) The series $\sum_{n=0}^{\infty} a_n$ is absolutely convergent.
(ii) $\sum_{n=0}^{\infty} a_n z^n$ converges uniformly on $\overline{D} = \{z : |z| \leq 1\}$ to a continuous function f on $\overline{D}$.
(iii) The set $\{\sum_{n=0}^{\infty} |a_n| x^n : 0 \leq x < 1\}$ is bounded.

Proof Since $|a_n z^n| \leq a_n$ for $z \in L_1$, the equivalence of (i) and (ii) follows from the complex version of Weierstrass' uniform M test (Exercise 6.5.3). If (i) holds, then f is bounded on $\overline{D}$, since $\sum_{n=0}^{\infty} |a_n| x^n \leq \sum_{n=0}^{\infty} |a_n|$, and so (iii) holds. Finally, suppose that (iii) holds, and that

$$M = \sup \left\{ \sum_{n=0}^{\infty} |a_n| x^n : 0 \leq x < 1 \right\}.$$

If $N \in \mathbf{N}$ then
$$\sum_{n=0}^{N} |a_n| = \lim_{x \nearrow 1} \sum_{n=0}^{N} |a_n| x^n \leq M,$$

so that $\sum_{n=0}^{\infty} |a_n| \leq M$, and (i) holds. □

What happens if $\sum_{n=0}^{\infty} a_n$ is conditionally convergent? First the radius of convergence is at least 1, since the sequence $(a_n)_{n=0}^{\infty}$ is bounded. If it were greater than 1, then $\sum_{n=0}^{\infty} a_n$ would converge absolutely. Consequently the radius of convergence is 1.

Theorem 6.6.5 (Abel's theorem) *Suppose that the series $\sum_{n=0}^{\infty} a_n$ is convergent, to s, say. Then $\sum_{n=0}^{\infty} a_n x^n \to s$ as $x \nearrow 1$.*

Here we only consider real values of x. A stronger result is obtained in Exercise 6.6.2.

Proof By replacing a_0 by $a_0 - s$, we can suppose that $s = 0$. Let $s_n = \sum_{j=0}^{n} a_j$, for $n \in \mathbf{Z}^+$. The sequence $(s_n)_{n=0}^{\infty}$ is bounded: let $M = \sup\{|s_n| :$

$n \in \mathbf{Z}^+\}$. If $0 \leq x < 1$, the series $\sum_{n=0}^{\infty} a_n x^n$ converges absolutely; let its sum be $f(x)$. The series $\sum_{n=0}^{\infty} x^n$ also converges absolutely, and so by Proposition 4.6.1, the convolution product $\sum_{n=0}^{\infty} c_n x^n$ converges absolutely to $f(x)/(1-x)$. But $c_n = s_n$, and so $f(x) = (1-x) \sum_{n=0}^{\infty} s_n x^n$.

Suppose that $0 < \epsilon < 1$. Let $\eta = \epsilon/2$. There exists n_0 such that $|s_n| < \eta$ for $n \geq n_0$, and so

$$\left| (1-x) \sum_{n=n_0}^{\infty} s_n x^n \right| \leq \eta(1-x) \sum_{n=n_0}^{\infty} x^n \leq \eta(1-x) \sum_{n=0}^{\infty} x^n = \eta.$$

On the other hand,

$$\left| (1-x) \sum_{n=0}^{n_0-1} s_n x^n \right| \leq (1-x) M n_0.$$

If $1 - \eta/(M+1)n_0 < x < 1$ then $|(1-x) \sum_{n=0}^{n_0-1} s_n x^n| < \eta$, and so

$$\left| \sum_{n=0}^{\infty} a_n x^n \right| = |f(x)| = \left| (1-x) \sum_{n=0}^{\infty} s_n x^n \right| < 2\eta = \epsilon.$$

$\square$

The next result involves a decreasing sequence of non-negative coefficients.

Proposition 6.6.6   *Suppose that $(a_n)_{n=0}^{\infty}$ is a decreasing null-sequence of positive numbers, and that $\sum_{n=0}^{\infty} a_n z^n$ has radius of convergence 1. Suppose that $0 < \delta \leq 1$. Then $\sum_{n=0}^{\infty} a_n z^n$ converges uniformly on the set*

$$P_\delta = \{z \in \mathbf{C} : |z| \leq 1, |z - 1| \geq \delta\}.$$

Proof   Let $t_n(z) = z^n$, for $z \in P_\delta$, so that $t_n \in C(P_\delta)$. Then

$$\left| \sum_{j=0}^{n} t_j(z) \right| = \left| \sum_{j=0}^{n} z^n \right| = \left| \frac{1 - z^{n+1}}{1 - z} \right| \leq \frac{2}{\delta},$$

so that $\left\| \sum_{j=0}^{n} t_j \right\| \leq 2/\delta$. The result now follows from Dirichlet's test for uniform convergence (Exercise 6.5.4).

$\square$

Suppose that the power series $\sum_{n=0}^{\infty} a_n z^n$ has positive radius of convergence R, and that $a_0 \neq 0$. The function $f(z) = \sum_{n=0}^{\infty} a_n z^n$ is continuous on $U_R = \{z : |z| < R\}$ and so there exists $0 < r \leq R$ such that if $|z| < r$ then $f(z) \neq 0$, and we can consider the function $1/f(z)$. Can it be expressed as a power series?

Theorem 6.6.7 *Suppose that the power series $\sum_{n=0}^{\infty} a_n z^n$ has positive radius of convergence R, and let $f(z) = \sum_{n=0}^{\infty} a_n z^n$ for $z \in U_R$. Suppose that $0 < S \le R$, and that f has no zeros in the disc $U_S = \{z : |z| < S\}$. Then there exists a power series $\sum_{n=0}^{\infty} c_n z^n$ with positive radius of convergence T such that, if we set $g(z) = \sum_{n=0}^{\infty} c_n z^n$ for $z \in U_T$, then $f(z)g(z) = 1$ for $|z| < \min(S, T)$.*

Proof By multiplying f by a_0^{-1}, we can suppose that $a_0 = 1$. (We do this to simplify the calculations.) Since the series $\sum_{n=0}^{\infty} a_n z^n$ converges absolutely for $|z| < R$, and since the function $\sum_{n=1}^{\infty} |a_n| t^n$ is continuous on $[0, R)$, there exists $t > 0$ such that $\sum_{n=1}^{\infty} |a_n| t^n \le 1$.

In order to see how to proceed, we consider the product of the two series. We require that $c_0 = 1$ and that $\sum_{j=0}^{n} a_j c_{n-j} = 0$ for $n \in \mathbf{N}$. Thus we require that

$$c_n = -\sum_{j=1}^{n} a_j c_{n-j} \quad \text{for } j \in \mathbf{N}.$$

This provides a recursive formula for the sequence $(c_n)_{n=0}^{\infty}$. We now show that the series $\sum_{n=0}^{\infty} c_n z^n$ has radius of convergence at least t. First we show, by induction, that $|c_n| t^n \le 1$ for all n. The result is true if $n = 0$. Suppose that it is true for $j < n$. Then

$$|c_n| t^n = |\sum_{j=1}^{n} (a_j t^j)(c_{n-j} t^{n-j})| \le \sum_{j=1}^{n} (|a_j| t^j)(|c_{n-j}| t^{n-j}) \le \sum_{j=1}^{n} |a_j| t^j \le 1,$$

establishing the claim. If $|z| < t$ then $\sum_{n=0}^{\infty} |c_n z^n| \le \sum_{n=0}^{\infty} (|z|/t)^n < \infty$, so that the series $\sum_{n=0}^{\infty} c_n z^n$ has positive radius of convergence T, with $T \ge t$. Finally, if $|z| < \min(S, T)$ then $f(z)g(z) = 1$, by Proposition 4.6.1. $\qquad \square$

Exercises

6.6.1 Suppose that f is a complex-valued function on a subset A of $\mathbf{C}$. Show that f is continuous on A if and only if its real and imaginary parts are continuous, and if and only if $\overline{f}$ is continuous. Show that $|f|$ is continuous if f is.

6.6.2 Suppose that the series $\sum_{n=0}^{\infty} a_n$ is convergent, to s, say. Suppose that $K > 0$. Let $W_K = \{z : |1 - z| \le K(1 - |z|)\}$ Sketch W_K. Show that $\sum_{n=0}^{\infty} a_n z^n \to s$ as $z \to 1$ in W_K.

6.6.3 State and prove Weierstrass' uniform M test for complex-valued functions.

6.6.4 *Dirichlet's test for uniform convergence: the complex case.* Suppose that $(f_j)_{j=0}^{\infty}$ is a decreasing sequence of non-negative real-valued functions on a set S which converges uniformly to 0 and that $(z_j)_{j=0}^{\infty}$ is a sequence of complex-valued functions on S for which the sequence of partial sums $(\sum_{j=0}^{n} z_j)_{n=0}^{\infty}$ is uniformly bounded: there exists M such that $|\sum_{j=0}^{n} z_j(s)| \leq M$ for all $n \in \mathbf{Z}^+$ and all $s \in S$. Use Abel's formula to show that $\sum_{j=0}^{\infty} a_j z_j$ converges uniformly.

7

Differentiation

7.1 Differentiation at a point

We now restrict attention to real-valued functions defined on an interval. Suppose that f is a real-valued function on an interval I, and that a is an interior point of I, so that there exists $\eta > 0$ such that $(a - \eta, a + \eta) \subseteq I$. Then f is *differentiable* at a, with *derivative* $f'(a)$, if whenever $\epsilon > 0$ there exists $0 < \delta \leq \eta$ such that if $0 < |x - a| < \delta$ then

$$\left| \frac{f(x) - f(a)}{x - a} - f'(a) \right| < \epsilon.$$

In other words, $(f(x) - f(a))/(x - a) \to f'(a)$ as $x \to a$. Thus if f is differentiable at a, then the derivative $f'(a)$ is uniquely determined. The derivative $f'(a)$ is also denoted by $\frac{df}{dx}(a)$.

Note that if $0 < |x - a| < \min(\delta, 1)$ then

$$|f(x) - f(a)| \leq \left| \frac{f(x) - f(a)}{x - a} \right| . |x - a| < \epsilon,$$

so that f is continuous at a.

This definition of the derivative involves division. It is convenient to have characterizations which avoid this.

Proposition 7.1.1 *Suppose that f is a real-valued function on an interval I, that a is an interior point of I, that $(a - \eta, a + \eta) \subseteq I$ and that $l \in \mathbf{R}$. The following are equivalent.*

(i) *f is differentiable at a, with derivative l.*
(ii) *There is a real-valued function r on $(-\eta, \eta) \setminus \{0\}$ such that*

$$f(a + h) = f(a) + lh + r(h) \text{ for } 0 < |h| < \eta$$

for which $r(h)/h \to 0$ as $h \to 0$.

(*iii*) *There is a real-valued function s on* $(-\eta, \eta)$ *such that*

$$f(a + h) = f(a) + (l + s(h))h \ \text{for} \ |h| < \eta$$

for which $s(0) = 0$ *and s is continuous at* 0.

Proof Conditions (i) and (ii) are equivalent, since

$$\frac{r(h)}{h} = \frac{f(a + h) - f(a)}{h} - l,$$

and (ii) and (iii) are equivalent, since $s(h) = r(h)/h$ for $h \neq 0$. $\square$

There are several closely related reasons for considering differentiability. Suppose that $b \in I$ and that $b \neq a$. Then the graph of the function $l_{a,b}$ defined by

$$l_{a,b}(x) = f(a) + \frac{f(b) - f(a)}{b - a}(x - a)$$

is a straight line which includes the line segment $[(a, f(a)), (b, f(b))]$. The quantity $(f(b) - f(a))/(b - a)$ is the *slope* of the line. Thus f is differentiable at a, with derivative $f'(a)$, if and only if the slope tends to $f'(a)$ as b tends to a. If so, then the graph of the function t_a defined by

$$t_a(x) = f(a) + f'(a)(x - a)$$

is the *tangent* to the graph of f at a.

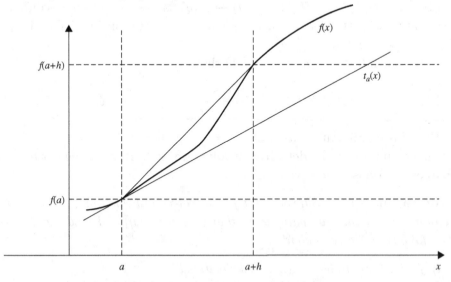

Figure 7.1. Differentiation, and the tangent.

If $|h| < \eta$, and we write

$$f(a + h) = t_a(a + h) + r(h) = f(a) + f'(a)h + r(h)$$

then $r(h)/h \to 0$ as $h \to 0$, so that $r(h) = o(|h|)$ and t_a is a linear approximation to f near a. Further, a small change h in the variable produces a small change approximately equal to $f'(a)h$ in the value of the function f, so that $f'(a)$ is the *rate of change* of f at a.

Let us give some easy examples.

Example 7.1.2   The function $f(x) = x^n$, with $n \in \mathbf{N}$, $n \geq 2$.

By the binomial theorem,

$$f(a + h) = a^n + na^{n-1}h + r(h), \text{ where } r(h) = h^2 q(h),$$

with q a polynomial in h of degree $n - 2$. Thus $r(h)/h \to 0$ as $h \to 0$, and so f is differentiable, with derivative na^{n-1}.

Example 7.1.3   The function $f(x) = 1/x$, on $(0, \infty)$, or on $(-\infty, 0)$.

If $0 < |h| < |a|$ then

$$f(a + h) - f(a) = -\frac{h}{a(a + h)}, \text{ so that } \frac{f(a + h) - f(a)}{h} \to -\frac{1}{a^2}$$

as $h \to 0$. Thus f is differentiable at a, with derivative $-1/a^2$.

Example 7.1.4   The real exponential function $\exp(x)$ on $\mathbf{R}$.

Since $\exp(a + h) = \exp(a)\exp(h)$, it follows that $\exp(a + h) = \exp(a) + \exp(a)h + s(h)h$, where

$$s(h) = \exp(a)\frac{\exp(h) - 1 - h}{h} = \exp(a)\left(\frac{h}{2!} + \frac{h^2}{3!} + \cdots\right),$$

so that if $|h| < 1$ then

$$|s(h)| \leq \frac{\exp(a)|h|}{2}\left(1 + |h| + |h|^2 + \cdots\right) = \frac{\exp(a)|h|}{2(1 - |h|)},$$

and $s(h) \to 0$ as $h \to 0$. Thus $\exp$ is differentiable, and the derivative at a is $\exp(a)$.

Here are some basic properties of differentiation.

Proposition 7.1.5   *Suppose that f and g are real-valued functions on an interval I, that $(a - \eta, a + \eta) \subseteq I$, and that f and g are differentiable at a. Suppose also that $\lambda, \mu \in \mathbf{R}$.*

(i) The derivative $f'(a)$ is unique.
(ii) $\lambda f + \mu g$ is differentiable at a, with derivative $\lambda f'(a) + \mu g'(a)$.

(*iii*) *The product fg is differentiable at a, with derivative $f'(a)g(a) + f(a)g'(a)$.*

(*iv*) *If f is an increasing function, then $f'(a) \geq 0$.*

(*v*) *If $f'(a) > 0$ then there exists $0 < \delta \leq \eta$ such that $f(x) < f(a) < f(y)$ for $a - \delta < x < a < y < a + \delta$.*

Proof   (i), (ii) and (iv) follow immediately from the definition.

(iii) If $0 < |h| < \eta$,

$$\frac{f(a+h)g(a+h) - f(a)g(a)}{h} =$$

$$= \left(\frac{f(a+h) - f(a)}{h}\right) g(a+h) + f(a) \left(\frac{g(a+h) - g(a)}{h}\right)$$

$$\to f'(a)g(a) + f(a)g'(a) \text{ as } h \to 0.$$

(v) There exists $0 < \delta < \eta$ such that

$$\left|\frac{f(a+h) - f(a)}{h} - f'(a)\right| < |f'(a)| \text{ for } 0 < |h| < \delta,$$

from which it follows that $(f(a+h) - f(a))/h > 0$ if $0 < h < \delta$ and $(f(a+h) - f(a))/h < 0$ if $0 > h > -\delta$. $\square$

It is tempting to suppose that if $f'(a) > 0$ then f must be an increasing function in some interval $(a - \delta, a + \delta)$. Exercise 7.1.3 shows that this is not the case.

Next we turn to the composition of two functions.

Theorem 7.1.6 (The chain rule)   *Suppose that f is a real-valued function on an open interval I, that g is a real-valued function on an open interval J, and that $f(I) \subseteq J$. Suppose that $a \in I$, that f is differentiable at a and that g is differentiable at $f(a)$. Then the composite function $g \circ f$ is differentiable at a, with derivative $(g \circ f)'(a) = g'(f(a))f'(a)$.*

Proof   First let us give an inadequate 'proof'. For small h,

$$\frac{g(f(a+h)) - g(f(a))}{h} = \frac{g(f(a+h)) - g(f(a))}{f(a+h) - f(a)} \cdot \frac{f(a+h) - f(a)}{h} \qquad (*).$$

Since f is continuous at a, $f(a+h) - f(a) \to 0$ as $h \to 0$, and so

$$\frac{g(f(a+h)) - g(f(a))}{f(a+h) - f(a)} \to g'(f(a)) \text{ as } h \to 0.$$

Since $(f(a+h) - f(a))/h \to f'(a)$ as $h \to 0$, the result follows.

What is wrong with this 'proof'? It may happen that $f(a + h) = f(a)$, in which case, the expression $(*)$ makes no sense. We must avoid dividing by 0.

We consider two possibilities. First, there exists $\delta > 0$ such that $(a - \delta, a + \delta) \subseteq I$ and $f(a + h) \neq f(a)$ for $0 < |h| < \delta$. In this case the preceding argument is valid.

Secondly, a is the limit point of a sequence $(a_n)_{n=1}^\infty$ in $I \setminus \{a\}$ for which $f(a_n) = f(a)$. In this case it follows that $f'(a) = 0$, and we must show that $g'(f(a)) = 0$. Let $b = f(a)$. We use Proposition 7.1.1. There exists $\eta > 0$ such that $(b - \eta, b + \eta) \subset J$ and a function t on $(-\eta, \eta)$, with $t(0) = 0$, such that $g(b + k) = g(b) + (g'(b) + t(k))k$ for $k \in (-\eta, \eta)$ and such that t is continuous at 0. Similarly, there exists $\delta > 0$ such that $(a - \delta, a + \delta) \subset I$ and a function s on $(-\delta, \delta)$, with $s(0) = 0$, such that $f(a + h) = b + (s(h)h$ for $h \in (-\delta, \delta)$ and such that s is continuous at 0. Since f is continuous, we can suppose that $f((a - \delta, a + \delta)) \subseteq (b - \eta, b + \eta)$. If $0 < |h| < \delta$ then

$$g(f(a + h)) = g(b + s(h)h) = g(b) + (g'(b) + t(s(h)h))s(h)h$$

so that
$$\frac{g(f(a + h)) - g(f(a))}{h} = (g'(b) + t(s(h)h))s(h) \to 0 \text{ as } h \to 0,$$

since $s(h) \to 0$ and $t(s(h)h) \to 0$ as $h \to 0$. $\qquad\square$

Corollary 7.1.7 *Suppose that g is a real-valued function on an open interval I, that $a \in I$, that $g(a) \neq 0$ and that g is differentiable at a. Then there exists $\delta > 0$ such that $(a - \delta, a + \delta) \subseteq I$ and $g(x) \neq 0$ for $a - \delta < x < a + \delta$. The function $1/g$ on $(a - \delta, a + \delta)$ is differentiable at a, with derivative $-g'(a)/(g(a))^2$.*

Further, if f is a real-valued function on I which is differentiable at a, then the function f/g on $(a - \delta, a + \delta)$ is differentiable at a, with derivative

$$\left(\frac{f}{g}\right)'(a) = \frac{f'(a)g(a) - f(a)g'(a)}{(g(a))^2}.$$

Proof We can suppose without loss of generality that $g(a) > 0$. Since g is continuous at a, there exists $\delta > 0$ such that $(a - \delta, a + \delta) \subseteq I$ and $|g(x) - g(a)| < |g(a)|$ for $|x - a| < \delta$. Then $g(x) > 0$ for $x \in (a - \delta, a + \delta)$. Let $h(y) = 1/y$ for $y \in (0, \infty)$. Then h is differentiable at $g(a)$, with derivative $-1/g(a)^2$. The first result therefore follows from the chain rule, applied to the functions g and h. The second result then follows from the product formula, applied to the functions f and $1/g$. $\qquad\square$

Suppose that f is a strictly increasing continuous function on an open interval I. Recall that $f(I)$ is an open interval, and that $f^{-1} : f(I) \to I$ is continuous (Corollary 6.4.3 and Proposition 6.4.5). Suppose that f is differentiable at $a \in I$. Then $f'(a) \geq 0$, but it can happen that $f'(a) = 0$ [for example, if $f(x) = x^3$ for $x \in \mathbf{R}$ then f is strictly increasing and continuous, and $f'(0) = 0$]. But if $f'(a) > 0$ then f^{-1} is differentiable at $f(a)$.

Theorem 7.1.8 *Suppose that f is a strictly increasing continuous function on an open interval I, that f is differentiable at $a \in I$, and that $f'(a) > 0$. Let $b = f(a)$. Then f^{-1} is differentiable at b and $(f^{-1})'(b) = 1/f'(a)$.*

Proof Suppose that $\epsilon > 0$. Since $(f(a + h) - f(a))/h \to f'(a)$ as $h \to 0$, since $f(a + h) - f(a) \neq 0$ for $h \neq 0$ and since $f'(a) \neq 0$, it follows that

$$\frac{h}{f(a + h) - f(a)} \to \frac{1}{f'(a)} \text{ as } h \to 0.$$

Thus there exists $\eta > 0$ such that $(a - \eta, a + \eta) \subseteq I$ and

$$\left| \frac{h}{f(a + h) - f(a)} - \frac{1}{f'(a)} \right| < \epsilon \text{ for } 0 < |h| < \eta.$$

By Proposition 6.4.5, the inverse mapping $f^{-1}; f(I) \to I$ is continuous. There therefore exists $\delta > 0$ such that $(b - \delta, b + \delta) \subseteq f(I)$ and such that $|f^{-1}(b + k) - f^{-1}(b)| < \eta$ for $|k| < \delta$. Suppose that $0 < |k| < \delta$; let $h = f^{-1}(b + k) - a$, so that $f^{-1}(b + k) = a + h$. Then $0 < |h| < \eta$ and $f(a + h) - f(a) = k$. Consequently

$$\left| \frac{f^{-1}(b + k) - f^{-1}(b)}{k} - \frac{1}{f'(a)} \right| = \left| \frac{h}{f(a + h) - f(a)} - \frac{1}{f'(a)} \right| < \epsilon.$$

$\square$

It is at times useful to consider one-sided derivatives, for example at the end points of intervals. Suppose that f is a real-valued function on an interval I, and that $[a, a + \eta) \subseteq I$. Then f is *differentiable on the right* at a, with *right-hand derivative* $f'(a+)$, if $(f(x) - f(a))/(x - a) \to f'(a+)$ as $x \searrow a$. If so, then f is continuous on the right. Differentiability on the left and the *left-hand derivative* $f'(a-)$ are defined similarly. Then f is differentiable at an interior point a if and only if it is differentiable on the right and on the left and $f'(a+) = f'(a-)$.

It is important to realize that differentiability is a very special property.

Example 7.1.9 A bounded continuous function s on $\mathbf{R}$ which is not differentiable at any point.

Let $f_0 = f$ be the saw-tooth function defined in Section 6.3:

$$f_0(x) = \begin{cases} \{x\} & \text{for } 2k \leq x < 2k+1, \\ 1 - \{x\} & \text{for } 2k+1 \leq x < 2k+2, \end{cases}$$

for $k \in \mathbf{Z}$. Thus f_0 is continuous, is periodic, with period 2 (that is, $f(x+2) = f(x)$ for all $x \in \mathbf{R}$), and is linear, with derivative ± 1, in each open interval $(k, k+1)$, with $k \in \mathbf{Z}$.

Next we define f_n, for $n \in \mathbf{N}$. We set $f_n(x) = f_0(6^n x)/2^n$. Thus f_0 is shrunk by a factor of $1/2^n$, but oscillates more rapidly. Let us list some of the properties of f_n. Suppose that $x \in \mathbf{R}$.

1. $0 \leq f_n(x) \leq 1/2^n$.
2. f_n is linear on intervals of length $1/6^n$, and has derivative $\pm 3^n$ on each interval. Thus there exists x_n such that $|x_n - x| = 1/6^{n+1}$ and $|f_n(x_n) - f_n(x)| = 3^n|x_n - x|$.
3. If $j < n$ then $|f_j(x_n) - f_j(x)| = 3^j|x_n - x|$.
4. f_j is periodic, with period $2/6^j$, so that if $j > n$ then $f_j(x) = f_j(x_n)$.

Now let $s_n(x) = \sum_{j=1}^{n} f_j(x)$. Then s_n is a continuous function on $\mathbf{R}$. By (i), and Weierstrass' uniform M test, $s_n(x)$ converges uniformly on $\mathbf{R}$ to a continuous function, $s(x)$ say, as $n \to \infty$. Further $|s(x) - s_n(x)| \leq 1/2^n$.

Suppose that $x \in \mathbf{R}$. We show that s is not differentiable at x. Suppose that $n \in \mathbf{N}$ and that x_n is defined as above. Then by (iv), $s(x_n) - s(x) = s_n(x_n) - s_n(x)$. Now

$$|s_n(x_n) - s_n(x)| \geq |f_n(x_n) - f_n(x)| - \sum_{j=1}^{n-1} |f_j(x_n) - f_j(x)|$$

$$= 3^n|x_n - x| - \sum_{j=1}^{n-1} 3^j|x_n - x| = \left(\frac{3^n + 3}{2}\right)|x_n - x|,$$

by (ii) and (iii). Thus $x_n \to x$ as $n \to \infty$, while

$$\left| \frac{s(x_n) - s(x)}{x_n - x} \right| \to \infty \text{ as } n \to \infty,$$

so that s is not differentiable at x.

Exercises

7.1.1 Suppose that $n \in \mathbf{N}$. Show that the function $f(x) = x^{1/n}$ on $I = (0, \infty)$ is differentiable at each point of I, and find its derivative.

7.1.2 Where is the function $f(x) = |x|$ differentiable? Let $(q_j)_{j=1}^{\infty}$ be an enumeration of the rational numbers in $(0, 1)$. Let

$$f(x) = \sum_{j=1}^{\infty} \frac{|x - q_j|}{2^j}, \text{ for } x \in (0, 1).$$

Show that f is a continuous function on $(0, 1)$. Show that f is not differentiable at the rational points of $(0, 1)$ and that f is differentiable at the irrational points of $(0, 1)$.

7.1.3 Let $x_n = 1/2^n$ and let $y_n = x_n + 1/5^n$, so that $y_1 > x_1 > y_2 > x_2 > \dots$. Define a real-valued function f by setting

$$f(x) = 1 - \frac{|x| - x_n}{y_n - x_n} \quad \text{if } x_n < |x| \leq y_n,$$

$$= \frac{|x| - y_{n+1}}{x_n - y_{n+1}} \quad \text{if } y_{n+1} < |x| \leq x_n,$$

$$= 0, \quad \text{otherwise.}$$

Sketch the graph of f. Let $g(x) = x + x^2 f(x)$. Show that f is differentiable at 0 and that $g'(0) = 1$. Suppose that $\delta > 0$. Show that there exist $0 < a < b < \delta$ such that $g(a) > g(b)$.

7.2 Convex functions

We now consider an important class of functions, with interesting continuity and differentiability properties.

Suppose that E is a real or complex vector space, and that $u, v \in E$. Let $\sigma : [0, 1] \to E$ be defined by

$$\sigma(t) = u + (v - u)t = (1 - t)u + tv \text{ for } 0 \leq t \leq 1.$$

Then $\sigma([0, 1])$ is the *straight line segment* $[u, v]$ between u and v. A subset C of E is *convex* if $[u, v] \subseteq C$, for each u, v in C. Thus a subset of $\mathbf{R}$ is convex if and only if it is an interval.

Suppose that f is a function on an interval I. f is said to be *convex* if the subset $\{(x, y) \in \mathbf{R}^2 : x \in I, y \geq f(x)\}$ of $\mathbf{R}^2$ is a convex set. Equivalently, if $x_0, x_1 \in I$, then the straight line segment $[(x_0, f(x_0)), (x_1, f(x_1))]$ in $\mathbf{R}^2$ lies above the graph $G_f = \{(x, f(x)) \in \mathbf{R}^2 : x \in I\}$. Since

$$[(x_0, f(x_0)), (x_1, f(x_1))] =$$
$$\{((1 - t)x_0 + tx_1, (1 - t)f(x_0) + tf(x_1)) : 0 \leq t \leq 1\},$$

this says that

$$(1-t)f(x_0) + tf(x_1) \geq f((1-t)x_0 + tx_1)$$

for all $x_0, x_1 \in I$ and all $0 \leq t \leq 1$.

We say that f is *strictly convex* if

$$(1-t)f(x_0) + tf(x_1) > f((1-t)x_0 + tx_1)$$

for distinct $x_0, x_1 \in I$ and all $0 < t < 1$. f is *concave* if $-f$ is convex; that is,

$(1-t)f(x_0) + tf(x_1) \leq f((1-t)x_0 + tx_1)$ for all $x_0, x_1 \in I$ and all $0 \leq t \leq 1$.

Strict concavity is defined similarly.

The next proposition provides some alternative characterizations of convexity.

Proposition 7.2.1   *Suppose that f is a real-valued function on an open interval I. The following are equivalent.*

(i) f is convex.
(ii) If $a, b, c \in I$ and $a < b < c$ then

$$\frac{f(b) - f(a)}{b - a} \leq \frac{f(c) - f(a)}{c - a}.$$

(iii) If $a, b, c \in I$ and $a < b < c$ then

$$\frac{f(c) - f(a)}{c - a} \leq \frac{f(c) - f(b)}{c - b}.$$

(iv) If $a, b, c \in I$ and $a < b < c$ then

$$\frac{f(b) - f(a)}{b - a} \leq \frac{f(c) - f(b)}{c - b}.$$

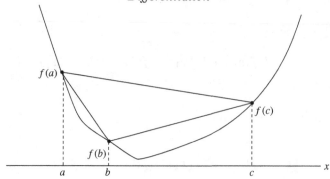

Figure 7.2. A convex function.

Proof   Let $t = (b-a)/(c-a)$, so that $0 < t < 1$, $1 - t = (c-b)/(c-a)$ and

$$b = \frac{c-b}{c-a}a + \frac{b-a}{c-a}c = (1-t)a + tc.$$

The proof is then simply a matter of using this equation in the definition of convexity, and rearranging the inequality. For example, if f is convex, then

$$f(b) \le \frac{c-b}{c-a}f(a) + \frac{b-a}{c-a}f(c),$$

so that

$$f(b) - f(a) \le \frac{(c-b)-(c-a)}{c-a}f(a) + \frac{b-a}{c-a}f(c) = \frac{b-a}{c-a}(f(c) - f(a)),$$

which gives (ii). Conversely, if (ii) holds, and if $x_0 < x_1$ and $0 < t < 1$ then setting $x_t = (1-t)x_0 + tx_1$,

$$\frac{f(x_t) - f(x_0)}{x_t - x_0} \le \frac{f(x_1) - f(x_0)}{x_1 - x_0}.$$

Since $x_t - x_0 = t(x_1 - x_0)$, this gives

$$f(x_t) \le f(x_0) + t(f(x_1) - f(x_0)) = (1-t)f(x_0) + tf(x_1).$$

The other equivalences are proved in a similar way.   □

Here are some basic properties of convex functions.

Proposition 7.2.2   *(i) If f and g are convex functions on an interval I and $a \ge 0$ then $f + g$ and af are convex.*
   (ii) If $(f_n)_{n=1}^{\infty}$ is a sequence of convex functions on an interval I, and if $f_n(x) \to f(x)$ as $n \to \infty$, for each $x \in I$, then f is convex.

(iii) If $\{f : f \in F\}$ is a family of convex functions on an interval I for which $g(x) = \sup\{f(x) : f \in F\}$ is finite for each $x \in I$ then g is convex.

(iv) If f, g are convex, non-negative increasing functions on an interval I then fg is convex.

(v) If f is a convex function on an interval I, and if ϕ is an increasing convex function on an interval J which contains $f(I)$, then $\phi \circ f$ is a convex function on I.

Proof (i) and (ii) follow immediately from the definitions.

We suppose that $x_0, x_1 \in I$ and that $0 < t < 1$, and we set $x_t = (1-t)x_0 + tx_1$.

(iii) Suppose that $\epsilon > 0$. There exists a function f in F such that $f(x_t) \geq g(x_t) - \epsilon$. Then

$$g(x_t) - \epsilon \leq f(x_t) \leq (1-t)f(x_0) + tf(x_1) \leq (1-t)g(x_0) + tg(x_1).$$

Since this holds for all $\epsilon > 0$, $g(x_t) \leq (1-t)g(x_0) + tg(x_1)$.

(iv) Since f and g are increasing,

$$(g(x_1) - g(x_0))(f(x_1) - f(x_0)) \geq 0.$$

Expanding and rearranging,

$$f(x_0)g(x_1) + f(x_1)g(x_0) \leq f(x_0)g(x_0) + f(x_1)g(x_1),$$

and so

$$
\begin{aligned}
f(x_t)g(x_t) &\leq \\
&\leq ((1-t)f(x_0) + tf(x_1))((1-t)g(x_0) + tg(x_1)) \\
&= (1-t)^2 f(x_0)g(x_0) + t(1-t)(f(x_0)g(x_1) + f(x_1)g(x_0)) + t^2(f(x_1)g(x_1) \\
&\leq (1-t)^2 f(x_0)g(x_0) + t(1-t)(f(x_0)g(x_0) + f(x_1)g(x_1)) + t^2(f(x_1)g(x_1) \\
&= (1-t)f(x_0)g(x_0) + tf(x_1)g(x_1).
\end{aligned}
$$

(v) Since ϕ is convex and increasing,

$$\phi(f(x_t)) \leq \phi((1-t)f(x_0) + tf(x_1)) \leq (1-t)\phi(f(x_0)) + t\phi(f(x_1)). \quad \square$$

We now turn to continuity and differentiability properties of convex functions.

Theorem 7.2.3 *Suppose that f is a convex function on an open interval I.*

(i) f is continuous on I.

(ii) f is differentiable on the right and on the left at each point a of I, and $f'(a-) \le f'(a+)$.

(iii) If $a < b$ then $f'(a+) \le f'(b-)$.

(iv) The mapping $a \to f'(a^+)$ is increasing, and is continuous on the right at each point a of I.

(v) The mapping $a \to f'(a^-)$ is increasing, and is continuous on the left at each point a of I.

Proof (i) is a consequence of (ii).

(ii) Suppose that $y < a < x$, with $x, y \in I$. By Proposition 7.2.1, the function $x \to (f(x) - f(a))/(x - a)$ is an increasing function on $I \cap (a, \infty)$, bounded below by $(f(a) - f(y))/(a - y)$. Thus $(f(x) - f(a))/(x - a)$ tends to a limit $f'(a+)$ as $x \searrow a$, and $(f(a) - f(y))/(a - y) \le f'(a+)$. Similarly, $(f(a) - f(y))/(a - y) \to f'(a-)$, and $f'(a-) \le f'(a+)$.

(iii) $f'(a+) \le (f(b) - f(a))/(b - a) \le f'(b-)$.

(iv) By (ii) and (iii), if $a < b$ then $f'(a+) \le f'(b-) \le f'(b+)$, so the mapping $x \to f'(x+)$ is increasing. Suppose that $a \in I$. Given $\epsilon > 0$, there exists $\delta > 0$ such that $(a, a + \delta) \subseteq I$ and

$$f'(a+) \le \frac{f(x) - f(a)}{x - a} < f'(a+) + \epsilon/2 \text{ for } x \in (a, a + \delta).$$

Choose $b \in (a, a + \delta)$. Since $(f(b) - f(x))/(b - x) \to (f(b) - f(a))/(b - a)$ as $x \searrow a$, there exists $0 < \eta < \delta$ such that

$$\frac{f(b) - f(x)}{b - x} < \frac{f(x) - f(a)}{x - a} + \epsilon/2 < f'(a+) + \epsilon \text{ for } x \in (a, a + \eta).$$

Thus if $x \in (a, a + \eta)$ then

$$f'(a+) \le f'(x+) \le \frac{f(b) - f(x)}{b - x} \le f'(a+) + \epsilon.$$

The proof of (v) is exactly similar. □

Corollary 7.2.4 *The set D of points of discontinuity of the mapping $a \to f'(a+)$ is countable. D is the set of points of discontinuity of the mapping $a \to f'(a-)$, and is the set of points at which f is not differentiable.*

Proof Since the mapping $a \to f'(a+)$ is increasing, D is countable, by Theorem 6.3.5. Suppose that $d \in D$. Since the mapping $y \to f'(y-)$ is continuous on the left at d,

$$f'(d-) = \lim_{y \nearrow d} f'(y-) \leq \lim_{y \nearrow d} f'(y+) < f'(d+),$$

so that f is not differentiable at d. Further, $f'(d+) \leq f'(z-)$ for $z > d$, so that $f'(d-) < \lim_{z \searrow d} f'(z-)$; d is a point of discontinuity of the mapping $a \to f'(a-)$ as well.

Conversely, if $c \notin D$ then

$$f'(c-) = \lim_{y \nearrow c} f'(y-) = \lim_{y \nearrow c} f'(y+) = f'(c+),$$

so that f is differentiable at c, and

$$f'(c-) = f'(c+) = \lim_{x \searrow c} f'(x-),$$

so that the mapping $x \to f'(x-)$ is continuous at c.   □

Exercises

7.2.1 Give an example of a convex function on $[0, 1]$ which is discontinuous at 0 and at 1.

7.2.2 A real-valued function on an interval I is *midpoint-convex* if

$$f((a+b)/2) \leq (f(a) + f(b))/2 \text{ for all } a, b \in I.$$

Suppose that f is a midpoint-convex function on I.

(a) Suppose that $c - h, c, c + h \in I$, where $h > 0$. Show that if $n \in \mathbf{N}$ then

$$\frac{f(c-h) - f(c)}{n+1} \leq f(c + h/n) - f(c) \leq \frac{f(c+h) - f(c)}{n},$$

$$\frac{f(c+h) - f(c)}{n+1} \leq f(c - h/n) - f(c) \leq \frac{f(c-h) - f(c)}{n}.$$

(b) Show that if f is bounded on I then f is continuous at c.

(c) Show that if f is bounded on I then f is convex.

7.2.3 State and prove results corresponding to Propositions 7.2.1 and 7.2.2 and Theorem 7.2.3 for strictly convex functions.

7.2.4 Suppose that f is a convex function on an interval I, that $x_1, \ldots, x_n$ are distinct points in I and that $p_1, \ldots p_n$ are positive numbers for which $\sum_{j=1}^{n} p_j = 1$. Show that

$$f\left(\sum_{j=1}^{n} p_j x_j\right) \leq \sum_{j=1}^{n} p_j f(x_j). \quad [\textit{Jensen's inequality}]$$

Show that the inequality is strict if f is strictly convex.

7.2.5 Suppose that f is a convex strictly increasing function on an open interval I. Show that the inverse function f^{-1} is concave and strictly increasing on $f(I)$.

7.3 Differentiable functions on an interval

Proposition 7.3.1 *Suppose that f is a real-valued function defined on an interval I and that f has a local maximum or local minimum at an interior point c of I. If f is differentiable at c then $f'(c) = 0$.*

Proof Suppose that f has a local maximum at c. Then

$$f'(c) = \lim_{x \searrow c} \frac{f(x) - f(c)}{x - c} \le 0 \text{ and } f'(c) = \lim_{x \nearrow c} \frac{f(x) - f(c)}{x - c} \ge 0,$$

and so $f'(c) = 0$. The proof when f has a local minimum at c is exactly similar. $\square$

A function on an open interval I which is differentiable at every point of the interval is said to be *differentiable on I*.

Theorem 7.3.2 (Rolle's theorem) *Suppose that f is a real-valued function, defined on a closed interval $[a, b]$, which is continuous on the closed interval $[a, b]$ and differentiable on the open interval (a, b). Suppose that $f(a) = f(b)$. Then there exists $c \in (a, b)$ such that $f'(c) = 0$.*

Proof If $f(x) = f(a)$ for all $x \in (a, b)$ then $f'(x) = 0$ for all $x \in (a, b)$. Otherwise f is not monotonic on $[a, b]$, and therefore, by Corollary 6.3.7, has a local maximum or local minimum at an interior point c of $[a, b]$. Then $f'(c) = 0$, by Proposition 7.3.1. $\square$

Corollary 7.3.3 *Suppose that $f'(x) \ne 0$, for each $x \in (a, b)$. Then f is strictly monotonic.*

Proof If not, f is not injective (Proposition 6.4.4), and so there exists $a \le a' < b' \le b$ for which $f(a') = f(b')$. But then there exists $a' < c < b'$ with $f'(c) = 0$, giving a contradiction. $\square$

As Exercise 7.5.6 shows, if f is differentiable on an interval then the derivative need not be continuous. Nevertheless, it satisfies an intermediate value property. This property is known as *Darboux continuity*.

Theorem 7.3.4 *Suppose that f is a real-valued function which is differentiable on the open interval (a, b) and that $f'(c) < f'(d)$ for some $a < c < d < b$. If $f'(c) < v < f'(d)$ there exists $c < e < d$ with $f'(e) = v$.*

Proof We apply a shear to the graph of f: let $h(x) = f(x) - vx$. Then $h'(c) = f'(c) - v < 0$ and $h'(d) = f'(d) - v > 0$. There exists $e \in [c, d]$ such that $h(e) = \inf\{h(x) : x \in [c, d]\}$. By Proposition 7.1.5 (v), there exists $0 < \delta < d - c$ such that $h(x) < h(c)$ for $c < x < c + \delta$ and $h(x) < h(d)$ for $d - \delta < x < d$, and so e must be an interior point of $[c, d]$. Thus h has a local minimum at e, and $h'(e) = f'(e) - v = 0$. □

We applied a shear to obtain this result. We do it again to obtain a mean-value theorem.

Theorem 7.3.5 (The mean-value theorem) *Suppose that f is a real-valued function which is continuous on the closed interval $[a, b]$ and differentiable on the open interval (a, b). Then there exists $a < c < b$ with $f'(c) = (f(b) - f(a))/(b - a)$.*

Proof Let $h_\lambda(x) = f(x) - \lambda x$. If we set $\lambda = (f(b) - f(a))/(b - a)$ then $h_\lambda(a) = h_\lambda(b)$, and so there exists $a < c < b$ such that $h'_\lambda(c) = f'(c) - \lambda = 0$. Thus $f'(c) = (f(b) - f(a))/(b - a)$. □

This theorem says that there is a point c in (a, b) at which the tangent to the graph of f is parallel to the chord joining $(a, f(a))$ and $(b, f(b))$.

The next corollary is 'obviously' true, but it is not a trivial result; it is however an immediate consequence of the mean-value theorem.

Corollary 7.3.6 *Suppose that f is a real-valued function which is continuous on the closed interval $[a, b]$ and differentiable on the open interval (a, b). If $f'(x) = 0$ for $a < x < b$ then f is constant on $[a, b]$: $f(a) = f(x) = f(b)$ for all $x \in [a, b]$.*

Here is a more sophisticated mean-value theorem.

Theorem 7.3.7 (Cauchy's mean-value theorem) *Suppose that f and g are real-valued functions which are continuous on the closed interval $[a, b]$ and differentiable on the open interval (a, b), and suppose that $g'(x) \neq 0$ for all $x \in (a, b)$. Then $g(a) \neq g(b)$, and there exists $c \in (a, b)$ such that*

$$\frac{f'(c)}{g'(c)} = \frac{f(b) - f(a)}{g(b) - g(a)}.$$

Proof By Corollary 7.3.3, g is strictly monotonic on (a, b), and so $g(a) \neq g(b)$. Let

$$\lambda = \frac{f(b) - f(a)}{g(b) - g(a)} \text{ and let } h_\lambda(x) = f(x) - \lambda g(x).$$

Then $h_\lambda(a) = h_\lambda(b)$, and so there exists $a < c < b$ such that $h'_\lambda(c) = f'(c) - \lambda g'(c) = 0$. Thus $f'(c)/g'(c) = (f(b) - f(a))/(g(b) - g(a))$. $\qquad\square$

Corollary 7.3.8 (L'Hôpital's rule) *Suppose that f and g are real-valued functions which are continuous on the closed interval $[a, b]$ and differentiable on the open interval (a, b), that $f(a) = g(a) = 0$ and that $g'(x) \neq 0$ for all $x \in (a, b)$. If $f'(x)/g'(x) \to l$ as $x \searrow a$ then $f(x)/g(x) \to l$ as $x \searrow a$.*

Proof Suppose that $\epsilon > 0$. There exists $0 < \delta \leq b - a$ such that $|f'(x)/g'(x) - l| < \epsilon$ for $a < x < a + \delta$. If $a < x < a + \delta$ there exists $a < c < x$ such that

$$\frac{f(x)}{g(x)} = \frac{f(x) - f(a)}{g(x) - g(a)} = \frac{f'(c)}{g'(c)},$$

and so

$$\left| \frac{f(x)}{g(x)} - l \right| = \left| \frac{f'(c)}{g'(c)} - l \right| < \epsilon. \qquad\square$$

Exercises

7.3.1 Suppose that f is continuous on $[a, b]$ and differentiable on (a, b) and that f has a derivative on the right at a [which we denote here by $f'(a)$] and a derivative on the left at b [which we denote here by $f'(b)$]. Show that if f' is continuous on $[a, b]$ and $\epsilon > 0$ then there exists $\delta > 0$ such that

$$\left| \frac{f(x) - f(y)}{x - y} - f'(x) \right| < \epsilon \text{ if } x, y \in [a, b] \text{ and } |x - y| < \delta.$$

7.3.2 Suppose that $a_0, \ldots, a_n \in \mathbf{R}$ and that

$$a_0 + \frac{a_1}{2} + \cdots + \frac{a_{n-1}}{n} + \frac{a_n}{n + 1} = 0.$$

Show that there exists $0 < c < 1$ such that

$$a_0 + a_1 c + \cdots + a_{n-1} c^{n-1} + a_n c^n = 0.$$

7.3.3 Suppose that f is continuous on $[a, b]$ and differentiable on (a, b). Show that f' is an increasing function on (a, b) if and only if f is convex.

7.3.4 Suppose that f is a real-valued function on an interval I which satisfies $|f(x) - f(y)| \leq |x - y|^2$ for all $x, y \in I$. Show that f is constant.

7.3.5 Suppose that f is a differentiable function on $\mathbf{R}$ and that $f'(x) \to l$ as $x \to +\infty$. Show that $f(x)/x \to l$ as $x \to +\infty$.

7.3.6 Suppose that f is continuous on $[a, b]$ and differentiable on (a, b), and that $f'(x) \to l$ as $x \searrow a$. Show that f is differentiable on the right at a and that $f'(a+) = l$.

7.3.7 Suppose that f is a differentiable function on $[a, b]$ and that f' is continuous on $[a, b]$. Let $N = \{x \in [a, b] : f'(x) = 0\}$. Suppose that $\epsilon > 0$. Show that there are finitely many disjoint intervals $I_1, \ldots, I_k$ in $[a, b]$ such that $N \subseteq \cup_{j=1}^{k} I_j$ and such that $|f'(x)| \leq \epsilon$ for $x \in \cup_{j=1}^{k} I_j$. Show that $f(N)$ is a closed subset of $\mathbf{R}$ with no interior points.

7.3.8 Suppose that a is an algebraic number which is not rational. Show that there exists a non-zero polynomial $p(x) = a_n x^n + \cdots + a_0$ with integer coefficients such that $p(a) = 0$, whereas $p(r) \neq 0$ for $r \in \mathbf{Q}$. Thus if $r = p/q$ then $q^n p(r)$ is a non-zero integer. Let

$$M = \sup\{|p'(x)| : a - 1 \leq x \leq a + 1\}.$$

Suppose that $r = p/q \in \mathbf{Q}$ and that $|r - a| \leq 1$. Use the mean-value theorem to show that $|r - a| \geq 1/Mq^n$. (This result is due to Liouville.) Let $x = \sum_{n=1}^{\infty} 10^{-n!}$. Show that x is not rational. Show that x is not algebraic.

7.4 The exponential and logarithmic functions; powers

We now consider how the results that we have obtained can be used to establish properties of some of the fundamental functions of analysis.

We have defined the real exponential function

$$\exp(x) = 1 + \frac{x}{1!} + \frac{x^2}{2!} + \cdots + \frac{x^n}{n!} + \cdots$$

for $x \in \mathbf{R}$, and have shown that $\exp(x + y) = \exp(x) \exp(y)$, and that $\exp$ is differentiable, with derivative $\exp'(x) = \exp(x)$. We set $e = \exp(1)$. The reader should use these results, and the results that have been proved, to justify the following statements.

1. $\exp$ is a non-negative strictly increasing function on $\mathbf{R}$.
2. $\exp$ is a strictly convex function on $\mathbf{R}$.
3. If $n \in \mathbf{Z}^+$ then $\exp(x)/x^n \to \infty$ as $x \to +\infty$.
4. If $n \in \mathbf{Z}^+$ then $x^n \exp(x) \to 0$ as $x \to -\infty$.
5. $\exp$ is a continuous bijection of $\mathbf{R}$ onto $(0, \infty)$ which is an isomorphism of the additive group $(\mathbf{R}, +)$ onto the multiplicative group $((0, \infty), \times)$.
6. The inverse mapping from $(0, \infty)$ to $\mathbf{R}$, which is called the *logarithmic function*, and denoted $\log x$, is differentiable, and $d\log x/dx = 1/x$, for $0 < x < \infty$.

7. If $x, y > 0$ then $\log xy = \log x + \log y$.
8. $\log x$ is a strictly increasing strictly concave function on $(0, \infty)$.
9. $\log 1 = 0$, $\log e = 1$, $\log x \to \infty$ as $x \to \infty$ and $\log x \to -\infty$ as $x \searrow 0$.
10. If $m \in \mathbf{N}$ then $\log x / x^{1/m} \to 0$ as $x \to \infty$ and $x^{1/m} \log x \to 0$ as $x \searrow 0$.
11. $1/x = \exp(-\log x)$, and if $x > 0$ and $n \in \mathbf{N}$ then

$$x^n = \exp(n \log x) \text{ and } x^{1/n} = \exp((\log x)/n).$$

Thus if $r = p/q \in \mathbf{Q}$ then $x^{p/q} = \exp((p/q) \log x)$.
This leads us to define $x^\alpha = \exp(\alpha \log x)$, for $x > 0$ and $\alpha \in \mathbf{R}$. x^α is x *raised to the power* α. Note that, with this terminology, $\exp(x) = \exp(x \log e) = e^x$. In future, we shall usually write e^x for $\exp x$.
12. If $x > 0$ and $\alpha, \beta \in \mathbf{R}$ then $x^{\alpha+\beta} = x^\alpha x^\beta$ and $x^0 = 1$.
13. For fixed $x > 0$, the function $\alpha \to x^\alpha$ from $\mathbf{R}$ to $(0, \infty)$ is continuous. Thus if $(r_n)_{n \in \mathbf{N}}$ is a sequence in $\mathbf{Q}$ and $r_n \to \alpha$ then $x^{r_n} \to x^\alpha$ as $n \to \infty$.
14. For fixed $x > 0$, the function $\alpha \to x^\alpha$ from $\mathbf{R}$ to $(0, \infty)$ is differentiable, with derivative $dx^\alpha/d\alpha = x^\alpha \log x$.
15. If $x > 1$ then the function $\alpha \to x^\alpha$ from $\mathbf{R}$ to $(0, \infty)$ is a strictly convex and strictly increasing bijection of $\mathbf{R}$ onto $(0, \infty)$.
16. If $0 < x < 1$ then the function $\alpha \to x^\alpha$ from $\mathbf{R}$ to $(0, \infty)$ is a strictly convex and strictly decreasing bijection of $\mathbf{R}$ onto $(0, \infty)$.
17. For fixed $\alpha \in \mathbf{R}$, the function $x \to x^\alpha$ from $(0, \infty) \to (0, \infty)$ is differentiable, with derivative $dx^\alpha/dx = \alpha x^{\alpha-1}$.
18. If $\alpha > 1$ then the function $x \to x^\alpha$ is a strictly increasing strictly convex bijection of $(0, \infty)$ onto $(0, \infty)$.
19. If $0 < \alpha < 1$ then the function $x \to x^\alpha$ is a strictly increasing strictly concave bijection of $(0, \infty)$ onto $(0, \infty)$.
20. If $\alpha < 0$ then the function $x \to x^\alpha$ is a strictly decreasing strictly convex bijection of $(0, \infty)$ onto $(0, \infty)$.
A strictly positive function f on an interval I is *logarithmically convex* if $\log f$ is convex. Since

$$(1 - \theta) \log f(x) + \theta \log f(y) = \log(f(x)^{1-\theta} f(y)^\theta),$$

f is logarithmically convex if and only if

$$f((1 - \theta)x + \theta y) \le f(x)^{1-\theta} f(y)^\theta$$

for $x, y \in I$ and $0 < \theta < 1$. Since $f = e^{\log f}$, it follows from Proposition 7.2.2 (v) that a logarithmically convex function is convex.

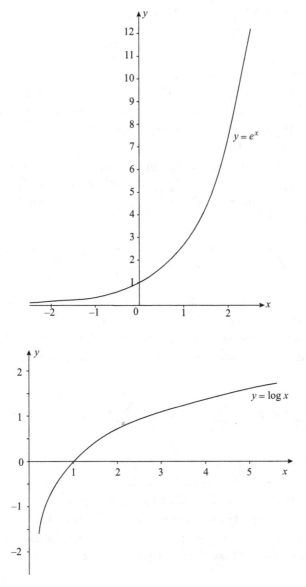

Figure 7.4. The exponential and logarithmic functions.

Exercises

7.4.1 Show that $n^{1/n} \to 1$ as $n \to \infty$.

7.4.2 Use the strict concavity of $\log x$ to prove the following generalization of the arithmetic mean-geometric mean inequality. Suppose that $x_1, \ldots, x_n$ are positive numbers and that $p_1, \ldots, p_n$ are strictly positive numbers with $\sum_{j=1}^{n} p_j = 1$. Show that

$$x_1^{p_1} x_2^{p_2} \ldots x_n^{p_n} \le p_1 x_1 + p_2 x_2 + \cdots + p_n x_n,$$

with equality if and only if $x_1 = x_2 + \cdots = x_n$.

7.4.3 Suppose that $1 < p, q < \infty$. If $1/p + 1/q = 1$, then p and q are called *conjugate indices*. Suppose that p and q are conjugate indices and that x and y are non-negative numbers. Show that $xy \le x^p/p + y^q/q$. When does equality hold?

Suppose now that $x_1, \ldots, x_n$, $y_1, \ldots, y_n$ are real numbers, and that $\sum_{j=1}^{n} |x_j y_j| \ne 0$. Let $S = (\sum_{j=1}^{n} |x_j|^p)^{1/p}$, $T = (\sum_{j=1}^{n} |y_j|^q)^{1/q}$, and let $a_j = x_j/S$, $b_j = y_j/T$ for $1 \le j \le n$. Show that $\sum_{j=1}^{n} |a_j b_j| \le 1$. Deduce that

$$\left| \sum_{j=1}^{n} x_j y_j \right| \le \sum_{j=1}^{n} |x_j y_j| \le \left(\sum_{j=1}^{n} |x_j|^p \right)^{1/p} \cdot \left(\sum_{j=1}^{n} |y_j|^q \right)^{1/q}.$$

(*Hölder's inequality*). Note that this generalizes Cauchy's inequality. When does equality hold? Extend this result to infinite sums.

7.4.4 Here is another version of Hölder's inequality. Suppose that $x_1, \ldots, x_n$, $y_1, \ldots, y_n$ are real numbers, and that $a_1, \ldots a_n$ are non-negative numbers. Show that

$$\left| \sum_{j=1}^{n} a_j x_j y_j \right| \le \sum_{j=1}^{n} a_j |x_j y_j| \le \left(\sum_{j=1}^{n} a_j |x_j|^p \right)^{1/p} \cdot \left(\sum_{j=1}^{n} a_j |y_j|^q \right)^{1/q}.$$

7.4.5 Show that the function $\log((1 + x)/(1 - x))$ is a strictly increasing bijection of $(-1, 1)$ onto $\mathbf{R}$. Show that it is convex on $(0, 1)$.

7.4.6 Suppose that $y > 0$. Show that $(y - 1)^2 > y(\log y)^2$.

7.4.7 Using the convexity of the function 4^{-x}, show that if $0 < x \le 1/2$ then $1 - x \ge 4^{-x}$. Let $(p_1, p_2, \ldots)$ be an enumeration of the primes in increasing order. Show that, for each $n \in \mathbf{N}$,

$$\sum_{j=1}^{n} \frac{1}{j} \le \prod_{j=1}^{n} (1 + \frac{1}{p_j} + \frac{1}{p_j^2} + \cdots) = \left(\prod_{j=1}^{n} (1 - \frac{1}{p_j}) \right)^{-1}.$$

Deduce that $\sum_{j=1}^{n} 1/j \le 4^{t_n}$, where $t_n = \sum_{j=1}^{n} 1/p_j$, and deduce that $\sum_{j=1}^{\infty} 1/p_j = \infty$.

7.4.8 Use the mean-value theorem to show that if $x > 0$ then

$$\frac{x}{1 + x} < \log(1 + x) < x.$$

Deduce that $(1 + x)^{1/x} \nearrow e$ as $x \searrow 1$.

7.4.9 Sketch the graph of the function $f(x) = x \log x$, for $x \in (0, \infty)$. Is it convex? Does it have any maxima or minima? What is $\lim_{x \searrow 0} f(x)$? What is $\lim_{x \searrow 0} f'(x)$?

7.4.10 Suppose that f is positive and differentiable on an open interval I. Show that $(\log f)'(x) = f'(x)/f(x)$. Let $g(x) = x^x$, for $x \in (0, \infty)$. Calculate $g'(x)$. Show that g is logarithmically convex. Sketch the graph of the function g, answering the same questions as in the previous exercise.

7.4.11 Investigate

$$\lim_{x \searrow 0} \frac{(1+x)^{1/x} - e}{x} \quad \text{and} \quad \lim_{x \to \infty} \frac{x(x^{1/x} - 1)}{\log x}.$$

7.5 The circular functions

Next we consider the cosine and sine functions. These functions arise in geometry and trigonometry, but we are not yet in a position to consider this aspect of things. Instead, we treat them in a purely analytic way. As we shall see later, they also have an important part to play in complex analysis, and this will also throw more light on them.

Each of the power series

$$\cos z = \sum_{k=0}^{\infty} (-1)^k \frac{z^{2k}}{(2k)!} = 1 - \frac{z^2}{2!} + \frac{z^4}{4!} - \cdots,$$

$$\sin z = \sum_{k=0}^{\infty} (-1)^k \frac{z^{2k+1}}{(2k+1)!} = z - \frac{z^3}{3!} + \frac{z^5}{5!} - \cdots$$

has infinite radius of convergence.

The cosine function cos is an *even* function ($\cos z = \cos(-z)$) and the sine function sin is an *odd* function ($\sin z = -\sin(-z)$).

Following custom, if $n \in \mathbf{N}$ we write $\cos^n z$ for $(\cos z)^n$ and $\sin^n z$ for $(\sin z)^n$. But $1/\cos z$ is denoted by $\sec z$ and $1/\sin z$ is denoted by $\operatorname{cosec} z$: $\cos^{-1}$ and $\sin^{-1}$ have quite different meanings (see Exercises 7.5.3 and 7.5.4).

We restrict attention to the real-valued functions cos and sin, defined on the real line $\mathbf{R}$.

Theorem 7.5.1 $\cos x$ *is differentiable, and* $\cos' x = -\sin x$. $\sin x$ *is differentiable, and* $\sin' x = -\cos x$.

Proof First we establish an elementary inequality. We prove this for complex numbers, since we shall need such an inequality in Volume III.

Lemma 7.5.2 *Suppose that $z, w \in \mathbf{C}$ and that $n \in \mathbf{N}$. Then*

$$|(z+w)^n - z^n - nwz^{n-1}| \le \frac{n(n-1)}{2}|w|^2(|z|+|w|)^{n-2}.$$

Proof The proof is trivially true if $n = 1$ or 2. Suppose that $n \ge 3$. By the binomial theorem,

$$(z+w)^n - z^n - nwz^{n-1} = w^2 \sum_{j=2}^{n} \binom{n}{j} w^{j-2} z^{n-j}$$

$$= \sum_{k=0}^{n-2} \frac{n(n-1)}{(k+2)(k+1)} \binom{n-2}{k} w^k z^{n-k-2},$$

so that

$$|(z+w)^n - z^n - nwz^{n-1}| \le \sum_{k=0}^{n-2} \frac{n(n-1)}{(k+2)(k+1)} \binom{n-2}{k} |w|^k |z|^{n-k-2}$$

$$\le \frac{n(n-1)}{2}(|z|+|w|)^{n-2}. \qquad \square$$

We now prove the theorem. Suppose that $h \ne 0$.

$$\frac{\cos(x+h) - \cos x}{h} + \sin x$$

$$= \sum_{k=0}^{\infty} (-1)^k \frac{(x+h)^{2k} - x^{2k} - 2khx^{2k-1}}{h(2k)!},$$

so that

$$\left| \frac{\cos(x+h) - \cos x}{h} - \sin x \right|$$

$$\le \sum_{k=0}^{\infty} \frac{|(x+h)^{2k} - x^{2k} - 2khx^{2k-1}|}{|h|(2k)!}$$

$$\le |h| \sum_{k=0}^{\infty} \frac{|(|x|+|h|)^{2k-2}}{(2k-2)!} \le |h| e^{|x|+|h|}.$$

A similar argument, left to the reader as an easy exercise, establishes the result for $\sin x$. $\qquad \square$

Corollary 7.5.3 $\cos^2 x + \sin^2 x = 1$, *so that* $-1 \le \cos x \le 1$ *and* $-1 \le \sin x \le 1$.

Proof Let $f(x) = \cos^2 x + \sin^2 x$. Then

$$f'(x) = 2\cos x(-\sin x) + 2\sin x \cos x = 0,$$

so that f is constant, by the mean-value theorem. Thus $f(x) = f(0) = 1$.
□

The alternating series test shows that $\sin x \geq x - x^3/3! = x(1 - x^2/6) > 0$ for $0 < x \leq 2$, so that $\cos x$ is strictly decreasing on $[0, 2]$. The alternating series test also shows that $\cos x \geq 1 - x^2/2 \geq 0$ for $0 \leq x \leq \sqrt{2}$, and that $\cos \sqrt{3} \leq 1 - 3/2 + 9/24 = -1/8$. Thus, by the intermediate-value theorem, there exists $\sqrt{2} < x_0 < \sqrt{3}$ such that $\cos x_0 = 0$. Since the function $\cos$ is strictly decreasing on $[0, 2]$, x_0 is unique. We set $\pi = 2x_0$.

Since $\sin' x = \cos x$ is positive on $(0, \pi/2)$, $\sin x$ is strictly increasing on $[0, \pi/2]$. Since

$$\sin^2(\pi/2) = \sin^2(\pi/2) + \cos^2(\pi/2) = 1,$$

$\sin 0 = 0$ and $\sin(\pi/2) = 1$. Since $\cos' x = -\sin x$ is negative and decreasing on $(0, \pi/2)$, $\cos x$ is decreasing and concave on $[0, \pi/2]$; since $\cos x$ is an even function, $\cos x$ is concave on $[-\pi/2, \pi/2]$. Similarly, $\sin x$ is convex on $[-\pi/2, 0]$ and concave on $[0, \pi/2]$.

In order to go further, we need the addition formula.

Theorem 7.5.4 *If $x, y \in \mathbf{R}$ then $\sin(x + y) = \sin x \cos y + \cos x \sin y$.*

Proof Since the series are absolutely convergent, we can expand the products as Cauchy products.

$$\sin x \cos y = \left(\sum_{j=0}^{\infty}(-1)^j \frac{x^{2j+1}}{(2j+1)!}\right)\left(\sum_{k=0}^{\infty}(-1)^k \frac{y^{2k}}{(2k)!}\right)$$

$$= \sum_{n=0}^{\infty}\left(\sum_{j+k=n}(-1)^n \frac{x^{2j+1}}{(2j+1)!}\cdot\frac{y^{2k}}{(2k)!}\right).$$

Similarly

$$\cos x \sin y = \sum_{n=0}^{\infty}\left(\sum_{j+k=n}(-1)^n \frac{x^{2j}}{(2j)!}\cdot\frac{y^{2k+1}}{(2k+1)!}\right).$$

Adding,

$$\sin x \cos y + \cos x \sin y = \sum_{l=0}^{\infty} \left(\sum_{j+k=2l+1} (-1)^l \frac{x^j}{j!} \cdot \frac{y^k}{k!} \right)$$

$$= \sum_{l=0}^{\infty} (-1)^l \frac{(x+y)^{2l+1}}{(2l+1)!} = \sin(x+y). \qquad \square$$

Corollary 7.5.5 $\cos(x+y) = \cos x \cos y - \sin x \sin y.$

Proof   Differentiate with respect to x, or with respect to y. $\qquad \square$

Corollary 7.5.6 $\sin(x + \pi/2) = \cos x$ *and* $\cos(x + \pi/2) = -\sin x.$

Proof   Put $y = \pi/2$. $\qquad \square$

Corollary 7.5.7 $\sin(x+\pi) = -\sin x$ *and* $\cos(x+\pi) = -\cos x.$
$\sin(x + 2\pi) = \sin x$ *and* $\cos(x + 2\pi) = \cos x.$

Thus the cosine and sine functions are periodic, with period 2π.

Proposition 7.5.8   *Suppose that* $(x, y) \in \mathbf{R}^2$ *and that* $x^2 + y^2 = r^2 > 0.$
Then there exists a unique $\theta \in (-\pi, \pi]$ *such that* $x = r \cos \theta$ *and* $y = r \sin \theta.$

Proof   Suppose first that y is non-negative. Since $\cos 0 = 1$ and $\cos \pi = -1$, and since $-1 \leq x/r \leq 1$, it follows from the intermediate value theorem that there exists $0 \leq \theta \leq \pi$ such that $x/r = \cos \theta$. θ is unique, since $\cos$ is a strictly decreasing function on $[0, \pi]$. Then $(y/r)^2 = 1 - (x/r)^2 = 1 - \cos^2 \theta = \sin^2 \theta$, and so $y = r \sin \theta$, since y and $\sin \theta$ are both non-negative.
   If $y < 0$ then there exists $0 < \phi < \pi$ such that $x = r \cos \phi$ and $-y = r \sin \phi$. Let $\theta = -\phi$. Then $x = r \cos \theta$ and $y = r \sin \theta$, and again, θ is uniquely determined. $\qquad \square$

We now consider the complex case. Inspection shows that

$$\cos z = \frac{e^{iz} + e^{-iz}}{2} \quad \text{and} \quad \sin z = \frac{e^{iz} - e^{-iz}}{2i},$$

and so we obtain *Euler's formula*

$$e^{iz} = \cos z + i \sin z.$$

In particular, if $x \in \mathbf{R}$ then $\cos x$ and $\sin x$ are the real and imaginary parts of e^{ix}.

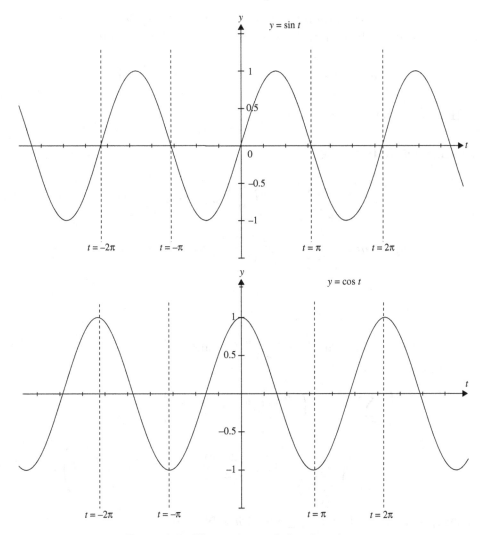

Figure 7.5. The cosine and sine functions.

Proposition 7.5.9 *The mapping*

$$t \to e^{it} = \cos t + i \sin t : \mathbf{R} \to \mathbf{T} = \{z : |z| = 1\}$$

is a continuous homomorphism of the additive group $(\mathbf{R}, +)$ *onto the multiplicative group* $(\mathbf{T}, .)$, *with kernel* $2\pi \mathbf{Z}$.

Proof The mapping is certainly continuous, and is a homomorphism into $(\mathbf{T}, .)$. It is surjective, by Proposition 7.5.8. Finally, $e^{it} = \cos t + i \sin t = 1$ if and only if $\cos t = 1$ and $\sin t = 0$, which happen if and only if $t = 2\pi k$, for some $k \in \mathbf{Z}$. $\square$

In fact, most of the properties of the real-valued functions cos and sin can be deduced from the equation $e^{it} = \cos t + i \sin t$. For example,

$$\cos^2 t + \sin^2 t = |e^{it}|^2 = e^{it}\overline{e^{it}} = e^{it}e^{-it} = 1.$$

Here are two more examples.

Example 7.5.10 If $n \in \mathbf{N}$ then

$$\cos nt = \sum_{0 \le 2k \le n} (-1)^k \binom{2n}{2k} \sin^{2k} t \cos^{2n-2k} t$$

and $\sin nt = \sum_{0 \le 2k < n} (-1)^k \binom{2n}{2k+1} \sin^{2k+1} t \cos^{2n-2k-1} t.$

For

$$e^{int} = (\cos t + i \sin t)^n = \sum_{j=0}^{n} \binom{n}{j} (i^j \sin^j t \cos^{n-j} t)$$

$$= \left(\sum_{0 \le 2k \le n} (-1)^k \binom{2n}{2k} \sin^{2k} t \cos^{2n-2k} t \right)$$

$$+ i \left(\sum_{0 \le 2k < n} (-1)^k \binom{2n}{2k+1} \sin^{2k+1} t \cos^{2n-2k-1} t \right).$$

Example 7.5.11 If $0 < |t| \le \pi$ and $n \in \mathbf{N}$ then

$$\sum_{j=-n}^{n} \cos jt = \frac{\sin(n+\frac{1}{2})t}{\sin t/2}.$$

For

$$\sum_{j=-n}^{n} \cos jt = \sum_{j=-n}^{n} e^{ijt} = e^{-int} \sum_{j=0}^{2n} e^{ijt}$$

$$= e^{-int} \frac{e^{i(2n+1)t} - 1}{e^{it} - 1} = \frac{e^{i(n+\frac{1}{2})t} - e^{-i(n+\frac{1}{2})t}}{e^{it/2} - e^{-it/2}} = \frac{\sin(n+\frac{1}{2})t}{\sin t/2}.$$

Exercises

7.5.1 Show that $2x/\pi < \sin x < x$ for $0 < x < \pi/2$.

7.5.2 Use the mean-value theorem to show that

$$0 < \frac{1}{\sin t} - \frac{1}{t} < t \text{ for } 0 < t < \pi/2.$$

7.5.3 Show that the function sin is a continuous strictly increasing bijection of $[-\pi/2, \pi/2]$ onto $[-1, 1]$. The inverse mapping is denoted by $\sin^{-1}$, or arcsin. Show that

$$\frac{d\sin^{-1}}{dx}(x) = \frac{1}{\sqrt{1-x^2}} \text{ for } -1 < x < 1.$$

7.5.4 Show that the function cos is a continuous strictly decreasing bijection of $[0, \pi]$ onto $[-1, 1]$. The inverse mapping is denoted by $\cos^{-1}$, or arccos. Show that

$$\frac{d\cos^{-1}}{dx}(x) = -\frac{1}{\sqrt{1-x^2}} \text{ for } -1 < x < 1.$$

7.5.5 Explain why

$$\frac{d(\sin^{-1} + \cos^{-1})}{dx}(x) = 0 \text{ for } -1 < x < 1.$$

7.5.6 Let $f(x) = \sin(1/x)$ for $x \neq 0$ and let $f(0) = 0$. Sketch the graph of f.

(a) For what values of α is the function $x^\alpha f(x)$ continuous on $\mathbf{R}$?
(b) For what values of α is the function $x^\alpha f(x)$ differentiable on $\mathbf{R}$?
(c) For what values of α is the function $x^\alpha f(x)$ continuously differentiable on $\mathbf{R}$?
(d) For what values of α is the function $x + x^\alpha f(x)$ strictly increasing on $\mathbf{R}$?

7.5.7 The *tangent function* tan is defined as $\tan x = \sin x / \cos x$ for $-\pi/2 < x < \pi/2$. Show that it is a strictly increasing differentiable mapping of $(-\pi/2, \pi/2)$ onto $\mathbf{R}$. Its inverse is denoted by $\tan^{-1}$, or arctan. What is the derivative of $\tan^{-1}$?

7.5.8 Investigate

$$\lim_{x\to 0} \frac{x(1 - \cos x)}{x - \sin x} \text{ and } \lim_{x\to 0} \frac{\tan^2 x - \sin^2 x}{(1 - \cos x)^2}.$$

7.5.9 Let $f(x) = x + 2x^2 \sin(1/x)$, for $x \neq 0$ and let $f(0) = 0$. Show that f is differentiable on $\mathbf{R}$, and calculate its derivative. Show that $f'(0) = 1$, but that there is no interval $(-\delta, \delta)$ on which f is monotonic.

7.6 Higher derivatives, and Taylor's theorem

Suppose that f is a differentiable function on an open interval I. Then, as with the functions exp, log, sin and cos, it may happen that f' is also differentiable. We then denote the derivative of f' by f'', or $f^{(2)}$, or d^2f/dx^2. The process may continue, and we obtain *higher derivatives* $f^{(n)}$, or d^nf/dx^n, of order n. If f has derivatives of all orders, we say the f is *infinitely differentiable*.

We have the following formula for products.

Proposition 7.6.1 (Leibniz' formula)   *Suppose that f, g are functions on an open interval which have derivatives of order n. Then*

$$(fg)^{(n)} = f^{(n)}g + nf^{(n-1)}g' + \cdots + \binom{n}{r}f^{(n-r)}g^{(r)} + \cdots + fg^{(n)}.$$

Proof   The proof is by induction on n. It is true for $n = 1$, by Proposition 7.1.5 (iii). Suppose that it is true for n. Using the result for $n = 1$,

$$(f^{(n-r)}g^{(r)})' = f^{(n-r+1)}g^{(r)} + f^{(n-r)}g^{(r+1)},$$

so that

$$(fg)^{(n+1)} = \sum_{r=0}^{n} \binom{n}{r} \left(f^{(n-r+1)}g^{(r)} + f^{(n-r)}g^{(r+1)} \right)$$

$$= \sum_{r=0}^{n+1} \left(\binom{n}{r} + \binom{n}{r-1} \right) f^{(n+1-r)}g^{(r)}$$

$$= \sum_{r=0}^{n+1} \binom{n+1}{r} f^{(n+1-r)}g^{(r)},$$

since

$$\binom{n}{r} + \binom{n}{r-1} = \binom{n+1}{r},$$

by de Moivre's formula. $\square$

If f is differentiable at a, then the function $t_a(x) = f(a) + (x-a)f'(a)$ provides a linear approximation to f; if $r_a = f - t_a$, then $r_a(x) = o(|x-a|)$. Suppose that f has higher derivatives, up to order n; can we obtain a better polynomial approximation? Let us consider a polynomial which has the same derivatives as f at a. Let

$$p_n(x) = f(a) + (x-a)f'(a) + \frac{(x-a)^2}{2!}f''(a) + \cdots + \frac{(x-a)^n}{n!}f^{(n)}(a).$$

Then p_n is a polynomial of degree at most n, $p_n(a) = f(a)$ and

$$p_n^{(s)}(x) =$$

$$f^{(s)}(a) + (x-a)f^{s+1}(a) + \frac{(x-a)^2}{2!}f^{(s+2)}(a) + \cdots + \frac{(x-a)^{n-s}}{(n-s)!}f^{(n)}(a),$$

so that $p_n^{(s)}(a) = f^{(s)}(a)$ for $1 \le s \le n$. Let $r_{n+1} = f - p_n$: r_{n+1} is the *remainder term*. Then $r_{n+1}(a) = 0$ and $r_{n+1}^{(s)}(a) = 0$ for $1 \le s \le n$, and we might hope that the remainder term is small, so that p_n is an even better approximation to f, near a.

Taylor's theorem provides information about the remainder term. We give two versions of this theorem here, and shall give another one in Theorem 8.7.3. The different versions each depend in detail upon the conditions that are placed on f and on its derivatives.

Theorem 7.6.2 (Taylor's theorem, with Lagrange's remainder)   *Suppose that f is a continuous function on $[a,b]$ which is n-times differentiable on $[a,b)$ (with one-sided derivatives at a). Then there exists $c \in (a,b)$ such that*

$$f(b) = f(a) + (b-a)f'(a) + \cdots + \frac{(b-a)^{n-1}}{(n-1)!}f^{(n-1)}(a) + \frac{(b-a)^n}{n!}f^{(n)}(c)$$

$$= p_{n-1}(b) + \frac{(b-a)^n}{n!}f^{(n)}(c).$$

Proof   The proof is just like the proof of the mean-value theorem. We shall assume that $a < b$; a similar proof applies if $b < a$. Let $h_\lambda(x) = f(x) - p_{n-1}(x) - \lambda(x-a)^n/n!$, where λ is a real number chosen so that $h_\lambda(b) = 0$. Then h_λ is continuous on $[a,b]$, $h_\lambda(a) = 0$, and $h_\lambda^{(s)}(a) = 0$ for $1 \le s < n$.

We need to show that there exists $a < c < b$ such that $\lambda = f^{(n)}(c)$. To do this, we repeatedly use Rolle's theorem. Since $h_\lambda(a) = h_\lambda(b) = 0$, there exists $a < c_1 < b$ such that $h_\lambda'(c_1) = 0$. Now h_λ' is continuous on $[a, c_1]$ and $h_\lambda'(a) = h_\lambda'(c_1) = 0$, and so, using Rolle's theorem again, there exists $a < c_2 < c_1$ such that $h_\lambda''(c_2) = 0$. Continuing in this way (that is, giving a proof by induction), we find that there exist $a < c_n < c_{n-1} < \cdots < c_1 < b$ such that $h_\lambda^{(n)}(c_n) = 0$. But $h_\lambda^{(n)} = f^{(n)} - \lambda$; setting $c = c_n$, we see that $\lambda = f^{(n)}(c)$. Thus

$$f(b) = p_{n-1}(b) + \frac{(b-a)^n}{n!}f^{(n)}(c). \qquad \square$$

For the second theorem, we impose slightly stronger conditions.

Theorem 7.6.3 (Taylor's theorem, with Cauchy's remainder)   *Suppose that f is a continuous function on $[a, b]$ which is n-times differentiable on $[a, b)$ (with one-sided derivatives at a), and for which the derivatives are bounded on $[a, b)$. Suppose that $k \in \mathbf{R}$ and that $k > 0$. Then there exists $c \in (a, b)$ such that*

$$f(b) = p_{n-1}(b) + \frac{(b-c)^{n-k}(b-a)^k}{k(n-1)!} f^{(n)}(c).$$

If we write $c = (1 - \theta_n)a + \theta_n b$, this becomes

$$f(b) = p_{n-1}(b) + \frac{(1 - \theta_n)^{n-k}(b-a)^n}{k(n-1)!} f^{(n)}(c).$$

Proof   Suppose that $a < b$; a similar proof holds if $b < a$. Let

$$h(x) = f(x) + \sum_{s=1}^{n-1} \frac{(b-x)^s}{s!} f^{(s)}(x)$$

for $a \le x < b$, and let $h(b) = f(b)$. Since the derivatives are bounded on $[a, b)$, the function h is continuous on $[a, b]$ and differentiable on (a, b). The idea behind this definition is that $h(a) = p_{n-1}(b)$ and

$$\frac{d}{dx}\left(\frac{(b-x)^s}{s!} f^{(s)}(x) \right) = -\frac{(b-x)^{s-1}}{(s-1)!} f^{(s)}(x) + \frac{(b-x)^s}{s!} f^{(s+1)}(x),$$

so that

$$h'(x) = \frac{(b-x)^{n-1} f^{(n)}(x)}{(n-1)!},$$

all the other terms cancelling in pairs. Let $g(x) = -(b-x)^k$, so that

$$g(b) - g(a) = (b-a)^k, \text{ and } g'(x) = k(b-x)^{k-1} \neq 0 \text{ for } x \in (a, b).$$

Thus by Cauchy's mean-value theorem there exists $a < c < b$ such that

$$\frac{h(b) - h(a)}{g(b) - g(a)} = \frac{h'(c)}{g'(c)}.$$

Thus

$$f(b) = h(b) = h(a) + (h(b) - h(a))$$

$$= h(a) + (g(b) - g(a))\frac{h'(c)}{g'(c)}$$

$$= p_{n-1}(b) + \frac{(b-c)^{n-1}(b-a)^k}{k(b-c)^{k-1}(n-1)!}f^{(n)}(c)$$

$$= p_{n-1}(b) + \frac{(b-c)^{n-k}(b-a)^k}{k(n-1)!}f^{(n)}(c). \qquad \Box$$

Suppose that f is infinitely differentiable. Then we can write $f(x) = p_n(x) + r_{n+1}(x)$ for each n. We might hope that $r_n(x) \to 0$ as $n \to \infty$, so that we can write

$$f(x) = f(a) + \sum_{j=1}^{\infty} \frac{(x-a)^j}{j!} f^{(j)}(a),$$

in which case the series on the right-hand side is called the *Taylor series* for f. The following example shows that this is not always the case. Let $f(x) = e^{-1/x^2}$ for $x \neq 0$ and let $f(0) = 0$. Then f is continuous on **R**. If $x \neq 0$ then $f'(x) = (2/x^3)e^{-1/x^2}$, and $f'(0) = \lim_{x \to 0} e^{-1/x^2}/x = 0$. An inductive argument then shows that there exists a sequence (s_j) of polynomials such that

$$f^{(j)}(x) = \frac{s_j(x)}{x^{3j}}e^{-1/x^2} \text{ for } x \neq 0,$$

$$\text{and } f^{(j+1)}(0) = \lim_{x \to 0} \frac{s_j(x)}{x^{3j+1}}e^{-1/x^2} = 0,$$

for all $j \in \mathbf{N}$. Thus $p_n(x) = 0$ for all n, and so $r_{n+1}(x) = f(x)$. In this case, the Taylor series gives us no useful information about f; the trouble is that f is *too* smooth at 0.

Let us give two applications of Taylor's theorem. Our first application is to the *Newton–Raphson method* of approximation. We consider a continuous function f on an interval $[a, b]$ with the following properties:

(i) f is twice differentiable on (a, b), and f'' is bounded on (a, b): there exists M such that $|f''(x)| \leq M$ for all $x \in (a, b)$;

(ii) there exists $m > 0$ such that $f'(x) \geq m$ for all $x \in (a, b)$;

(iii) $f(a) < 0$ and $f(b) > 0$.

Then f is strictly increasing on $[a, b]$, and so, by the intermediate value theorem, there exist a unique $c \in (a, b)$ with $f(c) = 0$. The Newton–Raphson method provides a sequence of successive approximations to c, and Taylor's theorem tells us that the approximation can improve extremely rapidly. We must start with a reasonably good approximation x_0. Let $K = M/2m$. Suppose that we have found $0 < h < K$, a_1 and b_1 such that

$$a < b_1 - h < a_1 < b_1 < a_1 + h \le b, \text{ and such that } f(a_1) < 0 < f(b_1).$$

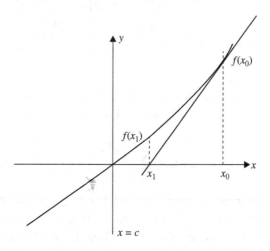

Figure 7.6. The Newton–Raphson method.

Let $\lambda = h/K$, so that $0 < \lambda < 1$ (the smaller λ is, the better the approximation will be). Then $c \in (a_1, b_1)$, and

$$[a_1, b_1] \subseteq (c - h, c + h) \subseteq (a, b).$$

Start by choosing $x_0 \in [a_1, b_1]$. Then $K|x_0 - c| < \lambda$. By Taylor's theorem, with Lagrange's remainder, there exists $y_0 \in (c, x_0)$ such that

$$0 = f(c) = f(x_0) + (c - x_0)f'(x_0) + (c - x_0)^2 f''(y_0)/2.$$

Hence

$$\frac{f(x_0)}{f'(x_0)} = (x_0 - c) - (c - x_0)^2 \frac{f''(y_0)}{2f'(x_0)}.$$

We set $x_1 = x_0 - f(x_0)/f'(x_0)$, so that

$$x_1 - c = (c - x_0)^2 \frac{f''(y_0)}{2f'(x_0)},$$

and so $|x_1 - c| \le Kh^2 = \lambda h$. Thus $x_1 \in (c - h, c + h)$, and $K|x_1 - c| \le \lambda^2$. Iterating the process, we obtain a sequence $(x_n)_{n=0}^{\infty}$ such that

$K|x_n - c| \leq \lambda^{2^n}$. This can lead to very rapid convergence; for example, if $K = 1$ and $\lambda = 1/10$, then $|x_3 - c| \leq 1/10^8$.

The proof of the existence of the nth root of a positive number that was given in Section 3.2 used the Newton–Raphson method.

The classic application of Taylor's theorem is to the binomial theorem.

Theorem 7.6.4 (The binomial theorem)   *Suppose that $\alpha \in \mathbf{R} \setminus \mathbf{N}$ and that $-1 < x < 1$. Let $f_\alpha(x) = (1 + x)^\alpha$. Then*

$$f_\alpha(x) = 1 + \alpha x + \sum_{j=2}^{\infty} \frac{\alpha(\alpha - 1) \dots (\alpha - j + 1)}{j!} x^j = 1 + \sum_{j=1}^{\infty} \binom{\alpha}{j} x^j,$$

the sum converging absolutely.

Proof   The proof is not quite straightforward. (It is unfortunate that Professor James Moriarty's treatise is not extant, as it would have thrown light on how the theorem was considered towards the end of the nineteenth century.) The ratio test shows that the series converges absolutely. Further,

$$f_\alpha^{(j)}(x) = \alpha(\alpha - 1) \dots (\alpha - j + 1)(1 + x)^{\alpha - j}.$$

Thus

$$f_\alpha(x) = 1 + \sum_{j=1}^{n-1} \binom{\alpha}{j} x^j + r_n(x).$$

We need to show that the remainder $r_n(x)$ tends to 0 as $n \to \infty$. The Lagrange form of the remainder is

$$r_n(x) = \binom{\alpha}{n}(1 + \theta_n x)^{\alpha - n} x^n = (1 + \theta_n x)^\alpha \binom{\alpha}{n} \left(\frac{x}{1 + \theta_n x} \right)^n,$$

where $0 < \theta_n < 1$. If $0 \leq x < 1$ then $\sup_n |x/(1 + \theta_n x)| < 1$, and so $r_n(x) \to 0$ as $n \to \infty$ (see Exercise 3.2.5). If $-1 < x \leq -1/2$, this argument does not work.

Instead, we use Cauchy's form of the remainder. Choose $k > |\alpha|$. We find that

$$r_n(x) = \frac{\alpha}{k}(1 - \theta_n)^{n-k} \binom{\alpha - 1}{n - 1}(1 + \theta_n x)^{\alpha - n} x^n.$$

Since $1 - \theta_n < 1 + \theta_n x$, it follows that if $n \geq \alpha$ then

$$|r_n(x)| \leq \left| \frac{1}{(1 - |x|)^{k - \alpha}} \binom{\alpha - 1}{n - 1} x^n \right|,$$

and so $r_n(x) \to 0$ as $n \to \infty$. $\qquad \square$

This is a remarkably technical proof. Another, easier, proof is given as an exercise in Section 8.7. We shall see in Volume III, that these proofs, and the proofs of other consequences of Taylor's theorem, are superseded by the complex version of Taylor's theorem.

Notice that if $\alpha = -\beta < 0$ and $0 < x < 1$ then

$$\frac{1}{(1-x)^\beta} = (1 + (-x))^\alpha = 1 + \sum_{j=1}^{\infty} \left| \binom{\alpha}{j} \right| x^j,$$

so that all the summands are positive. In particular,

$$\frac{1}{(1-x)^{1/2}} = 1 + \frac{x}{2} + \sum_{j=2}^{\infty} \frac{1.3.\ldots.(2j-1)}{j!2^j} x^j = 1 + \frac{x}{2} + \sum_{j=2}^{\infty} \binom{2j}{j} \frac{x^j}{2^{2j}}.$$

If f is differentiable at a, then $f(x) - f(a) - (x-a)f'(a) = o(|x-a|)$; for this, we do not need to suppose that f is differentiable at any point other than a. There is a corresponding result for n-times differentiable functions; this is due to W. H. Young. (We shall not use this result later, and it may be omitted.)

Theorem 7.6.5 *Suppose that f is $(n-1)$-times differentiable in an interval I and that $f^{(n-1)}$ is differentiable at an interior point a of I. Let*

$$p_n(x) = f(a) + (x-a)f'(a) + \frac{(x-a)^2}{2!} f''(a) + \cdots + \frac{(x-a)^n}{n!} f^{(n)}(a),$$

and let $r_{n+1}(x) = f(x) - p_n(x)$. Then $r_{n+1}(x) = o(|x-a|^n)$.

Proof Let $u(x) = r_{n+1}(x)/(x-a)^n$, for $x \neq a$. Then we must show that $u(x) \to 0$ as $x \to a$. Suppose that $\epsilon > 0$. Let

$$v_\epsilon(x) = r_{n+1}(x) + \epsilon(x-a)^n.$$

Then $v_\epsilon(x) = (u(x) + \epsilon)(x-a)^n$ for $x \neq a$, and v_ϵ is n-times differentiable at a;

$$v_\epsilon(a) = 0, \ v_\epsilon^{(s)}(a) = 0 \text{ for } 1 \le s \le n-1 \text{ and } v_\epsilon^{(n)}(a) = n!\epsilon > 0.$$

By Proposition 7.1.5 (v), there exists $\delta > 0$ such that $[a, a+\delta) \subseteq I$ and $v_\epsilon^{(n-1)}(x) > 0$ for $a < x < a+\delta$. By Corollary 7.3.6, $v_\epsilon^{(n-2)}$ is strictly increasing on $[a, a+\delta)$, and so $v_\epsilon^{(n-2)}(x) > 0$ for $a < x < a+\delta$. Iterating the argument, it follows that $v_\epsilon(x) = (u(x) + \epsilon)(x-a)^n > 0$ for $a < x < a+\delta$. Thus $u(x) > -\epsilon$ for $a < x < a+\delta$. Applying the same argument to $w_\epsilon(x) = -r_{n+1}(x) + \epsilon(x-a)^n$, it follows that there exists $\delta' > 0$ such that $[a, a+\delta') \subseteq I$ and $u(x) < \epsilon$ for $a < x < a+\delta'$. Consequently, $u(x) \to 0$ as $x \searrow a$. Similarly $u(x) \to 0$ as $x \nearrow a$. $\square$

Exercises

7.6.1 Suppose that f is differentiable in an open interval I, and that f is twice differentiable at $a \in I$. Show that

$$\frac{f(a+h) + f(a-h) - 2f(a)}{h^2} \to 0 \text{ as } h \to 0.$$

[Hint: L'Hôpital's rule.]

7.6.2 Suppose that f is $2k$-times differentiable on an open interval I, that $f^{(j)}(a) = 0$ for $1 \le j \le 2k-1$ and that $f^{(2k)}(a) < 0$. Show that f has a local maximum at a.

7.6.3 Suppose that f is twice differentiable on an open interval I, that $a, b, c \in I$, with $a < b < c$. Show that there exists $d \in (a, b)$ such that

$$\frac{f(c) - f(a)}{c - a} - \frac{f(b) - f(a)}{b - a} = \tfrac{1}{2}(c - b)f''(d).$$

7.6.4 Apply the Newton–Raphson method to the function $f(x) = x^2 - 2$, starting with $x_0 = 3/2$, to obtain rational approximations to $\sqrt{2}$. How good is the approximation after three iterations?

7.6.5 Let $f(x) = \log(1 + x)$, for $-1 < x < 1$, and let $r_n(x)$ be the nth remainder in the Taylor series expansion of f. Show that $r_n(x) \to 0$ as $n \to \infty$, and determine the infinite Taylor series for f.

7.6.6 Let $f(x) = \tan^{-1}(x)$. Apply the Newton–Raphson method when $0 < |x_0| < 1$, when $x_0 = 1$ and when $|x| > 1$. When (x_n) converges, how fast does it converge?

7.6.7 Suppose that f is a convex increasing function on the closed interval $[a, b]$ which is differentiable on the open interval (a, b), and for which $f(a) < 0 < f(b)$. Suppose that $x_0 \in (a, b)$ and that $f(x_0) > 0$. Show that the sequence $(x_n)_{n=0}^{\infty}$ defined by the Newton–Raphson method is decreasing, that $x_n > b$ and that $f(x_n) \ge 0$. Suppose that $x_n \to c$. Show that $f(c) = 0$, and that there exists $0 < \lambda < 1$ such that $x_n - c \le \lambda^n(x_0 - c)$. What happens if $f(x_0) < 0$?

7.6.8 Apply the Newton–Raphson method to the function $f(x) = x^n$, where $n \ge 2$, starting with $x_0 > 0$. Calculate x_n. Why is the convergence slower than that described in the text?

7.6.9 Apply the Newton–Raphson method to the function $f(x) = x + x^{\alpha+1}$, where $x > 0$ and $0 < \alpha < 1$, starting with $x_0 > 0$. Calculate x_n. Why is the convergence slower than that described in the text?

7.6.10 Suppose that $(a_j)_{j=1}^\infty$ is a sequence of positive terms, and that there exists $a > 0$ such that $a_{j+1}/a_j = 1 - a/j + r_j$, where $r_j/j \to 0$ as $j \to \infty$. Show that $\sum_{j=1}^\infty a_j$ converges if $a > 1$ and that $\sum_{j=1}^\infty a_j$ diverges if $a < 1$. (Consider $b_j = 1/j^s$, where s is between 1 and a. This extends D'Alembert's ratio test.)

8
Integration

8.1 Elementary integrals

We now turn to integration, which we develop as the 'area under the curve'. We establish the existence and properties of the Riemann integral; this is an integral whose development is quite straightforward, and which is good for many of the needs of analysis. It has some shortcomings: it can only be applied to a restricted class of functions, and it is not easy to obtain good results about limits of integrals. For this, a more sophisticated integral, the Lebesgue integral, is needed; we shall consider this in Volume III.

As with all theories of integration, we proceed by approximation. To begin with, we restrict attention to bounded real-valued functions on a finite interval $[a, b]$. The easiest functions to start with are the *step functions* – functions which take constant values v_j on a finite set $\{I_j : 1 \leq j \leq k\}$ of disjoint subintervals of $[a, b]$. The graph of such a function is a *bar graph*, and we define the elementary integral of such a function to be $\sum_{j=1}^{k} v_j l(I_j)$, where $l(I_j)$ is the length of the interval I_j. Note that v_j can be positive or negative, so that the integral can take positive and negative values.

The idea of the Riemann integral of a function f is to approximate f from above and below by step functions. If the integrals of the approximations from above and from below approach a common limit, then we take this limit to be the Riemann integral of f.

In order to carry out this programme, we need to set up the appropriate machinery. A *dissection D* of $[a, b]$ is a finite subset of $[a, b]$ which contains both a and b. We arrange the elements of D, the *points of dissection of D*, in increasing order: $a = x_0 < x_1 < \cdots < x_k = b$. The dissection splits $[a, b]$ into k disjoint intervals $I_1, \ldots I_k$. We need to decide what to do with the endpoints; we adopt the convention that $I_1 = [x_0, x_1]$ and that $I_j = (x_{j-1}, x_j]$ for $2 \leq j < k$.

We order the dissections of $[a, b]$ by inclusion: we say that D_2 *refines* D_1 if $D_1 \subseteq D_2$, and write $D_1 \le D_2$. This is a partial order on the set Δ of all dissections of $[a, b]$, and Δ is a lattice: $D_1 \vee D_2 = D_1 \cup D_2$ and $D_1 \wedge D_2 = D_1 \cap D_2$. Δ has a least element $\{a, b\}$, but has no greatest element.

Suppose that D is a dissection, with intervals $I_1, \ldots, I_k$. We denote the indicator function of I_j by χ_j: $\chi_j(x) = 1$ if $x \in I_j$, and $\chi_j(x) = 0$ otherwise. Similarly, we write $\chi_{[a,b]}$ for the indicator function of $[a, b]$. We denote the linear span of $\{\chi_j : 1 \le j \le k\}$ by E_D; thus a function $f \in E_D$ is of the form $f = \sum_{j=1}^{k} v_j \chi_j$, where $v_1, \ldots, v_k$ are real numbers. The elements of E_D are the step functions on $[a, b]$ whose points of discontinuity are contained in D; note that, according to our convention, step functions are continuous on the left. E_D is a k-dimensional vector space of functions.

If D_2 refines D_1, then $E_{D_1} \subseteq E_{D_2}$, and so the set of spaces $\{E_D : D \in \Delta\}$ also forms a lattice:

$$E_{D_1} \wedge E_{D_2} = E_{D_1} \cap E_{D_2} = E_{D_1 \wedge D_2}$$

$$\text{and } E_{D_1} \vee E_{D_2} = \text{span}\left(E_{D_1} \cup E_{D_2}\right) = E_{D_1 \vee D_2}.$$

The union $E_\Delta = \cup\{E_D : D \in \Delta\}$ is the infinite-dimensional vector space of all (left-continuous) step functions.

We now wish to define the *elementary integral* of a step function f. If $f = \sum_{j=1}^{k} v_j \chi_j$, we want to define $\int_a^b f(x)\, dx$ to be $\sum_{j=1}^{k} v_j l(I_j)$, where $l(I_j) = x_j - x_{j-1}$ is the length of I_j. But the representation is not unique, and we need to show that the integral is well-defined.

Proposition 8.1.1 *Suppose that D and D' are dissections of $[a, b]$, and that $f \in E_D \cap E_{D'}$, with representations $f = \sum_{j=1}^{k} v_j \chi_j$ and $f = \sum_{j=1}^{k'} v'_j \chi'_j$. Then $\sum_{j=1}^{k} v_j l(I_j) = \sum_{j=1}^{k'} v'_j l(I'_j)$.*

Proof We use the lattice property of Δ. Let $D'' = D \cup D'$. Let $D = \{x_0, \ldots, x_k\}$ and $D'' = \{x''_0, \ldots, x''_{k''}\}$. Then there exist $0 = r_0 < r_1 < \cdots < r_k = k''$ such that $x_j = x''_{r_j}$ for $0 \le j \le k$. Thus $l(I_j) = \sum_{r=r_{j-1}+1}^{r_j} l(I''_r)$. We can write $f = \sum_{r=1}^{k''} v''_r \chi''_r$, where $v_j = v''_r$ for $r_{j-1} < r \le r_j$. Consequently,

$$\sum_{j=1}^{k} v_j l(I_j) = \sum_{j=1}^{k} \left(\sum_{r=r_{j-1}+1}^{r_j} v''_r l(I''_r) \right) = \sum_{r=1}^{k''} v''_r l(I''_r).$$

Similarly, $\sum_{j=1}^{k'} v_j' l(I_j') = \sum_{r=1}^{k''} v_r'' l(I_r'')$, so that

$$\sum_{j=1}^{k} v_j l(I_j) = \sum_{j=1}^{k'} v_j' l(I_j').$$

$\square$

We can therefore define the elementary integral as

$$\int_a^b f(x)\,dx = \sum_{j=1}^{k} v_j l(I_j).$$

Proposition 8.1.2 *Suppose that f and g are step functions and that $c \in \mathbf{R}$. Then $f + g$ and cf are step functions, and*
(i) $\int_a^b (f(x) + g(x))\,dx = \int_a^b f(x)\,dx + \int_a^b g(x)\,dx$.
(ii) $\int cf(x)\,dx = c\int_a^b f(x)\,dx$.
(iii) *If $f(x) \leq g(x)$ for all x then $\int_a^b f(x)\,dx \leq \int_a^b g(x)\,dx$.*

Proof (i) Since Δ is a lattice, there exists $D \in \Delta$, with intervals $I_1, \ldots, I_k$, such that $f, g \in E_D$. Then we can write $f = \sum_{j=1}^{k} v_j \chi_j$ and $g = \sum_{j=1}^{k} w_j \chi_j$. Then $f + g = \sum_{j=1}^{k} (v_j + w_j)\chi_j$ is a step function and

$$\int_a^b (f(x) + g(x))\,dx = \sum_{j=1}^{k} (v_j + w_j)l(I_j) = \sum_{j=1}^{k} v_j l(I_j) + \sum_{j=1}^{k} w_j l(I_j)$$

$$= \int_a^b f(x)\,dx + \int_a^b g(x)\,dx.$$

The proofs of (ii) and (iii) are just as easy, and are left as exercises for the reader. $\square$

8.2 Upper and lower Riemann integrals

We now consider a bounded function f on $[a, b]$, with $m \leq f(x) \leq M$ for all $x \in [a, b]$. We try to integrate it by approximating from above and below by step functions. Let

$$U_f = \{g : g \in E_\Delta \text{ and } g \geq f\}$$

be the set of step functions which are greater than or equal to f. U_f is non-empty, since $M\chi_{[a,b]} \in U_f$. If $g \in U_f$, $g \geq m\chi_{[a,b]}$, and so $\int_a^b g(x)\,dx \geq$

$m(b - a)$. Thus the set $\{\int_a^b g(x)\,dx : g \in U_f\}$ is bounded below. We define the *upper Riemann integral* of f to be

$$\overline{\int_a^b} f(x)\,dx = \inf\{\int_a^b g(x)\,dx : g \in U_f\}.$$

Similarly we set

$$L_f = \{h : h \in E_\Delta \text{ and } h \leq f\}$$

and define the *lower Riemann integral* of f to be

$$\underline{\int_a^b} f(x)\,dx = \sup\{\int_a^b h(x)\,dx : h \in L_f\}.$$

Proposition 8.2.1   *Suppose that f is a bounded function on $[a,b]$. Then $\underline{\int_a^b} f(x)\,dx \leq \overline{\int_a^b} f(x)\,dx$.*

Proof   If $h \in L_f$ and $g \in U_f$ then $h \leq f \leq g$, so that

$$\int_a^b h(x)\,dx \leq \int_a^b g(x)\,dx.$$

Taking the supremum over L_f, we see that

$$\underline{\int_a^b} f(x)\,dx \leq \int_a^b g(x)\,dx,$$

so that, taking the infimum over U_f,

$$\underline{\int_a^b} f(x)\,dx \leq \inf\{\int_a^b g(x)\,dx : g \in U_f\} = \overline{\int_a^b} f(x)\,dx. \qquad \square$$

Suppose that D is a dissection, with intervals $I_1, \ldots, I_k$, and that f is a bounded function on $[a,b]$. Let $M(I_j) = \sup\{f(x) : x \in I_j\}$, and let $M_D(f) = \sum_{j=1}^k M_j \chi_j$. Then $M_D(f)$ is the least element of $E_D \cap U_f = \{g \in E_D, g \geq f\}$. We set

$$S_D = S_D(f) = \sum_{j=1}^k M(I_j) l(I_j) = \int_a^b M_D(f)(x)\,dx.$$

Then $S_D = \inf\{\int_a^b g(x)\,dx : g \in U_f \cap E_D\}$, so that

$$\overline{\int_a^b} f(x)\,dx = \inf\{\inf\{\int_a^b g(x)\,dx : g \in U_f \cap E_D\} : D \in \Delta\}$$
$$= \inf\{S_D : D \in \Delta\}.$$

Similarly, we define $m(I_j) = \inf\{f(x) : x \in I_j\}$ and $m_D(f) = \sum_{j=1}^k m(I_j)\chi_j$, and set

$$s_D = s_D(f) = \sum_{j=1}^k m_j l(I_j) = \int_a^b m_D(f)(x)\,dx.$$

Then $s_D = \sup\{\int_a^b g(x)\,dx : g \in L_f \cap E_D\}$, so that

$$\underline{\int_a^b} f(x)\,dx = \sup\{\sup\{\int_a^b g(x)\,dx : g \in U_f \cap E_D\} : D \in \Delta\}$$
$$= \sup\{s_D : D \in \Delta\}.$$

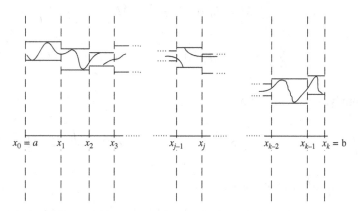

Figure 8.2. Upper and lower sums S_D and s_D.

Note that if D' refines D then $S_{D'} \le S_D$ and $s_{D'} \ge s_D$.

In fact, we do not need to consider all the dissections to determine the upper and lower Riemann integrals. If D is a dissection, with intervals $I_1, \ldots, I_k$, we define the *mesh size* $\delta(D)$ to be $\max\{l(I_j) : 1 \le j \le k\}$.

Theorem 8.2.2 *Suppose that $(D_r)_{r=1}^\infty$ is a sequence of dissections of $[a, b]$, and that $\delta(D_r) \to 0$ as $r \to \infty$. If f is a bounded function on $[a, b]$ then $S_{D_r}(f) \to \overline{\int_a^b} f(x)\,dx$ as $r \to \infty$.*

Proof Suppose that $\epsilon > 0$. Then there exists a dissection D of $[a, b]$, with points of dissection $a = x_0 < x_1 < \cdots < x_k = b$ such that

$S_D < \overline{\int_a^b} f(x)\,dx + \epsilon/2$. The idea of the proof is to choose r large enough so that the set D is contained in a set of intervals of D_r of small total length. Let $\eta = \epsilon/2(k+1)(M-m+1)$. There exists r_0 such that $\delta(D_r) < \eta$ for $r \geq r_0$. Suppose that $r \geq r_0$. Let $D' = D \vee D_r$. Then $S_{D'} \leq S_D$. Let $\{J_1, \ldots, J_q\}$ be the intervals of the dissection D_r, and let $K_1, \ldots, K_s$ be the intervals of the dissection D'. We divide $\{1, \ldots, q\}$ into two disjoint subsets. Let $p \in B$ if J_p contains one or more elements of D, and let $p \in G$ otherwise. (B is the set of bad indices, and G is the set of good indices.) Then $|B| \leq k+1$. If $p \in B$, then J_p is the disjoint union $\cup_{r \in S_p} K_r$ of finitely many of intervals in D'. Since $m \leq f(x) \leq M$,

$$Ml(J_p) \geq M(J_p)l(J_p) \geq \sum_{r \in S_p} M(K_r)l(K_r) \geq m \sum_{r \in S_p} l(K_r) = ml(J_p).$$

If $p \in G$, then $J_p = K_r$ for some $r \in \{1, \ldots, s\}$, so that $M(J_p) = M(K_r)$. Thus

$$S_{D_r} - S_{D'} = \sum_{p \in B} \left(M_p l(J_p)l(J_p) - \sum_{r \in S_p} M(K_r)l(K_r)) \right)$$

$$\leq \sum_{p \in B} (M-m)l(J_p) \leq (M-m)(k+1)\delta(D_r) < \epsilon/2.$$

Consequently, if $r \geq r_0$ then

$$\overline{\int_a^b} f(x)\,dx \leq S_{D_r} \leq S_{D'} + \epsilon/2 \leq S_D + \epsilon/2 < \overline{\int_a^b} f(x)\,dx + \epsilon$$

so that $S_{D_r} \to \overline{\int_a^b} f(x)\,dx$ as $r \to \infty$. $\qquad\square$

We can for example take D_r to be the dissection dividing $[a, b]$ into r intervals of equal length. Alternatively, we can repeatedly bisect the intervals, so that D_r is a dissection dividing $[a, b]$ into 2^r intervals of equal length; in this case, $(D_r)_{r=1}^{\infty}$ is an increasing sequence of dissections, so that $(S_{D_r})_{r=1}^{\infty}$ is a decreasing sequence, converging to $\overline{\int_a^b} f(x)\,dx$ as $r \to \infty$.

8.3 Riemann integrable functions

We say that a bounded function f on $[a, b]$ is *Riemann integrable* if its upper and lower integrals are equal. The common value is then the *Riemann integral* $\int_a^b f(x)\,dx$. In this expression, f is called the *integrand*. First, we must check that this extends the elementary integral of step functions.

Proposition 8.3.1 *If f is a step function then it is Riemann integrable, and the Riemann integral is the same as the elementary integral.*

Proof Let E be the elementary integral. Since f is in both U_f and L_f,

$$E \leq \underline{\int_a^b} f(x)\,dx \leq \overline{\int_a^b} f(x)\,dx \leq E,$$

and so all the quantities are equal. $\square$

Proposition 8.3.2 *Suppose that f is a bounded function on $[a,b]$. Then f is Riemann integrable if and only if given $\epsilon > 0$ there exist step functions g and h with $h \leq f \leq g$ and $\int_a^b g(x)\,dx - \int_a^b h(x)\,dx < \epsilon$.*

Proof This follows immediately from the definition. $\square$

Proposition 8.3.3 *Suppose that f is a bounded function on $[a,b]$. Then f is Riemann integrable if and only if given $\epsilon > 0$ there exists a dissection D such that $S_D - s_D < \epsilon$.*

Proof The condition is clearly sufficient. If f is Riemann integrable and $\epsilon > 0$ then there exist dissections D_1 and D_2 such that $s_{D_1} + \epsilon/2 > \int_a^b f(x)\,dx > S_{D_2} - \epsilon/2$. Let $D = D_1 \vee D_2$. Then

$$S_D \leq S_{D_2} \leq s_{D_1} + \epsilon \leq s_D + \epsilon. \qquad \square$$

We can express this proposition in terms of the oscillation of f. Suppose that f is a bounded real-valued function on a non-empty set S. The *oscillation* $\Omega = \Omega(f, S)$ of f on S is defined as

$$\Omega(f, S) = \sup\{|f(s) - f(t)| : s, t \in S\} = \sup_{s \in S} f(s) - \inf_{s \in S} f(s).$$

Corollary 8.3.4 *Suppose that f is a bounded function on $[a,b]$. Then f is Riemann integrable if and only if given $\epsilon > 0$ there exists a dissection $D = \{a = x_0 < \cdots < x_k = b\}$ of $[a,b]$, with intervals $I_1, \ldots, I_k$ such that*

$$\sum_{j=1}^{k} \Omega(f, I_j)(x_j - x_{j-1}) < \epsilon.$$

Proof For $S_D - s_D = \sum_{j=1}^{k} \Omega(f, I_j)(x_j - x_{j-1})$. $\square$

Corollary 8.3.5 *Suppose that f is a bounded function on $[a,b]$. Then f is Riemann integrable if and only if given $\epsilon > 0$ there exist a dissection*

$D = \{a = x_0 < \cdots < x_k = b\}$ of $[a,b]$ and a partition $G \cup B$ of $\{1, \ldots, k\}$ such that

$$\Omega(f, I_j) \leq \epsilon \text{ for } j \in G \quad \text{and} \quad \sum_{j \in B} l(I_j) < \epsilon,$$

where $I_1, \ldots I_k$ are the intervals of the dissection.

Proof Suppose that the condition is satisfied, and that $\epsilon > 0$. Let $\eta = \epsilon/(b - a + \Omega(f, [a, b]))$. Then

$$\sum_{j=1}^{k} \Omega(f, I_j)(x_j - x_{j-1}) =$$

$$= \sum_{j \in G} \Omega(f, I_j)(x_j - x_{j-1}) + \sum_{j \in B} \Omega(f, I_j)(x_j - x_{j-1})$$

$$\leq (\sup_{j \in G} \Omega(f, I_j)) \sum_{j \in G}(x_j - x_{j-1}) + \Omega(f, [a, b]) \sum_{j \in B}(x_j - x_{j-1})$$

$$\leq (b - a)\eta + \Omega(f, [a, b])\eta = \epsilon,$$

so that f is Riemann integrable. Suppose conversely that f is Riemann integrable. By the previous corollary there exists a dissection D with

$$\sum_{j=1}^{k} \Omega(f, I_j)(x_j - x_{j-1}) < \min(\epsilon, \epsilon^2).$$

Let $G = \{j \in D : \Omega(f, I_j) \leq \epsilon\}$ and let $B = \{j \in D : \Omega(f, I_j) > \epsilon\}$. Then

$$\epsilon \sum_{j \in B} l(I_j) \leq \sum_{j \in B} \Omega(f, I_j)l(I_j) < \epsilon^2,$$

which give the result. $\square$

Many important functions are Riemann integrable.

Theorem 8.3.6 *(i) A continuous function on $[a, b]$ is Riemann integrable.*
(ii) A monotonic function on $[a, b]$ is Riemann integrable.

Proof In both cases, we use Proposition 8.3.3.

(i) We use the fact that f is uniformly continuous. Suppose that $\epsilon > 0$. There exists $\delta > 0$ such that if $|x - y| < \delta$ then $|f(x) - f(y)| < \epsilon/(b - a)$. Choose N so that $(b - a)/N < \delta$, and let D_N be the dissection of $[a, b]$ into N intervals $I_1, \ldots, I_N$ of equal length. Then $l(I_j) = (b - a)/N < \delta$, so that

$$M_j = \sup\{f(x) : x \in I_j\} < \inf\{f(x) : x \in I_j\} + \epsilon/(b - a) = m_j + \epsilon/(b - a)\},$$

for $1 \leq j \leq N$, and so $S_{D_N} < s_{D_N} + \epsilon$.

(ii) Without loss of generality we can suppose that f is increasing. Suppose that $\epsilon > 0$. Choose N so that $N > (f(b) - f(a))(b - a)/\epsilon$. Let D_N be the dissection of $[a, b]$ into N intervals of equal length, as before, and let $a = x_0 < x_1 < \cdots < x_N = b$ be the points of dissection. Then $m_j \geq f(x_{j-1})$ and $M_j = f(x_j)$, so that

$$S_D = \sum_{j=1}^{N} f(x_j) \frac{b-a}{N} = \sum_{j=1}^{N} f(x_{j-1}) \frac{b-a}{N} + (f(b) - f(a)) \frac{b-a}{N} < s_D + \epsilon.$$

□

As an easy example, let us calculate $\int_0^a x \, dx$, where $a > 0$. Then

$$S_{D_N} = \sum_{j=1}^{N} \left(\frac{aj}{N} \right) \left(\frac{a}{N} \right) = \frac{a^2}{N^2} \sum_{j=1}^{N} j = \frac{a^2}{2} \left(1 + \frac{1}{N} \right),$$

and

$$s_{D_N} = \sum_{j=1}^{N} \left(\frac{a(j-1)}{N} \right) \left(\frac{a}{N} \right) = \frac{a^2}{N^2} \sum_{j=1}^{N} (j-1) = \frac{a^2}{2} \left(1 - \frac{1}{N} \right),$$

from which it follows that $\int_0^a x \, dx = a^2/2$.

We can also characterize Riemann integrability in terms of a sequence of dissections.

Proposition 8.3.7 *Suppose that $(D_r)_{r=1}^{\infty}$ is a sequence of dissections of $[a, b]$, and that $\delta(D_r) \to 0$ as $r \to \infty$. If f is a bounded function on $[a, b]$, then f is Riemann integrable if and only if $S_{D_r} - s_{D_r} \to 0$ as $r \to \infty$. If so, then $\int_a^b f(x) \, dx = \lim_{r \to \infty} S_{D_r} = \lim_{r \to \infty} s_{D_r}$.*

Proof This follows immediately from the definition and Theorem 8.2.2. □

Corollary 8.3.8 *Suppose that f is a bounded function on $[a, b]$, and that $J \in \mathbf{R}$. Then the following are equivalent.*
(i) f is Riemann integrable, and $\int_a^b f(x) \, dx = J$.
(ii) If $(D_r)_{r=1}^{\infty}$ is a sequence of dissections of $[a, b]$ with $\delta(D_r) \to 0$ as $r \to \infty$, if $I_{r,1}, \ldots, I_{r,q_r}$ are the intervals of the dissection D_r, and if $y_{r,p} \in I_{r,p}$, for $1 \leq p \leq q_r$, then

$$\sum_{p=1}^{q_r} f(y_{r,p}) l(I_{r,p}) \to J \text{ as } r \to \infty.$$

It is important that (ii) must hold for *every* choice of $y_{r,p} \in I_{r,p}$, and not just for one particular choice.

Proof If f is Riemann integrable, and $\int_a^b f(x)\,dx = J$, then

$$s_{D_r} \le \sum_{p=1}^{q_r} f(y_{r,p}) l(I_{r,p}) \le S_{D_r},$$

and so (ii) follows from the proposition and the sandwich principle.

Conversely, suppose that (ii) holds. For each $r \in \mathbf{N}$ and each p with $1 \le p \le q_r$ there exist $y_{r,p}$ and $z_{r,p}$ in $I_{r,p}$ for which $f(y_{r,p}) - f(z_{r,p}) \ge \Omega(f, I_{r,p})/2$. Then

$$0 \le \sum_{p=1}^{q_r} \Omega(f, I_{r,p}) l(I_{r,p}) \le 2 \sum_{p=1}^{q_r} (f(y_{r,p}) - f(z_{r,p})) l(I_{r,p}).$$

But

$$\sum_{p=1}^{q_r} (f(y_{r,p}) - f(z_{r,p})) l(I_{r,p}) \to 0 \text{ as } r \to \infty$$

and so

$$\sum_{p=1}^{q_r} \Omega(f, I_{r,p}) l(I_{r,p}) \to 0 \text{ as } r \to \infty.$$

Thus f is Riemann integrable, by Corollary 8.3.4. Further, since

$$s_{D_r} \le \sum_{p=1}^{q_r} f(y_{r,p}) l(I_{r,p}) \le S_{D_r},$$

it follows from the sandwich principle that

$$J = \lim_{r \to \infty} \left(\sum_{p=1}^{q_r} f(y_{r,p}) l(I_{r,p}) \right) = \int_a^b f(x)\,dx. \qquad \square$$

Let us consider some examples.

Example 8.3.9 A bounded function which is not Riemann integrable.

Let $f(x) = 1$ if x is rational, and $f(x) = 0$ if x is irrational. If $g = \sum_{j=0}^{k} v_j \chi_j \in U_f$ then each I_j contains a rational number, and so $v_j \ge 1$. Thus $\int_a^b g(x)\,dx \ge b - a$, and so $\overline{\int_a^b} f(x)\,dx \ge b - a$. Since $\chi_{[a,b]} \in U_f$, $\overline{\int_a^b} f(x)\,dx = b - a$. Similarly, If $h = \sum_{j=0}^{k} w_j \chi_j \in L_f$ then each I_j contains an irrational number, and so $w_j \le 0$. Thus $\int_a^b h(x)\,dx \le 0$, and so $\underline{\int_a^b} f(x)\,dx \le 0$. Since $0 \in L_f$, $\underline{\int_a^b} f(x)\,dx = 0$. Thus f is not Riemann integrable.

Example 8.3.10   A Riemann integrable function on $[0, 1]$ which is discontinuous at the rational points of $[0, 1]$.

If $r \in [0, 1]$ is rational, and $r = p/q$ in lowest terms, let $g(r) = 1/q$, and if x is irrational, let $g(x) = 0$. Then g is discontinuous at every rational number. Suppose that $\epsilon > 0$. Then there exists q_0 such that $1/q_0 < \epsilon$. Then in a closed interval $[a, b]$ there are only finitely many rational numbers $r = p/q$ with $q \leq q_0$, so that $L = \{x \in [a, b] : g(x) > \epsilon\}$ is finite. We can include L in a finite set of intervals of total length less than ϵ: there exists a dissection

$$D = \{0 = x_0 < y_0 < x_1 < y_1 < \cdots < x_k < y_k = 1\}$$

such that

$$L \subseteq [x_0, y_0] \cup (x_1, y_1] \cup \cdots \cup (x_k, y_k],$$

and $\sum_{i=0}^{k} y_k - x_k < \epsilon$. If we take $G = \{x_i : 1 \leq i \leq k\}$ and $B = \{y_i : 0 \leq i \leq k\}$ then $\Omega(g, I_j) \leq \epsilon$ for $j \in G$, and $\sum_{j \in B} l(I_j) < \epsilon$, so that g is Riemann integrable, by Corollary 8.3.5. Further, $S_D < \epsilon(b - a) + \epsilon$, so that $\int_a^b g(x)\, dx \leq 0$. Since g is non-negative, $\int_a^b g(x)\, dx = 0$.

Example 8.3.11   A function which is constant on a dense open subset of $[0, 1]$, but which is not Riemann integrable.

Let $C^{(\epsilon)}$ be a fat Cantor set. $C^{(\epsilon)}$ is a perfect subset of $[0, 1]$ with empty interior. Let $I_{C^{(\epsilon)}}$ be the indicator function of $C^{(\epsilon)}$. Then $I_{C^{(\epsilon)}}$ is zero on the dense open subset $[0, 1] \setminus C^{(\epsilon)}$ of $[0, 1]$. Since $C^{(\epsilon)}$ has an empty interior, $\int_0^1 I_{C^{(\epsilon)}}(x)\, dx = 0$. On the other hand, if D is a dissection of $[0, 1]$, with intervals $I_1, \ldots, I_k$, and if $G = \{j : I_j \cap C^{(\epsilon)} = \emptyset\}$, then $\sum_{j \in G} l(I_j) \leq \epsilon$, and so $S_D(I_{C^{(\epsilon)}}) \geq 1 - \epsilon$. Thus $\overline{\int_0^1} I_{C^{(\epsilon)}}(x)\, dx \geq 1 - \epsilon$.

Exercises

8.3.1 Suppose that f is a bounded function on $[a, b]$ which is continuous except at finitely many points of $[a, b]$. Show that f is Riemann integrable.

8.3.2 Suppose that f is a Riemann integrable function on $[a, b]$. Suppose that $\epsilon > 0$. Show that there exist $a \leq a_1 < b_1 \leq b$ such that

$$\sup\{f(x) : x \in [a_1, b_1]\} - \inf\{f(x) : x \in [a_1, b_1]\} < \epsilon.$$

Show that f has a point of continuity in $[a, b]$.
Suppose that $f(x) > 0$ for all $x \in [a, b]$. Show that $\int_a^b f(x)\, dx > 0$.

8.3.3 Suppose that f is an integrable function on $[a, b]$ and that ϕ is uniformly continuous on $f([a, b])$. Show that $\phi \circ f$ is Riemann integrable.

8.3.4 Suppose that f is a bounded on $[a, b]$. Show that

$$\overline{\int_a^b} f(x)\, dx = \inf\{\int_a^b g(x)\, dx : g \text{ continuous}, g \geq f\}.$$

8.3.5 Suppose that f is a bounded increasing function on $[a, b]$. Show that

$$\overline{\int_a^b} f(x)\, dx =$$

$$\inf\{\int_a^b g(x)\, dx : g \text{ continuous and strictly increasing}, g \geq f\}.$$

8.4 Algebraic properties of the Riemann integral

Here are some straightforward results about upper and Riemann integrals.

Proposition 8.4.1 *Suppose that f and g are bounded functions on $[a, b]$, and that $c \geq 0$.*

(i) $\underline{\int_a^b} f(x) + g(x)\, dx \geq \underline{\int_a^b} f(x)\, dx + \underline{\int_a^b} g(x)\, dx.$

(ii) $\overline{\int_a^b} f(x) + g(x)\, dx \leq \overline{\int_a^b} f(x)\, dx + \overline{\int_a^b} g(x)\, dx.$

(iii) $\underline{\int_a^b} cf(x)\, dx = c\underline{\int_a^b} f(x)\, dx$ *and* $\overline{\int_a^b} cf(x)\, dx = c\overline{\int_a^b} f(x)\, dx.$

(iv) $\underline{\int_a^b} (-f(x))\, dx = -\overline{\int_a^b} f(x)\, dx$ *and* $\overline{\int_a^b} (-f(x))\, dx = -\underline{\int_a^b} f(x)\, dx.$

(v) *If $f(x) \leq g(x)$ for all $x \in [a, b]$ then*

$$\underline{\int_a^b} f(x)\, dx \leq \underline{\int_a^b} g(x)\, dx \text{ and } \overline{\int_a^b} f(x)\, dx \leq \overline{\int_a^b} g(x)\, dx.$$

Proof If $h \in L_f$ and $k \in L_g$ then $h + k \in L_{f+g}$. Thus

$$\underline{\int_a^b} f(x) + g(x)\, dx \geq \int_a^b h(x) + k(x)\, dx = \int_a^b h(x)\, dx + \int_a^b k(x)\, dx.$$

Taking the suprema over L_f and L_g, we obtain the first result. The rest are just as easy. □

Corollary 8.4.2 *(i) If f and g are Riemann integrable and $c \in \mathbf{R}$ then $f + g$ and cf are Riemann integrable, and*

$$\int_a^b (f(x) + g(x))\, dx = \int_a^b f(x)\, dx + \int_a^b g(x)\, dx,$$

$$\int_a^b cf(x)\, dx = c \int_a^b f(x)\, dx.$$

(ii) If $f(x) \leq g(x)$ for all $x \in [a,b]$ then $\int_a^b f(x)\, dx \leq \int_a^b g(x)\, dx$.

Proof (i) We have

$$\overline{\int_a^b} (f(x) + g(x))\, dx \geq \overline{\int_a^b} f(x)\, dx + \overline{\int_a^b} g(x)\, dx$$

$$= \underline{\int_a^b} f(x)\, dx + \underline{\int_a^b} g(x)\, dx$$

$$\geq \underline{\int_a^b} (f(x) + g(x))\, dx \geq \overline{\int_a^b} (f(x) + g(x))\, dx,$$

and so they are all equal. Scalar multiplication is even easier.

(ii) This follows directly from (v). □

When f is continuous, we can say more.

Proposition 8.4.3 *Suppose that f is a non-negative continuous function on $[a,b]$ and that $\int_a^b f(x)\, dx = 0$. Then $f(x) = 0$ for all $x \in [a,b]$.*

Proof Suppose not, and suppose that $f(c) > 0$ for some $c \in [a,b]$. There exists $\delta > 0$ such that if $x \in (c - \delta, c + \delta) \cap [a,b]$ then $|f(x) - f(c)| < f(c)/2$. Choose $\max(a, c - \delta) < x_1 < x_2 < \min(b, c + \delta)$. Then $f(x) > f(c)/2$ for $x \in (x_1, x_2]$. Let $h(x) = (f(c)/2)\chi_{(x_1, x_2]}$. Then $f(x) \geq h(x)$ for all $x \in [a,b]$, so that

$$\int_a^b f(x)\, dx \geq \int_a^b h(x)\, dx = (x_2 - x_1)f(c)/2 > 0. \qquad \square$$

Corollary 8.4.4 *Suppose that f and g are continuous functions on $[a,b]$ and that $f(x) \geq g(x)$ for all $x \in [a,b]$. If $\int_a^b f(x)\, dx = \int_a^b g(x)\, dx$, then $f(x) = g(x)$ for all $x \in [a,b]$.*

Proof The function $h = f - g$ is continuous and non-negative, and $\int_a^b h(x)\, dx = 0$. Thus $f(x) - g(x) = 0$ for all $x \in [a,b]$. □

Recall that if f is a real-valued function on a set S then $f^+(s) = (f(s))^+ = \max(f(s), 0)$, $f^-(s) = (f(s))^- = \max(-f(s), 0)$ and $|f|(s) = |f(s)|$.

Theorem 8.4.5 *Suppose that f and g are Riemann integrable functions on $[a, b]$. Then $f^+, f^-, |f|, f^2$ and fg are all Riemann integrable.*

Proof We use Corollary 8.3.5. Since $\Omega(f^+, I) \leq \Omega(f, I)$, it follows from Corollary 8.3.5 that f^+ is Riemann integrable. Similarly, f^- is Riemann integrable, and so therefore is $|f| = f^+ + f^-$.

Next we consider f^2. Let $M = \sup\{f(x) : x \in [a, b]\}$. Suppose that $\epsilon > 0$. Let $\eta = \epsilon/(2M + 1)$. By Corollary 8.3.5, there exist a dissection $D = \{a = x_0 < \cdots < x_k = b\}$ of $[a, b]$ and a partition $G \cup B$ of $\{1, \ldots, k\}$ such that

$$\Omega(f, I_j) \leq \eta \text{ for } j \in G \quad \text{and} \quad \sum_{j \in B} l(I_j) < \eta,$$

where $I_1, \ldots I_k$ are the intervals of the dissection. Then $\sum_{j \in B} l(I_j) < \epsilon$. Since $|s^2 - t^2| = |s + t|.|s - t|$, it follows that $\Omega(f^2, I_j) \leq 2M\eta < \epsilon$ for $j \in G$, and so f^2 is Riemann integrable.

Finally, since $fg = \frac{1}{2}((f + g)^2 - f^2 - g^2)$, fg is Riemann integrable. [This last trick is called *polarization*.] □

Corollary 8.4.6 *If f is Riemann integrable on $[a, b]$ then*

$$\left| \int_a^b f(x) \, dx \right| \leq \int_a^b |f(x)| \, dx.$$

Proof For $-|f| \leq f \leq |f|$. □

Exercises

8.4.1 Give an example of a function on $[0, 1]$ which is not Riemann integrable, but for which $|f|$ and f^2 are Riemann integrable.

8.4.2 Suppose that f and g Riemann integrable on $[a, b]$. By considering the function $(f + \lambda g)^2$, for suitable λ, or otherwise, establish *Schwarz's inequality*:

$$\int_a^b f(x)g(x) \, dx \leq \left(\int_a^b f(x)^2 \, dx \right)^{1/2} \left(\int_a^b g(x)^2 \, dx \right)^{1/2}.$$

8.4.3 Suppose that f and g are Riemann integrable on $[a, b]$, and that p and q are conjugate indices. Show that $|f|^p$ and $|g|^q$ are Riemann integrable, and establish *Hölder's inequality for integrals*:

$$\left| \int_a^b f(x)g(x)\, dx \right| \le \int_a^b |f(x)g(x)|\, dx$$

$$\le \left(\int_a^b |f(x)|^p\, dx \right)^{1/p} \left(\int_a^b |g(x)|^q\, dx \right)^{1/q}.$$

8.5 The fundamental theorem of calculus

We have introduced the Riemann integral as the measure of an area under a curve. It also acts as the inverse of differentiation.

Proposition 8.5.1 *Suppose that f is a bounded function on $[a, b]$ and that $a < c < b$. Then f is Riemann integrable on $[a, b]$ if and only if it is Riemann integrable on $[a, c]$ and $[c, b]$. If so $\int_a^b f(x)\, dx = \int_a^c f(x)\, dx + \int_c^b f(x)\, dx$.*

Proof This is an easy exercise for the reader. □

If $a < b$ and f is Riemann integrable on $[a, b]$, we write $\int_b^a f(x)\, dx = -\int_a^b f(x)\, dx$. Thus the formula above can also be written as $\int_a^c f(x)\, dx = \int_a^b f(x)\, dx + \int_b^c f(x)\, dx$.

Theorem 8.5.2 (The fundamental theorem of calculus) *(i) Suppose that f is Riemann integrable on $[a, b]$. Set $F(t) = \int_a^t f(x)\, dx$, for $a \le t \le b$. F is continuous on $[a, b]$. If f is continuous at t then F is differentiable at t, and $F'(t) = f(t)$. (If $t = a$ or b, then F has a one-sided derivative.)*
(ii) Suppose that f is differentiable on $[a, b]$ (with one sided derivatives at a and b). If f' is Riemann integrable then $f(x) = f(a) + \int_a^x f'(t)\, dt$ for $a \le x \le b$.

Proof (i) The function f is bounded, and so there exists M such that $|f(x)| \le M$ for all $x \in [a, b]$. Then if $a \le t < s \le b$,

$$|F(t) - F(s)| = \left| \int_t^s f(x)\, dx \right| \le \int_t^s |f(x)|\, dx \le M(s - t),$$

from which it follows that F is continuous.

Suppose that f is continuous at t. Suppose that $\epsilon > 0$. There exists $\delta > 0$ such that if $|s - t| < \delta$ and $s \in [a, b]$ then $|f(s) - f(t)| < \epsilon$. Now

$$\int_t^s f(x) - f(t)\, dx = F(s) - F(t) - f(t)(s - t),$$

so that if $0 < |s - t| < \delta$ and $s \in [a, b]$ then

$$|F(s) - F(t) - f(t)(s - t)| \leq \left| \int_t^s f(x) - f(t)\, dx \right|$$

$$\leq \int_t^s |f(x) - f(t)|\, dx < \epsilon |s - t|,$$

since $|x - t| \leq |s - t| < \delta$ for $x \in [t, s]$. Thus F is differentiable at t, with derivative $f(t)$.

(ii) Let $(D_r)_{r=1}^{\infty}$ be a sequence of dissections of $[a, x]$, with $\delta(D_r) \to 0$. Suppose that D_r has points of dissection $a = x_{r,0} < \cdots < x_{r,k_r} = x$. By the mean-value theorem, for each $1 \leq j \leq k_r$ there exists $y_{r,j} \in [x_{r,j-1}, x_{r,j}]$ such that $f(x_{r,j}) - f(x_{r,j-1}) = f'(y_{r,j})(x_{r,j} - x_{r,j-1})$. Thus

$$\sum_{j=1}^{k_r} f'(y_{r,j})(x_{r,j} - x_{r,j-1}) = \sum_{j=1}^{k_r} (f(x_{r,j}) - f(x_{r,j-1})) = f(x) - f(a).$$

The result now follows from Corollary 8.3.8.   □

Thus if f is a continuous function on $[a, b]$, the integral enables us to solve the first-order differential equation $F'(x) = f(x)$, with boundary condition $F(a) = 0$. Any function F which satisfies $F' = f$ is called a *primitive*, or *anti-derivative*, of f. If F and G are primitives of f, then $(F - G)'(x) = F'(x) - G'(x) = 0$ on $[a, b]$, and so $F = G + c$, where c is a constant.

It is important to note that in Part (ii) of the theorem, we require f' to be Riemann integrable. In general, the primitive of a function is well-behaved, whereas the derivative, where it exists, need not be. The fundamental theorem of calculus allows us to calculate many integrals without difficulty. Here are some examples.

1. Suppose that $a < b$, that $k \neq 0$ and that $c > 0$. Then

$$\int_a^b e^{kx}\, dx = \int_a^b \frac{d}{dt}\left(\frac{e^{kt}}{k}\right) dt = \frac{1}{k}(e^{kb} - e^{ka}),$$

$$\int_a^b c^x\, dx = \int_a^b \frac{d}{dt}\left(\frac{c^t}{\log c}\right) dt = \frac{c^b - c^a}{\log c};$$

$$\int_a^b \cos x\, dx = \int_a^b \frac{d}{dt} \sin t\, dt = \sin b - \sin a,$$

$$\int_a^b \sin x \, dx = \int_a^b \frac{d}{dt}(-\cos t) \, dt = \cos a - \cos b,$$

$$\int_a^b \frac{dx}{1+x^2} = \int_a^b \frac{d}{dt}\tan^{-1}t \, dt = \tan^{-1}b - \tan^{-1}a.$$

2. If $0 < a < b$ then

$$\int_a^b \frac{dx}{x} = \int_a^b \frac{d}{dt}\log t \, dt = \log b - \log a = \log\frac{b}{a}.$$

3. If $-1 \le a < b \le 1$ then

$$\int_a^b \frac{dx}{\sqrt{1-x^2}} = \int_a^b \frac{d}{dt}\sin^{-1}t \, dt = \sin^{-1}b - \sin^{-1}a = \cos^{-1}a - \cos^{-1}b.$$

We use the fundamental theorem of calculus to obtain the following *change of variables* formula.

Theorem 8.5.3   *Suppose that g is a differentiable increasing function on $[a,b]$, that g' is Riemann integrable and that f is continuous on $[g(a), g(b)]$. Then*

$$\int_{g(a)}^{g(b)} f(y) \, dy = \int_a^b f(g(x))g'(x) \, dx.$$

Proof   Let F be a primitive for f, so that F is continuously differentiable on $[g(a), g(b)]$. Thus $F \circ g$ is differentiable on $[a,b]$, and its derivative $F'(g(x))g'(x) = f(g(x))g'(x)$ is Riemann integrable. Hence

$$F(g(b)) - F(g(a)) = (F \circ g)(b) - (F \circ g)(a) = \int_a^b f(g(x))g'(x) \, dx.$$

But

$$F(g(b)) - F(g(a)) = \int_{g(a)}^{g(b)} F'(y) \, dy = \int_{g(a)}^{g(b)} f(y) \, dy. \qquad \square$$

The next result, which is occasionally useful, concerns certain infinite Taylor series.

Proposition 8.5.4   *Suppose that f is an infinitely differentiable function on $[a,b)$ (with one-sided derivatives at a) and that $f^{(n)}(a) \ge 0$ for all $n \in \mathbf{N}$. Suppose that the Taylor series*

$$f'(a) + \sum_{j=1}^{\infty} \frac{f^{(j+1)}(a)}{j!}(x-a)^j$$

for $f'(x)$ converges to $f'(x)$ for each $x \in (a, b)$. Then

$$f(x) = f(a) + \sum_{j=1}^{\infty} \frac{f^{(j)}(a)}{n!}(x-a)^j$$

for each $x \in (a, b)$. If f is bounded then $f(b-) = \lim_{x \nearrow b} f(x)$ exists, and

$$f(b-) = f(a) + \sum_{j=1}^{\infty} \frac{f^{(j)}(a)}{j!}(b-a)^j.$$

Proof Let

$$s_n(x) = f'(a) + \sum_{j=1}^{n} \frac{f^{(j+1)}(a)}{j!}(x-a)^j, \quad u_n(x) = f(a) + \sum_{j=1}^{n} \frac{f^{(j)}(a)}{j!}(x-a)^j.$$

If $a < t < x$ then

$$0 \le f'(t) - s_n(t) = \sum_{j=n+1}^{\infty} \frac{f^{(j+1)}(a)}{j!}(t-a)^j$$

$$\le \sum_{j=n+1}^{\infty} \frac{f^{(j+1)}(a)}{j!}(x-a)^j = f'(x) - s_n(x)$$

so that

$$0 \le f(x) - u_{n+1}(x) = \int_a^x (f'(t) - s_n(t))\, dt \le (x-a)(f'(x) - s_n(x)).$$

Since $s_n(x) \to f'(x)$ as $n \to \infty$, it follows that $u_n(x) \to f(x)$ as $n \to \infty$.

Since $f'(x) \ge 0$ for $x \in [a, b)$, f is an increasing function, so that if f is bounded then $f(b-) = \lim_{x \nearrow b} f(x)$ exists. The sequence $(u_n(b))_{n=1}^{\infty}$ is increasing. Since each of the polynomials u_n is continuous, it follows that $f(b-) \ge u_n(b)$, for $n \in \mathbf{N}$. But

$$f(b-) = \sup_{a \le x < b} f(x) = \sup_{a \le x < b} \sup_{n \in \mathbf{N}} u_n(x) \le \sup_{n \in \mathbf{N}} u_n(b),$$

and so $u_n(b) \to f(b-)$ as $n \to \infty$. $\qquad \square$

Here is a particular application.

Corollary 8.5.5 *If $0 \le x \le 1$ then*

$$\sin^{-1}(x) = x + \sum_{j=1}^{\infty} \binom{2j}{j} \frac{x^{2j+1}}{2^{2j}(2j+1)}.$$

Proof For

$$\frac{d}{dx}\sin^{-1}x = (1-x^2)^{-1/2} = 1 + \sum_{j=1}^{\infty}\binom{2j}{j}\frac{x^{2j}}{2^{2j}}.$$

□

Exercises

8.5.1 Suppose that f is a Riemann integrable function on $[a, b]$. Show that the set of points of continuity of f is dense in $[a, b]$.

8.5.2 It is not easy to give conditions on a differentiable function f to ensure that f' is Riemann integrable. One sufficient condition is that f' is continuous. Use part (i) of the fundamental theorem of calculus to prove part (ii), in the case where f' is a continuous function on $[a, b]$.

8.5.3 Let

$$L(x) = \int_1^x \frac{dt}{t} \text{ for } 0 < x < \infty.$$

Show that L is a continuous strictly increasing mapping of $(0, \infty)$ onto **R**. Let E be the inverse function. Taking these as the definitions of the logarithmic and exponential functions, establish the basic results (i)–(x) of Section 7.4.

8.5.4 Suppose that f is a continuous convex function on $[a, b]$. Show that

$$f(b) - f(a) = \int_a^b f'(x+)\,dx = \int_a^b f'(x-)\,dx.$$

8.5.5 Suppose that f is a continuous periodic function on **R**, with period 1. Suppose that $a > 0$. Show that $\int_0^1 f(x+a) - f(x)\,dx = 0$. Deduce that there exist $0 \le x_1 < x_2 < 1$ such that $f(x_1 + a) = f(x_1)$ and $f(x_2 + a) = f(x_2)$.

8.5.6 Show that if $-1 < x < 1$ then

$$\tan^{-1}(x) = x - \frac{x^3}{3} + \frac{x^5}{5} - \cdots + \frac{(-1)^{n-1}x^{2n-1}}{2n-1} + \int_0^x \frac{(-t^2)^n}{1+t^2}\,dt,$$

and show that

$$\tan^{-1}(x) = \sum_{n=0}^{\infty}(-1)^n\frac{x^{2n+1}}{(2n+1)}.$$

Why is this the Taylor series expansion of $\tan^{-1}(x)$?

8.5.7 Show that $\tan(2x) = 2\tan(x)/(1-\tan^2(x))$. Deduce that

$$\pi/4 = 4\tan^{-1}(1/5) - \tan^{-1}(1/239).$$

Use this to calculate π to five decimal places.

$$[4/239 = 0.001673640....]$$

8.6 Some mean-value theorems

Here is an easy mean-value theorem.

Theorem 8.6.1 *Suppose that f is a continuous function on $[a, b]$. Then there exists $a < c < b$ such that $\int_a^b f(x)\, dx = (b - a)f(c)$.*

Proof Let $F(t) = \int_a^t f(x)\, dx$, for $a \leq t \leq b$. Then $\int_a^b f(x)\, dx = F(b) - F(a)$ and $F'(t) = f(t)$, and so the result follows from the mean-value theorem.

$\square$

The proof of the next theorem is considerably harder; it is similar to the proof of Dirichlet's test.

Theorem 8.6.2 (Bonnet's mean-value theorem) *Suppose that f is a Riemann integrable function on $[a, b]$ and that ϕ is a decreasing non-negative function on $[a, b]$. Then there exists $a < c < b$ such that*

$$\int_a^b \phi(x)f(x)\, dx = \phi(a) \int_a^c f(x)\, dx.$$

Proof Let $F(c) = \int_a^c f(x)\, dx$ for $a \leq c \leq b$, and let

$$\Lambda = \sup\{F(t) : t \in [a, b]\}, \quad \lambda = \inf\{F(t) : t \in [a, b]\}.$$

Since F is a continuous function on $[a, b]$, it is sufficient, by the intermediate value theorem, to show that

$$\phi(a)\lambda \leq \int_a^b f(x)\phi(x)\, dx \leq \phi(a)\Lambda.$$

Suppose that $\epsilon > 0$. There exists a dissection D, with points of dissection $a = x_0 < \cdots < x_k = b$, such that $S_D(f) < s_D(f) + \epsilon$ and

$$\int_a^b \phi(x)f(x)\, dx - \epsilon < \sum_{j=1}^k \phi(x_{j-1})f(x_{j-1})(x_j - x_{j-1}) < \int_a^b \phi(x)f(x)\, dx + \epsilon.$$

Let $a_j = f(x_{j-1})(x_j - x_{j-1})$, and let $b_j = a_1 + \cdots + a_j$, for $1 \le j \le k$; let $b_0 = 0$. Then

$$\sum_{j=1}^{k} \phi(x_{j-1}) f(x_{j-1})(x_j - x_{j-1}) = \sum_{j=1}^{k} \phi(x_{j-1}) a_j = \sum_{j=1}^{k} \phi(x_{j-1})(b_j - b_{j-1})$$

$$= \sum_{j=1}^{k-1} (\phi(x_{j-1}) - \phi(x_j)) b_j + \phi(x_{k-1}) b_k.$$

Now

$$\lambda - \epsilon \le F(x_j) - \epsilon \le \sum_{i=1}^{j} M_i(f)(x_i - x_{i-1}) - \epsilon \le \sum_{i=1}^{j} m_i(f)(x_i - x_{i-1}) \le b_j$$

and

$$b_j \le \sum_{i=1}^{j} M_i(f)(x_i - x_{i-1}) \le \sum_{i=1}^{j} m_i(f)(x_i - x_{i-1}) + \epsilon \le F(x_j) + \epsilon \le \Lambda + \epsilon.$$

Since ϕ is non-negative decreasing and $\lambda - \epsilon \le b_j \le \Lambda + \epsilon$, it follows that

$$\phi(a)(\lambda - \epsilon) = \sum_{j=1}^{k-1} (\phi(x_{j-1}) - \phi(x_j))(\lambda - \epsilon) + \phi(x_{k-1})(\lambda - \epsilon)$$

$$\le \sum_{j=1}^{k-1} (\phi(x_{j-1}) - \phi(x_j)) b_j + \phi(x_{k-1}) b_j$$

$$= \sum_{j=1}^{k} \phi(x_{j-1}) f(x_{j-1})(x_j - x_{j-1}) < \int_{a}^{b} \phi(x) f(x) \, dx + \epsilon.$$

A similar argument shows that

$$\phi(a)(\Lambda + \epsilon) > \int_{a}^{b} \phi(x) f(x) \, dx - \epsilon,$$

and so

$$\phi(a)(\lambda - \epsilon) - \epsilon \le \int_{a}^{b} \phi(x) f(x) \, dx \le \phi(a)(\Lambda + \epsilon) + \epsilon.$$

Since this holds for all $\epsilon > 0$, the result follows. $\qquad\qquad\qquad\square$

There is also a version for increasing functions.

Corollary 8.6.3 *Suppose that f is a Riemann integrable function on* $[a, b]$
and that ϕ *is an increasing non-negative function on* $[a, b]$. *Then there exists*
$a < c < b$ *such that*

$$\int_a^b \phi(x)f(x)\,dx = \phi(b)\int_c^b f(x)\,dx.$$

Proof Consider the function g on $[-b, -a]$ defined by $g(x) = f(-x)$.   □

We can also drop the non-negativity condition.

Corollary 8.6.4 (Du Bois–Reymond's mean-value theorem)   *Suppose that*
f is a Riemann integrable function on $[a, b]$ *and that* ψ *is a monotonic*
function on $[a, b]$. *Then there exists* $a < c < b$ *such that*

$$\int_a^b \psi(x)f(x)\,dx = \psi(a)\int_a^c f(x)\,dx + \psi(b)\int_c^b f(x)\,dx.$$

Proof If ψ is decreasing, set $\phi(x) = \psi(x) - \psi(b)$. If ψ is increasing, set
$\phi(x) = -\psi(x) + \psi(b)$.   □

Exercises

8.6.1 Suppose that $0 < a < b$. Show that

$$\left| \int_a^b \frac{\sin u}{u}\,du \right| \le \pi.$$

8.6.2 Suppose that $0 < a < b$ and that $K > 0$. Show that

$$\left| \int_a^b \frac{\sin Ku}{u}\,du \right| \le \pi.$$

8.6.3 Suppose that $0 < s < t \le \pi/2$ and that $K > 0$. Show that

$$\left| \frac{1}{2\pi} \int_s^t \frac{\sin Ku}{\sin u}\,du \right| < 1.$$

8.6.4 Suppose that ϕ is a non-negative increasing function on $(0, t]$, where
$0 < t < \pi/2$, that $\phi(u) \to 0$ as $u \searrow 0$, and that $K > 0$. Show that

$$\left| \frac{1}{2\pi} \int_0^t \frac{\phi(u)\sin Ku}{\sin u}\,du \right| \le \phi(t).$$

8.7 Integration by parts

Let us apply the fundamental theorem of calculus to the product of two functions.

Theorem 8.7.1 (Integration by parts) *Suppose that f is continuous on $[a, b]$ that g is continuous and differentiable on $[a, b]$, and that g' is Riemann integrable. Let F be a primitive for f. Then*

$$\int_a^b f(x)g(x)\,dx = F(b)g(b) - F(a)g(a) - \int_a^b F(x)g'(x)\,dx.$$

Proof Since $F' = f$, the function $F(x)g(x)$ has derivative $f(x)g(x) + F(x)g'(x)$, which is Riemann integrable. Applying the fundamental theorem of calculus,

$$F(b)g(b) = F(a)g(a) + \int_a^b (F(x)g(x))'\,dx$$

$$= F(a)g(a) + \int_a^b f(x)g(x)\,dx + \int_a^b F(x)g'(x)\,dx,$$

from which the result follows.   □

The difference $F(b)g(b) - F(a)g(a)$ is frequently written as $[F(x)g(x)]_a^b$.

As an example, let us calculate $\int_0^a x \sin x\,dx$, where $a > 0$. Set $f(x) = \sin x$ and $g(x) = x$. Then we can take $F(x) = -\cos x$, and so

$$\int_0^\pi x \sin x\,dx = (-\cos a).a - (-1).0 + \int_0^a \cos x\,dx = \sin a - a \cos a.$$

Although Theorem 8.7.1 is very easy, it is also extremely powerful. It provides a continuous analogue of the argument used to establish Dirichlet's test. To illustrate this, let us use the integration by parts formula to prove a version of Bonnet's mean-value theorem (where rather stronger conditions are imposed).

Theorem 8.7.2 *Suppose that f is a Riemann integrable function on $[a, b]$ and that ϕ is a decreasing, continuous, differentiable, non-negative function on $[a, b]$, and that ϕ' is Riemann integrable. Then there exists $a < c < b$ such that*

$$\int_a^b \phi(x)f(x)\,dx = \phi(a) \int_a^c f(x)\,dx.$$

Proof Let $F(c) = \int_a^c f(x)\,dx$ for $a \le c \le b$, and let $\Lambda = \sup\{F(t) : t \in [a, b]\}$, $\lambda = \inf\{F(t) : t \in [a, b]\}$. As in Theorem 8.6.2, since F is a continuous function on $[a, b]$, it is sufficient, by the intermediate value theorem, to show that

$$\phi(a)\lambda \le \int_a^b f(x)\phi(x)\,dx \le \phi(a)\Lambda.$$

Integrating by parts,

$$\int_a^b f(x)\phi(x)\,dx = F(b)\phi(b) - \int_a^b F(x)\phi'(x)\,dx.$$

Thus, since $\phi' \le 0$,

$$\lambda\phi(a) = \lambda\phi(b) - \lambda\int_a^b \phi'(x)\,dx \le F(b)\phi(b) - \int_a^b F(x)\phi'(x)\,dx$$

$$\le \Lambda\phi(b) - \Lambda\int_a^b \phi'(x)\,dx = \Lambda\phi(a). \qquad \square$$

Integration by parts enables us to give another version of Taylor's theorem with remainder.

Theorem 8.7.3 (Taylor's theorem with integral remainder)   *Suppose that f is k times continuously differentiable on $[a, b]$. Then*

$$f(b) = f(a) + \sum_{j=1}^{n-1} \frac{(b-a)^j}{j!} f^{(j)}(a) + \frac{1}{(n-1)!}\int_a^b (b-x)^{n-1} f^{(n)}(x)\,dx.$$

Proof   By induction on n. It is true for $n = 0$, by the fundamental theorem of calculus. Suppose that it is true for n, and that f is $(n+1)$-times continuously differentiable. Then it is n times differentiable, and so by the inductive hypothesis

$$f(b) = f(a) + \sum_{j=1}^{n-1} \frac{(b-a)^j}{j!} f^{(j)}(a) + \frac{1}{(n-1)!}\int_a^b (b-x)^{n-1} f^{(n)}(x)\,dx.$$

Now $-(b-x)^n/n$ is a primitive for $(b-x)^{n-1}$, and so, integrating by parts,

$$\int_a^b (b-x)^{n-1} f^{(n)}(x)\,dx =$$

$$0 - \left(\frac{-(b-a)^n}{n} f^{(n)}(a)\right) - \left(\int_a^b \frac{-(b-x)^n}{n} f^{(n+1)}(x)\,dx\right).$$

Thus

$$\frac{1}{(n-1)!} \int_a^b (b-x)^{n-1} f^{(n)}(x)\, dx =$$

$$\frac{1}{n!}(b-a)^n f^{(n)}(a) + \frac{1}{n!} \int_a^b (b-x)^n f^{(n+1)}(x)\, dx,$$

and so

$$f(b) = f(a) + \sum_{j=1}^{n-1} \frac{(b-a)^j}{j!} f^{(j)}(a) + \frac{1}{(n-1)!} \int_a^b (b-x)^{n-1} f^{(n)}(x)\, dx. \qquad \square$$

This form of Taylor's theorem differs from the earlier ones, in that it gives the remainder $r_n(b)$ explicitly as a function of b.

Exercises

8.7.1 Show that

$$\int_0^{\pi/2} x^n \sin x\, dx = n \left(\frac{\pi}{2}\right)^{n-1} - n(n-1) \int_0^{\pi/2} x^{n-2} \sin x\, dx \text{ for } n \geq 2.$$

8.7.2 Suppose that f is continuous and differentiable on $[a, b]$, and that f' is Riemann integrable. Show that

$$\int_a^b f(x)\, dx = (b-a)f(b) - \int_a^b (x-a)f'(x)\, dx.$$

Suppose that $m, n \in \mathbf{N}$ and that $a \leq m, n \leq b$. Establish *Euler's summation formula*:

$$\sum_{j=m+1}^n f(j) = \int_m^n f(x)\, dx + \int_m^n (x - [x])f'(x)\, dx.$$

(Here $[x]$ is the integral part of x; the least integer not greater than x.)

8.7.3 Use Taylor's theorem with integral remainder to give another proof of the binomial theorem.

8.8 Improper integrals and singular integrals

So far, we have considered integrals of bounded functions defined on a bounded closed interval. How can we deal with unbounded functions, or different sorts of interval? There are various limiting processes that we can use; the resulting integrals are called *improper integrals* and *singular integrals*.

First we consider a function f which is defined on a semi-infinite interval $[a, \infty)$ and whose restriction to each finite subinterval $[a, b]$ is Riemann integrable. If $\int_a^b f(x)\, dx$ tends to a limit l as $b \to \infty$, we set $l = \int_a^\infty f(x)\, dx$; this is an *improper integral*. It may also happen that $\int_a^b f(x)\, dx \to \infty$ as $b \to \infty$; in this case we write $\int_a^\infty f(x)\, dx = \infty$. For example, if f is non-negative, then the function $I(b) = \int_a^b f(x)\, dx$ is an increasing function on $[a, \infty)$, and so either $I(b)$ tends to a finite limit as $b \to \infty$, which is the integral $\int_a^\infty f(x)\, dx$, or $I(b) \to \infty$ as $b \to \infty$, in which case $\int_a^\infty f(x)\, dx = \infty$.

Let us give some examples.

First,

$$\int_0^\infty (1 + x^2)^{-1}\, dx = \lim_{b \to \infty} \int_0^b (1 + x^2)^{-1}\, dx = \lim_{b \to \infty} \tan^{-1}(b) = \pi/2.$$

Secondly, the function $\operatorname{sinc} x = (\sin x)/x$ is an important function in the theory of signal processing. What can we say about $\int_0^\infty \operatorname{sinc} x\, dx$? There is no problem at 0, since $\operatorname{sinc} x \to 1$ as $x \to 0$. Let

$$I_n = \left| \int_{(n-1)\pi}^{n\pi} \operatorname{sinc} x\, dx \right| = \int_{(n-1)\pi}^{n\pi} \frac{|\sin x|}{x}\, dx.$$

Then $I_1 > I_2 > \cdots$, and $I_n \to 0$ as $n \to \infty$. By the alternating series test, the limit

$$\lim_{n \to \infty} \int_0^{n\pi} \operatorname{sinc} x\, dx = \lim_{n \to \infty} \sum_{j=1}^n (-1)^{j+1} I_j$$

exists. Further $\int_{\lfloor b \rfloor \pi}^{b\pi} \operatorname{sinc} x\, dx \to 0$ as $b \to \infty$, so that

$$\int_0^\infty \operatorname{sinc} x\, dx = \lim_{b \to \infty} \int_0^b \operatorname{sinc} x\, dx$$

exists. But note that

$$|I_n| \geq \int_{(n-2/3)\pi}^{(n-1/3)\pi} |\operatorname{sinc} \pi x|\, dx \geq \frac{\pi}{6n},$$

so that $\int_0^\infty |\operatorname{sinc} x|\, dx = \infty$.

This last phenomenon shows that we must proceed with some care. For example, let $f(x) = (\sin x)/\sqrt{x}$. Then arguments just like those for sinc show that the improper integral $\int_0^\infty f(x)\, dx$ exists. But

$$\lim_{b \to \infty} \int_0^b f^2(x)\, dx = \lim_{b \to \infty} \int_0^b \frac{\sin^2 x}{x}\, dx = \infty,$$

since

$$\int_{(n-1)\pi}^{n\pi} \frac{\sin^2 x}{x}\, dx \geq \frac{1}{n\pi} \int_{(n-1)\pi}^{n\pi} \sin^2 x\, dx = \frac{1}{2n}$$

and $\sum_{n=1}^{\infty}(1/2n) = \infty$.

Theorem 8.8.1 (The integral test) *Suppose that f is a decreasing non-negative function on $[0, \infty)$ and that $f(x) \to 0$ as $x \to \infty$.*

(i) The series $\sum_{j=0}^{\infty} f(j)$ converges if and only if $\lim_{n\to\infty} \int_0^n f(x)\, dx$ exists, and then

$$\sum_{j=1}^{\infty} f(j) \leq \int_0^{\infty} f(x)\, dx \leq \sum_{j=0}^{\infty} f(j)$$

(ii) If

$$C_n = \sum_{j=0}^{n-1} f(j) - \int_0^n f(x)\, dx \quad \text{and} \quad D_n = \sum_{j=0}^{n} f(j) - \int_0^n f(x)\, dx,$$

so that $C_n \leq D_n$, then $(C_n)_{n=1}^{\infty}$ is an increasing sequence and $(D_n)_{n=1}^{\infty}$ is a decreasing sequence, and the two sequences converge to a common limit G.

Proof (i) Let us set $g(x) = f(\lfloor x \rfloor)$ and $h(x) = f(\lfloor x \rfloor + 1)$, so that $h \leq f \leq g$. Now

$$\int_0^n g(x)\, dx = \sum_{j=0}^{n-1} f(j) \quad \text{and} \quad \int_0^n h(x)\, dx = \sum_{j=1}^{n} f(j),$$

and so

$$\sum_{j=1}^{n} f(j) \leq \int_0^n f(x)\, dx \leq \sum_{j=0}^{n-1} f(j).$$

Thus either the sum and the integral both converge, or they both diverge.
(ii) Since

$$C_{n+1} - C_n = \int_n^{n+1} f(n) - f(x)\, dx \geq 0,$$

$(C_n)_{n=1}^{\infty}$ is an increasing sequence; similarly

$$D_{n+1} - D_n = \int_n^{n+1} f(n+1) - f(x)\, dx \leq 0,$$

so that $(D_n)_{n=1}^{\infty}$ is a decreasing sequence. Finally, $D_n - C_n = f(n)$, so that C_n and D_n converge to a common limit G. $\qquad \square$

If we set $f(x) = \log(1+x)$, we see that $\sum_{j=1}^{n}(1/j) - \log(n+1)$ increases to a constant γ and that $\sum_{j=1}^{n}(1/j) - \log n$ decreases to γ, where $0 < \gamma < 1$. The number γ is called *Euler's constant*; its value is $0.577\cdots$. It is not known if γ is rational, or if it is algebraic; but every instinct suggests that it is transcendental. It is sometimes called *Mascheroni's constant*, since in 1790 Mascheroni calculated it to 32 decimal places, although in fact only the first 19 were correct; in 1878, J.C. Adams calculated it to 260 decimal places. With the use of computers, γ is now known to 10^{10} decimal places.

As another example, let us consider $a_n = 1/(n \log n)$, for $n \geq 2$. Consider $f(x) = 1/(x \log x)$, for $x \geq e$. Then

$$\int_e^n \frac{dx}{x \log x} = \log(\log n) \to \infty,$$

as $n \to \infty$, and so $\sum_{n=2}^{\infty} 1/(n \log n)$ diverges. [Of course no harm is done by starting at 2, and at e.]

As a second example of an improper integral, let us consider a function f which is defined on $\mathbf{R}$ and whose restriction to each finite subinterval $[a, b]$ is Riemann integrable. There are then two ways of proceeding. First we may require that the two improper integrals $\int_0^\infty f(x)\, dx$ and $\int_{-\infty}^0 f(x)\, dx$ (defined in the obvious way) both exist, and then define $\int_{-\infty}^\infty f(x)\, dx$ to be their sum. In this case, we again call the resulting value the *improper* integral. Alternatively, we may simply require that $\lim_{b\to\infty} \int_{-b}^b f(x)\, dx$ exists. In this case, the limit is called the *Cauchy principal value* of the integral or the *singular integral* of the function f, and denote it by $(PV) \int_{-\infty}^\infty f(x)\, dx$. For example, if $f(x) = x/(1+x^2)$ then $\int_0^\infty f(x)\, dx = \infty$ and $\int_{-\infty}^0 f(x)\, dx = -\infty$, so that the improper integral does not exist. On the other hand, f is an odd function, and so the Cauchy principal value is 0. Great caution is needed in handling singular integrals.

Next, it may happen that f is defined on an interval $(a, b]$, that f is not bounded, but that f is bounded and Riemann integrable on every interval $[c, b]$, for $a < c < b$. If $\int_c^b f(x)\, dx$ tends to a limit l as $c \searrow a$, then we set $l = \int_a^b f(x)\, dx$; this is a *singular integral*. For example, let $f(x) = x^{\alpha-1}$, where $x \in (0, 1]$ and $0 < \alpha < 1$. This is unbounded as $x \searrow 0$, but

$$\int_0^1 f(x)\, dx = \lim_{\epsilon\searrow 0} \int_\epsilon^1 f(x)\, dx = \lim_{\epsilon\searrow 0}(1 - \epsilon^\alpha)/\alpha = 1/\alpha.$$

It may also happen that f is defined on a set $[a, b] \backslash \{c\}$, where c is an interior point of the interval $[a, b]$, and that f is bounded and Riemann integrable on each of the intervals $[a, d]$ (with $a < d < c$) and $[e, b]$ (with $c < e < b$), while f is unbounded.

Again we can proceed in two ways. First we can require that the two improper integrals $\int_a^c f(x)\,dx$ and $\int_c^b f(x)\,dx$ both exist, and then define the improper integral $\int_a^b f(x)\,dx$ to be their sum. Alternatively if

$$\lim_{\epsilon \searrow 0} \left(\int_a^{c-\epsilon} f(x)\,dx + \int_{c+\epsilon}^b f(x)\,dx \right)$$

then we again define the limit to be the *Cauchy principal value* of the integral. or the *singular integral* of the function f, and denote it by $(PV) \int_a^b f(x)\,dx$. For example if $f(x) = 1/x$ on $[-1,1] \setminus \{0\}$ then the singular integral $\int_0^1 f(x)\,dx = \infty$, and the improper integral $\int_{-1}^1 f(x)\,dx$ does not exist, whereas the singular integral does exist, with value 0.

It is also possible to consider multiple singularities, and to consider improper singular integrals. In each case, 'caution' should be your watchword: results for Riemann integrable functions do not always extend to improper integrals and singular integrals, and each case should be treated on its merits.

Finally, let us mention that we can extend all these results to complex-valued functions; we simply consider, and integrate, the real and imaginary parts separately.

Exercises

8.8.1 Suppose that f is a real-valued function defined on $[0, \infty)$ which is Riemann integrable on $[0, b]$ for each $0 < b < \infty$. f is said to be *absolutely integrable* if the improper integral $\int_0^\infty |f(x)|\,dx$ exists and is finite. Show that if f is absolutely integrable then the improper integral $\int_0^\infty f(x)\,dx$ exists. [If the improper integral $\int_0^\infty f(x)\,dx$ exists, but f is not absolutely integrable, then f is said to be *conditionally integrable*.] (As with sequences, absolutely integrable functions are relatively well behaved, whereas conditionally integrable functions need to be handled with care.)

8.8.2 Suppose that f and g are real-valued functions defined on $[0, \infty)$ which are Riemann integrable on $[0, b]$ for each $0 < b < \infty$. Suppose also that p and q are conjugate indices for which the improper integrals $\int_0^\infty |f(x)|^p\,dx$ and $\int_0^\infty |g(x)|^q\,dx$ are finite. Show that fg is absolutely integrable, and establish *Hölder's inequality for improper integrals*:

$$\left| \int_0^\infty f(x)g(x)\,dx \right| \le \int_0^\infty |f(x)g(x)|\,dx$$
$$\le \left(\int_0^\infty |f(x)|^p\,dx \right)^{1/p} \cdot \left(\int_0^\infty |g(x)|^q\,dx \right)^{1/q}.$$

8.8.3 Prove carefully that

$$\int_0^\infty \frac{\cos x}{1+x}\,dx = \int_0^\infty \frac{\sin x}{(1+x)^2}\,dx.$$

Show that the first integrand is conditionally integrable, and that the second is absolutely integrable.

8.8.4 (*Euler's summation formula*) Suppose that f is differentiable on $[0, \infty)$.

(a) Suppose that the improper integral $\int_0^\infty f(x)\,dx$ and the improper sum $\sum_{j=1}^\infty f(j)$ both exist. Show that

$$\sum_{j=1}^\infty f(j) = \int_0^\infty f(x)\,dx + \int_0^\infty (x - \lfloor x \rfloor) f'(x)\,dx.$$

(b) Suppose that f is decreasing and that $f(x) \to 0$ as $x \to \infty$. Show that the improper integral $\int_0^\infty (x - \lfloor x \rfloor) f'(x)\,dx$ exists and that

$$\sum_{j=1}^{\lfloor X \rfloor} f(j) - \int_0^X f(x)\,dx \to \int_0^\infty (x - \lfloor x \rfloor) f'(x)\,dx.$$

8.8.5 Suppose that f is a monotonic function on $[0, \pi/2]$. Show that $\int_0^{\pi/2} f(x) \sin nx\,dx \to 0$ as $n \to \infty$.

8.8.6 Establish the identity

$$(\sin 2nx - \sin(2n - 2)x) \cos x = (\cos 2nx + \cos(2n - 2)x) \sin x.$$

Show that

$$\int_0^{\pi/2} \frac{\sin 2nx}{\sin x} \cos x\,dx = \pi/2.$$

Show that $1/\tan x - 1/x$ is a bounded monotonic function on $(0, \pi/2]$. Show that

$$\int_0^{\pi/2} \frac{\sin nx}{x}\,dx \to \int_0^\infty \frac{\sin x}{x}\,dx \text{ as } n \to \infty.$$

Show that

$$\int_0^\infty \frac{\sin x}{x}\,dx = \frac{\pi}{2}.$$

(This is an ingenious proof. We shall see in Volume III that complex analysis avoids the need for such ingenuity.)

8.8.7 Show that

$$1 + \frac{1}{2} + \frac{1}{3} + \cdots + \frac{1}{n} = \int_0^1 \frac{1 - x^n}{1 - x} \, dx = \int_0^n \frac{1 - (1 - t/n)^n}{t} \, dt.$$

8.8.8 The *gamma function* is defined as

$$\Gamma(x) = \int_0^\infty t^{x-1} e^{-t} \, dt \text{ for } 0 < x < \infty.$$

Interpret this as an improper integral, and prove carefully that $\Gamma(x + 1) = x\Gamma(x)$. Deduce that $\Gamma(n + 1) = n!$, for $n \in \mathbf{N}$.

8.8.9 Show that

$$\gamma = \lim_{n \to \infty} \left(\int_0^1 \frac{1 - (1 - t/n)^n}{t} \, dt - \int_1^n \frac{(1 - t/n)^n}{t} \, dt \right).$$

Can we take the limit inside the integral? Yes, it is possible to prove this directly, but the limiting process becomes much clearer when these integrals are treated as Lebesgue integrals.

8.8.10 Prove the following continuous versions of Dirichlet's test and Abel's test.

(a) *Dirichlet's test.* Suppose that ϕ is a decreasing non-negative function on $[0, \infty)$ and that $\phi(x) \to 0$ as $x \to \infty$. Suppose that f is a function on $[0, \infty)$ for which the Riemann integral $F(x) = \int_0^x f(t) \, dt$ exists for all $x \in [0, \infty)$, and for which F is bounded on $(0, \infty)$. Show that the improper Riemann integral $\int_0^\infty \phi(t) f(t) \, dt$ exists.

(b) *Abel's test.* Suppose that ϕ is a decreasing non-negative function on $[0, \infty)$. Suppose that f is a function on $[0, \infty)$ for which the improper Riemann integral $F(x) = \int_0^\infty f(t) \, dt$ exists. Show that the improper Riemann integral $\int_0^\infty \phi(t) f(t) \, dt$ also exists.

9

Introduction to Fourier series

9.1 Introduction

Recall that a function f defined on $\mathbf{R}$ is *t-periodic* (where $t > 0$) if $f(s+t) = f(s)$ for all $s \in \mathbf{R}$: t is called a *period* of f. If f has period t_0 and $\alpha > 0$ then the function $f(\alpha t)$ has period t_0/α. Thus, by scaling, we can, and shall, restrict attention to 2π-periodic functions. For example, the functions $\cos nt$ and $\sin nt$, for $n \in \mathbf{Z}^+$, are examples of 2π-periodic functions. More generally, a function of the form

$$p(t) = a_0/2 + \sum_{j=1}^{m} a_j \cos jt + \sum_{j=1}^{n} b_j \sin jt,$$

where $a_j, b_j \in \mathbf{R}$, is called a *real trigonometric polynomial*. Trigonometric polynomials are 2π-periodic. The question that Fourier asked, and began to answer, is 'If f is a 2π-periodic function, can it be expressed as a limit of trigonometric polynomials?' This question has led to an enormous amount of mathematics, which has many fundamental applications to the physical sciences. We shall however restrict our attention to the mathematical analysis of Fourier's question.

Suppose that f is a 2π-periodic function. If f is Riemann integrable over the interval $[-\pi, \pi]$, we say that f is *locally Riemann integrable*. Note that this implies that f is Riemann integrable over any bounded interval, and that

$$\int_{-\pi}^{\pi} f(t)\, dt = \int_{t_0-\pi}^{t_0+\pi} f(t)\, dt \text{ for any } t_0 \in \mathbf{R}.$$

The set of all locally Riemann integrable 2π-periodic functions forms a vector space, which we denote by $\mathbf{V}$. An element of $\mathbf{V}$ is bounded. If $f \in \mathbf{V}$ we set

$$\|f\|_1 = \frac{1}{2\pi} \int_{-\pi}^{\pi} |f(t)|\, dt \text{ and } \|f\|_\infty = \sup_{t \in [-\pi,\pi]} |f(t)|.$$

The function $\|.\|_\infty$ is a *norm* on **V**: it satisfies

$$\|f + g\|_\infty \leq \|f\|_\infty + \|g\|_\infty,$$
$$\|\alpha f\|_\infty = |\alpha|.\|f\|_\infty$$

and $\|f\|_\infty = 0$ if and only if $f = 0$,

for $f, g \in \mathbf{V}$ and $\alpha \in \mathbf{R}$. The function $\|.\|_1$ is a *semi-norm* on **V**: it satisfies the first two conditions, but not the third. If $f \in \mathbf{V}$ and $\|f\|_1 = 0$, we say that f is a *trivial function*.

In this chapter, we use the results of analysis that we have obtained so far to obtain results concerning the Fourier analysis of functions in **V**. A more advanced theory requires the theory of Lebesgue integration, and we shall consider this in Volume III.

Suppose that $p(t) = a_0/2 + \sum_{j=1}^m a_j \cos jt + \sum_{j=1}^n b_j \sin jt$ is a real trigonometric polynomial function. Can we find the coefficients a_j and b_j from the knowledge of p? Here, and elsewhere, orthogonality relations play an essential role. Since

$$\cos a \cos b = \tfrac{1}{2}(\cos(a + b) + \cos(a - b)),$$
$$\sin a \sin b = \tfrac{1}{2}(\cos(a + b) - \cos(a - b))$$
$$\text{and } \sin a \cos b = \tfrac{1}{2}(\sin(a + b) + \sin(a - b)),$$

and since

$$\int_{-\pi}^{\pi} \cos mt \, dt = \int_{-\pi}^{\pi} \sin nt \, dt = 0$$

for $m, n \in \mathbf{Z}$, $m \neq 0$, it follows that

$$\int_{-\pi}^{\pi} \cos mt \cos nt \, dt = \int_{-\pi}^{\pi} \sin mt \sin nt \, dt = \int_{-\pi}^{\pi} \sin nt \cos pt \, dt = 0$$

for $m, n, p \in \mathbf{Z}$, $m \neq n$. Since

$$\cos^2 t = \tfrac{1}{2}(1 + \cos 2t) \text{ and } \sin^2 t = \tfrac{1}{2}(1 - \cos 2t)$$

it follows that

$$\int_{-\pi}^{\pi} \cos^2 mt \, dt = \int_{-\pi}^{\pi} \sin^2 nt \, dt = \pi, \text{ for } m, n \in \mathbf{Z}, m \neq 0.$$

Hence

$$a_j = \frac{1}{\pi} \int_{-\pi}^{\pi} p(t) \cos jt \, dt \text{ for } j \in \mathbf{Z}^+,$$

$$b_j = \frac{1}{\pi} \int_{-\pi}^{\pi} p(t) \sin jt \, dt \text{ for } j \in \mathbf{N}.$$

Note that this justifies the fact that the constant term appears as $a_0/2$, rather than a_0, in the definition of a trigonometric polynomial.

This suggests that we consider similar integrals for more general functions than trigonometric polynomial functions. Suppose that $f \in \mathbf{V}$. We set

$$a_j(f) = \frac{1}{\pi} \int_{-\pi}^{\pi} f(t) \cos jt \, dt \text{ for } j \in \mathbf{Z}^+$$

$$b_j(f) = \frac{1}{\pi} \int_{-\pi}^{\pi} f(t) \sin jt \, dt \text{ for } j \in \mathbf{N}.$$

The numbers $a_j(f)$ are the *Fourier cosine coefficients* of f and the numbers $b_j(f)$ are the *Fourier sine coefficients* of f. The *Fourier series* of f is then the formal expression

$$f(t) \sim a_0(f)/2 + \sum_{j=1}^{\infty} a_j(f) \cos jt + \sum_{j=1}^{\infty} b_j(f) \sin jt.$$

We can write this in another form. Let $A_0(f) = a_0(f)$ and $A_j(f) = (a_j(f)^2 + b_j(f)^2)^{\frac{1}{2}}$, for $j \in \mathbf{N}$. There exists $\phi_j(f) \in (-\pi, \pi]$ such that $\cos \phi_j(f) = a_j(f)/A_j(f)$ and $\sin \phi_j(f) = -b_j(f)/A_j(f)$. Then

$$a_j(f) \cos jt + b_j(f) \sin jt = A_j(f) \cos j(t + \phi_j(f)),$$

so that

$$f(t) \sim A_0(f)/2 + \sum_{j=1}^{\infty} A_j(f) \cos j(t + \phi_j(f)).$$

The quantities $A_j(f) \cos j(t + \phi_j(f))$ are the *harmonics* of f. $A_j(f)$ is the *amplitude* of the harmonic and $\phi_j(f)$ is its *phase*. The process of calculating the harmonics is called *harmonic analysis*.

Note that we use the symbol $\sim$ rather than an equality sign. As we shall see in Section 9.6, the sum need not converge, even when f is a continuous function. Our aim will be to see when it does converge to $f(t)$, and to see if there are other ways to approximate f, using the Fourier coefficients.

Let us remark that we can consider the cosine and sine series separately. Let

$$f_e(t) = \tfrac{1}{2}(f(t) + f(-t)) \text{ and } f_o(t) = \tfrac{1}{2}(f(t) - f(-t)).$$

Then f_e is an even function ($f_e(t) = f_e(-t)$), f_o is an odd function ($f_o(t) = -f_o(-t)$), and $f = f_e + f_o$. Further,

$$a_j(f_e) = a_j(f), \qquad b_j(f_e) = 0,$$
$$a_j(f_o) = 0, \qquad b_j(f_e) = b_j(f),$$

so that

$$f_e(t) \sim a_0(f)/2 + \sum_{j=1}^{\infty} a_j(f) \cos jt \text{ and } f_o(t) \sim \sum_{j=1}^{\infty} b_j(f) \sin jt.$$

Note that if f is an even function then

$$a_j(f) = \frac{2}{\pi} \int_0^{\pi} f(t) \cos jt \, dt \text{ for } j \in \mathbf{Z}^+,$$

and that if f is an odd function then

$$b_j(f) = \frac{2}{\pi} \int_0^{\pi} f(t) \sin jt \, dt \text{ for } j \in \mathbf{N}.$$

Suppose that f is any Riemann integrable function on the interval $[0, \pi]$. We can extend f to an even 2π-periodic function by setting $f(t) = f(-t)$ for $t \in [-\pi, 0]$, and setting $f(2\pi k + t) = f(t)$ for $t \in [-\pi, \pi]$ and $k \in \mathbf{Z}$. Note that the extension is continuous on $\mathbf{R}$ if f is continuous on $[0, \pi]$. Thus Fourier cosine series become a tool to consider functions defined on the interval $[0, \pi]$.

9.2 Complex Fourier series

We have seen that Euler's formulae

$$e^{it} = \cos t + i \sin t, \quad \cos t = \frac{1}{2}(e^{it} + e^{-it}), \quad \sin t = \frac{1}{2i}(e^{it} - e^{-it}),$$

are useful, when manipulating formulae involving sine and cosine functions. There are however stronger underlying reasons for considering the complex case. We consider the doubly infinite sequence $(\gamma_n)_{n=-\infty}^{\infty}$ of 2π-periodic functions defined by

$$\gamma_n(t) = e^{int}.$$

The subset $\mathbf{T} = \{z \in \mathbf{C} : |z| = 1\}$ of $\mathbf{C}$ is a group under multiplication. Each function γ_n is a 2π-periodic continuous homomorphism of the additive group $(\mathbf{R}, +)$ into $\mathbf{T}$ (which is surjective if $n \neq 0$), and in fact every such homomorphism is of this form (Exercise 9.2.1). Further, the set $\{\gamma_n : n \in \mathbf{Z}\}$ is a group under point-wise multiplication.

We therefore consider complex-valued 2π-periodic functions. These are functions of a real variable t: if $f = u + iv$, where u and v are the real and imaginary parts of f, then f is continuous, or 2π-periodic, or Riemann integrable over a bounded interval $[a, b]$ if and only if u and v are, and the integral $\int_a^b f(t)\, dt$ is defined as

$$\int_a^b f(t)\, dt = \int_a^b u(t)\, dt + i \int_a^b v(t)\, dt.$$

We therefore consider the complex vector space, which we again denote by $\mathbf{V}$, of complex-valued locally Riemann integrable 2π-periodic functions, and define $\|.\|_1$ and $\|.\|_\infty$, as in the real case. If $f \in \mathbf{V}$ and $\|f\|_1 = 0$, we again say that f is a *trivial function*.

A function of the form $\sum_{j=-n}^{n} c_j \gamma_j$ (where each $c_j \in \mathbf{C}$) is called a *complex trigonometric polynomial*. Fourier's question then becomes 'If $f \in \mathbf{V}$, can f be expressed as a limit of complex trigonometric polynomials?'

Suppose that f and g are in $\mathbf{V}$. We set

$$\langle f, g \rangle = \frac{1}{2\pi} f(t)\overline{g(t)}\, dt.$$

Note that this definition involves a complex conjugate. The function $(f, g) \to \langle f, g \rangle$ is an example of a *complex semi-inner product*; we shall study these further in Volume II. Let us list some of its properties, which follow immediately from the definition.

- $\langle f, f \rangle = \frac{1}{2\pi}|f(t)|^2\, dt \geq 0.$
- $\langle g, f \rangle = \overline{\langle f, g \rangle}.$
- $\langle \alpha_1 f_1 + \alpha_2 f_2, g \rangle = \alpha_1 \langle f_1, g \rangle + \alpha_2 \langle f_2, g \rangle.$
- $\langle f, \beta_1 g_1 + \beta_2 g_2 \rangle = \overline{\beta_1} \langle f, g_1 \rangle + \overline{\beta_2} \langle f, g_2 \rangle.$

The functions $(\gamma_n)_{n=-\infty}^{\infty}$ then form an *orthonormal set*:

$$\langle \gamma_m, \gamma_n \rangle = \begin{cases} 1 & \text{if } m = n, \\ 0 & \text{if } m \neq n. \end{cases}$$

If $p = \sum_{j=-n}^{n} c_j \gamma_j$ is a complex trigonometric polynomial, then it follows from the orthogonality relations that $c_j = \langle p, \gamma_j \rangle$ for $-n \leq j \leq n$. We consider similar semi-inner products for functions in $\mathbf{V}$. If $f \in \mathbf{V}$, we define its *complex Fourier coefficients* $(\hat{f}_n)_{n=-\infty}^{\infty}$ as $\hat{f}_n = \langle f, \gamma_n \rangle$, so that

$$\hat{f}_n = \frac{1}{2\pi}\int_{-\pi}^{\pi} f(t)\overline{\gamma_n(t)}\, dt = \frac{1}{2\pi}\int_{-\pi}^{\pi} f(t)\gamma_{-n}(t)dt = \frac{1}{2\pi}\int_{-\pi}^{\pi} f(t)e^{-int}\, dt.$$

In particular, $\hat{f}_0 = \frac{1}{2\pi}\int_{-\pi}^{\pi} f(t)\,dt$ is the average value of f over the interval $[-\pi, \pi]$. We then write

$$f \sim \sum_{-\infty}^{\infty} \hat{f}_n \gamma_n = \sum_{-\infty}^{\infty} \langle f, \gamma_n \rangle \, \gamma_n.$$

We can define the cosine Fourier coefficients and the sine Fourier coefficients for complex valued functions. If $f \in \mathbf{V}$, it is easy to pass between the complex Fourier coefficients and the cosine and sine Fourier coefficients. Let us define the *reversal* $R(f)$ of $f \in \mathbf{V}$ by setting $R(f)(t) = f(-t)$. To avoid too many superscripts, we set $C(f) = \bar{f}$ and $S(f) = R(\bar{f})$. We then have the following identities.

Proposition 9.2.1   *Suppose that $f \in \mathbf{V}$ and that $n \in \mathbf{N}$.*

1. $\hat{f}_0 = a_0(f)/2.$
2. $\hat{f}_n = a_n(f) + ib_n(f)$ and $\hat{f}_{-n} = a_n(f) - ib_n(f).$
3. If f is an even function then $\hat{f}_n = \hat{f}_{-n} = a_n(f)$, and if f is an odd function then $\hat{f}_0 = 0$ and $\hat{f}_n = -\hat{f}_{-n} = ib_n(f).$
4. $a_n(f) = \frac{1}{2}(\hat{f}_n + \hat{f}_{-n})$ and $b_n(f) = \frac{i}{2}(\hat{f}_{-n} - \hat{f}_n).$
5. $\widehat{C(f)}_n = \hat{f}_{-n}.$
6. $\widehat{R(f)}_n = \hat{f}_{-n}.$
7. $\widehat{S(f)}_n = \hat{f}_n.$
8. If f is real-valued, then $\hat{f}_{-n} = \overline{\hat{f}_n}.$

Proof   The reader should verify these identities. $\square$

Exercises

9.2.1 Suppose that γ is a continuous 2π-periodic homomorphism of $(\mathbf{R}, +)$ into the multiplicative group $\mathbf{T}$. There exists $0 < \delta < \pi$ such that if $|t| < \delta$ then $|\gamma(t) - 1| < 1.$
   (a) Suppose that $k > 2\pi/\delta$. Show that there exists $n \in \mathbf{Z}$, with $|n| < k/6$, such that $\gamma(2\pi/k) = e^{2\pi in/k}.$
   (b) Show that n does not depend upon k.
   (c) Show that if $q \in \mathbf{Q}$ then $\gamma(2\pi iq) = e^{2\pi inq}.$
   (d) Use continuity to show that $\gamma = \gamma_n.$
9.2.2 Verify the identities of Proposition 9.2.1.

9.3 Uniqueness

The size of a function controls the size of its Fourier coefficients. We establish two fundamental inequalities.

Theorem 9.3.1 (Bessel's inequality)   *If $f \in V$, then*

$$\sum_{n=-\infty}^{\infty} |\hat{f}_n|^2 \leq \langle f, f \rangle = \frac{1}{2\pi} \int_{-\pi}^{\pi} |f(t)|^2 \, dt.$$

Proof   Let $p_n = \sum_{j=-n}^{n} \hat{f}_j \gamma_j$. Then $\langle f, \gamma_k \rangle = \langle p_n, \gamma_k \rangle$ for $k \in \mathbf{Z}$, so that $\langle f, p_n \rangle = \langle p_n, p_n \rangle$, and similarly $\langle p_n, f \rangle = \langle p_n, p_n \rangle$. Hence

$$0 \leq \langle f - p_n, f - p_n \rangle = \langle f, f \rangle - \langle f, p_n \rangle - \langle p_n, f \rangle + \langle p_n, p_n \rangle$$

$$= \langle f, f \rangle - \langle p_n, p_n \rangle = \langle f, f \rangle - \sum_{j=-n}^{n} |\hat{f}_j|^2.$$

Since this holds for all $n \in \mathbf{N}$, the result follows.   □

In fact, equality holds; we prove this (Parseval's equation: Corollary 9.4.7) later.

Proposition 9.3.2   *If $f \in V$ then*

$$|\hat{f}_n| \leq \frac{1}{2\pi} \int_{-\pi}^{\pi} |f(t)| \, dt \leq \sup_{t \in \mathbf{R}} |f(t)| \, dt.$$

Proof   For

$$|\hat{f}_n| = \left| \frac{1}{2\pi} \int_{-\pi}^{\pi} f(t) e^{-int} \, dt \right| \leq \frac{1}{2\pi} \int_{-\pi}^{\pi} |f(t) e^{-int}| \, dt = \frac{1}{2\pi} \int_{-\pi}^{\pi} |f(t)| \, dt.$$

□

Corollary 9.3.3   *If f is a trivial function, then $\hat{f}_n = 0$ for all $n \in \mathbf{Z}$.*

More importantly, the converse is true.

Theorem 9.3.4   *If $f \in V$ and $\hat{f}_n = 0$ for all $n \in \mathbf{Z}$ then f is trivial.*

Proof   Let $f = u + iv$. Since $a_0(f) = 2\hat{f}_0$, and $a_n(f) = \frac{1}{2}(\hat{f}_n + \hat{f}_{-n})$ and $b_n(f) = \frac{i}{2}(\hat{f}_n - \hat{f}_{-n})$ for $n \in \mathbf{N}$, the Fourier cosine and sine coefficients of f are all zero, and so therefore are the Fourier cosine and sine coefficients of u and v. Consequently, if p is a real trigonometric polynomial, then $\frac{1}{2\pi} \int_{-\pi}^{\pi} u(t)p(t) \, dt = 0$ and $\frac{1}{2\pi} \int_{-\pi}^{\pi} v(t)p(t) \, dt = 0$. Suppose that

$\frac{1}{2\pi} \int_{-\pi}^{\pi} |f(t)|\, dt > 0$. Then one of

$$\frac{1}{2\pi} \int_{-\pi}^{\pi} u^{+}(t)\, dt, \quad \frac{1}{2\pi} \int_{-\pi}^{\pi} u^{-}(t)\, dt, \quad \frac{1}{2\pi} \int_{-\pi}^{\pi} v^{+}(t)\, dt, \quad \frac{1}{2\pi} \int_{-\pi}^{\pi} v^{-}(t)\, dt$$

is non-zero. Suppose that $\frac{1}{2\pi} \int_{-\pi}^{\pi} u^{+}(t)\, dt > 0$. (The argument in the other cases is essentially the same.) By considering a lower sum for the Riemann integral of u^{+}, we see that there exists an interval $[t_0 - \eta, t_0 + \eta]$ in $[-\pi, \pi]$ and $\lambda > 0$ such that $u(t) \geq \lambda$ for $t \in [t_0 - \eta, t_0 + \eta]$. The idea now is to find a real trigonometric polynomial which is large on the interval $[t_0 - \eta/2, t_0 + \eta/2]$, positive on the interval $[t_0 - \eta, t_0 + \eta]$ and bounded in modulus by 1 for other values of t in $[-\pi, \pi]$. Let $\alpha = \cos \eta/2 - \cos \eta$: then $\alpha > 0$. Let $l(t) = 1 + \cos(t - t_0) - \cos \eta$. Then

$$l(t) \geq 1 + \alpha \text{ for } t \in [t_0 - \eta/2, t_0 + \eta/2],$$

$$l(t) \geq 1 \text{ for } t \in [t_0 - \eta, t_0 + \eta]$$

and $|l(t)| \leq 1$ for other values of t in $[-\pi, \pi]$.

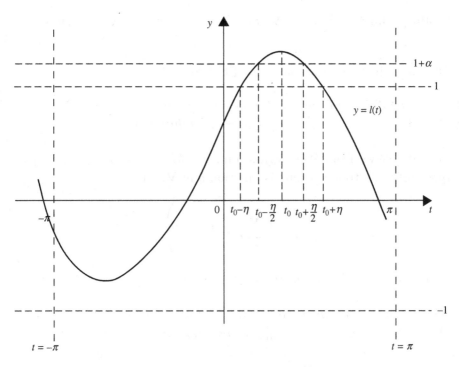

Figure 9.3a. The function $l(t)$.

Let $M = \sup_{t \in \mathbf{R}} u^+(t)$. Thus if $k \in \mathbf{N}$ then

$$\frac{1}{2\pi} \int_{-\pi}^{\pi} f(t)(l(t))^k \, dt = I_1 + I_2 + I_3,$$

where

$$I_1 = \frac{1}{2\pi} \int_{-\pi}^{t_0 - \eta} u^+(t)(l(t)^k \, dt \geq -\frac{M(t_0 - \eta + \pi)}{2\pi},$$

$$I_2 = \frac{1}{2\pi} \int_{t_0 - \eta}^{t_0 + \eta} u^+(t)(l(t))^k \, dt \geq \frac{1}{2\pi} \int_{t_0 - \eta/2}^{t_0 + \eta/2} u^+(t)(l(t)^k \, dt \geq \eta\lambda(1 + \alpha)^k,$$

$$I_3 = \frac{1}{2\pi} \int_{t_0 + \eta}^{\pi} u^+(t)(l(t))^k \, dt \geq -\frac{M(\pi - (t_0 + \eta))}{2\pi}.$$

Thus

$$\frac{1}{2\pi} \int_{-\pi}^{\pi} u^+(t)(l(t))^k \, dt \geq \eta\lambda(1 + \alpha)^k - M,$$

which is positive for large enough k. Since l^k is a trigonometric polynomial, we obtain a contradiction. $\qquad\square$

Corollary 9.3.5 *If $f, g \in \mathbf{V}$ and $\hat{f}_n = \hat{g}_n$ for all $n \in \mathbf{Z}$ then $f - g$ is trivial.*

Here is an important application of the corollary.

Theorem 9.3.6 *Suppose that f is a continuous function in $\mathbf{V}$ and that $\sum_{n=-\infty}^{\infty} |\hat{f}_n| < \infty$. Then $\sum_{j=-n}^{n} \hat{f}_j \gamma_j \to f$ uniformly as $n \to \infty$.*

Proof It follows from Weierstrass' uniform M test that $\sum_{j=-n}^{n} \hat{f}_j \gamma_j$ converges uniformly to a continuous function g in $\mathbf{V}$. Then

$$\hat{g}_k = \lim_{n \to \infty} \frac{1}{2\pi} \int_{-\pi}^{\pi} \left(\sum_{j=-n}^{n} \hat{f}_j(t)\gamma_j(t) \right) \overline{\gamma_k(t)} \, dt = \hat{f}_k.$$

It therefore follows from the corollary above that $f = g$. $\qquad\square$

This result is useful when we consider the indefinite integral of a function in $\mathbf{V}$. If $f \in \mathbf{V}$, the function $t \to \int_0^t f(s) \, ds$ is not necessarily periodic. Instead, we consider the function $F(t) = \int_0^t f(s) \, ds - \hat{f}_0 t$, which is a continuous element of $\mathbf{V}$.

Theorem 9.3.7 *Suppose that $f \in V$. Let $F(t) = \int_0^t f(s)\,ds - \hat{f}_0 t$. If $n \neq 0$ then $\hat{F}_n = i\hat{f}_n/n$. Further, $\sum_{n=-\infty}^{\infty} |\hat{F}_n| < \infty$, so that $\sum_{j=-n}^{n} \hat{F}_j \gamma_j$ converges uniformly to F as $n \to \infty$.*

Proof Suppose that $\epsilon > 0$. There exists a continuous function g in V with $\frac{1}{2\pi} \int_{-\pi}^{\pi} |f(s) - g(s)| < \epsilon/4\pi$ and with $\hat{g}_0 = \hat{f}_0$. Let $G(t) = \int_0^t g(s)\,ds - \hat{g}_0 t$. If $t \in [-\pi, \pi]$, $|G(t) - F(t)| \leq \int_0^t |f(s) - g(s)|\,ds < \epsilon/2$, so that $|\hat{G}_n - \hat{F}_n| < \epsilon/2$ for all $n \in \mathbf{N}$. If $n \in \mathbf{Z}$, and $n \neq 0$, we integrate by parts.

$$\hat{G}_n = \frac{1}{2\pi} \int_{-\pi}^{\pi} G(s) e^{-ins}\,ds = \frac{i}{2\pi n} \int_{-\pi}^{\pi} (g(s) - \hat{f}_0) e^{-ins}\,ds = \frac{i}{n}\hat{g}_n.$$

But $|\hat{g}_n - \hat{f}_n| \leq \frac{1}{2\pi} \int_{-\pi}^{\pi} |f(s) - g(s)|\,ds < \epsilon/4\pi$, and so $|\hat{F}_n - i\hat{f}_n/n| < \epsilon$. Since this holds for all $\epsilon > 0$, $\hat{F}_n = i\hat{f}_n/n$.

It now follows from the Cauchy–Schwarz inequality, and Bessel's inequality, that

$$\sum_{n=-\infty}^{\infty} |\hat{F}_n| \leq \left(|\hat{F}_0|^2 + \sum_{n=1}^{\infty} |\hat{f}_n|^2 + |\hat{f}_{-n}|^2 \right)^{\frac{1}{2}} \left(1 + 2\sum_{n=1}^{\infty} \frac{1}{n^2} \right)^{\frac{1}{2}} < \infty.$$

Thus $\sum_{j=-n}^{n} \hat{F}_j \gamma_j$ converges uniformly to F as $n \to \infty$, by Theorem 9.3.6. $\square$

Corollary 9.3.8 *If f is a continuously differentiable function in V, then $\sum_{j=-n}^{n} |\hat{f}_j| < \infty$ and $\sum_{j=-n}^{n} \hat{f}_j \gamma_j$ converges uniformly to f as $n \to \infty$.*

Let us give two examples.

Example 9.3.9 Suppose that $0 < \delta \leq \pi/2$. Let $I_\delta(t) = \pi/\delta$ if $0 \leq |t| \leq \delta$, let $I_\delta(t) = 0$ if $\delta < |t| \leq \pi$ and let $I_\delta(t + 2k\pi) = I_\delta(t)$ for $k \in \mathbf{Z}$. Then

$$I_\delta(t) \sim 1 + \sum_{n=1}^{\infty} \frac{2\sin n\delta}{n\pi} \cos nt.$$

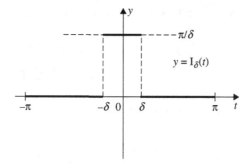

Figure 9.3b. The function $I_\delta(t)$.

Certainly $a_0(I_\delta) = 1$. (This is the reason for the choice of the constant π/δ.) Further

$$a_n(I_\delta) = \frac{2}{\delta} \int_0^\delta \cos nt = \frac{2 \sin n\delta}{n\delta}.$$

Thus

$$\sum_{n=1}^N a_n(I_\delta) \cos nt = \frac{2}{\delta} \sum_{n=1}^N \frac{\sin n\delta \cos nt}{n} = \frac{1}{\delta} \sum_{n=1}^N \frac{\sin(n(t+\delta)) - \sin(n(t-\delta))}{n}.$$

Do these sums converge, as $N \to \infty$? If $0 < \alpha < \pi$ then

$$\left| \sum_{n=1}^N \sin n\alpha \right| \le \left| \sum_{n=1}^N e^{in\alpha} \right| = \left| \frac{e^{i(N+1)\alpha} - e^{i\alpha}}{e^{i\alpha} - 1} \right| = \frac{2}{|e^{i\alpha/2} - e^{-i\alpha/2}|} = \frac{1}{\sin \alpha/2}.$$

Thus if $|t| \le \pi$, and if $|t-\delta| > \eta$ and $|t+\delta| > \eta$, then

$$\left| \sum_{n=1}^N (\sin(n(t+\delta)) - \sin(n(t-\delta))) \right| \le 2 \sin \eta/2.$$

It therefore follows from the uniform version of Dirichlet's test that the Fourier cosine series converges to a continuous function on $[-\pi, \pi] \setminus \{\delta, -\delta\}$, and converges uniformly on

$$[-\pi, \pi] \setminus ((\delta - \eta, \delta + \eta) \cup (-\delta - \eta, -\delta + \eta)).$$

Does the Fourier series converge to I_δ? We shall consider this question in Section 9.6.

Example 9.3.10 Suppose that $0 < \delta \le \pi/2$. Let

$$J_\delta(t) = \begin{cases} \frac{\pi}{\delta} \left(1 - \frac{t}{2\delta}\right) & \text{if } 0 \le |t| \le 2\delta, \\ 0 & \text{if } 2\delta < |t| \le \pi, \end{cases}$$

and let $J_\delta(t + 2k\pi) = J_\delta(t)$ for $k \in \mathbf{Z}$. Then

$$J_\delta(t) \sim 1 + \sum_{n=1}^\infty \frac{2 \sin^2 n\delta}{n^2 \delta^2} \cos nt.$$

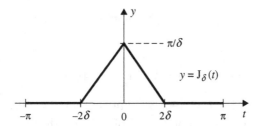

Figure 9.3c. The function $J_\delta(t)$.

Once again, $a_0(J_\delta) = 1$. If $n > 0$ then, integrating by parts, and using the identity $2\sin^2 a = 1 - \cos 2a$, it follows that

$$a_n(J_\delta) = \frac{2}{\delta}\int_0^{2\delta}(1 - \frac{t}{2\delta})\cos nt \, dt$$

$$= \frac{1}{n\delta^2}\int_0^{2\delta}\sin nt \, dt$$

$$= \frac{1}{n^2\delta^2}(1 - \cos 2n\delta) = \frac{2\sin^2 n\delta}{n^2\delta^2}.$$

In this case, all the Fourier coefficients are non-negative, and $\sum_{n=0}^{\infty} a_n(J_\delta) < \infty$, and so the Fourier series converges uniformly to J_δ.

When $\delta = \pi/2$ then $J_\delta(0) = 2$, $a_{2k-1}(J_\delta) = 8/(2k-1)^2\pi^2$, and $a_{2k} = 0$. Hence

$$2 = 1 + \sum_{k=1}^{\infty}\frac{8}{(2k-1)^2\pi^2},$$

so that

$$\sum_{k=1}^{\infty}\frac{1}{(2k-1)^2} = \frac{\pi^2}{8}.$$

Since

$$\sum_{n=1}^{\infty}\frac{1}{n^2} = \sum_{k=1}^{\infty}\frac{1}{(2k-1)^2} + \sum_{k=1}^{\infty}\frac{1}{(2k)^2} = \frac{\pi^2}{8} + \frac{1}{4}\sum_{n=1}^{\infty}\frac{1}{n^2},$$

it follows that

$$\sum_{n=1}^{\infty}\frac{1}{n^2} = \frac{\pi^2}{6}.$$

This famous equation was proved by Euler, and was one of his early triumphs. But Euler did not know about Fourier series: we give another proof, due to him, in Section 10.8.

Exercises

9.3.1 Show that
$$1 + \sum_{k=1}^{\infty}(-1)^k \left(\frac{1}{(4k-1)^2} + \frac{1}{(4k+1)^2} \right) = \frac{\sqrt{2}.\pi^2}{16}.$$

9.3.2 Let $f(t) = t^2$ for $t \in [-\pi, \pi]$, and extend by periodicity. Calculate the Fourier coefficients of f, and obtain another proof of the equation $\sum_{n=1}^{\infty} 1/n^2 = \pi^2/6$.

9.3.3 Let $f(t) = 1$ for $|t| \leq \pi/2$, let $f(t) = -1$ for $\pi/2 < |t| \leq \pi$, and extend by periodicity. Calculate the Fourier coefficients of f.

9.3.4 Let $f(t) = \pi - 2|t|$ for $t \in [-\pi, \pi]$. Use Example 9.3.10 to calculate the Fourier coefficients of f.

9.3.5 Suppose that f is a continuously differentiable function in $\mathbf{V}$. Obtain an upper bound for $\sum_{n=-\infty}^{\infty} |\hat{f}_n|$.

9.3.6 Suppose that $(b_n)_{n=1}^{\infty}$ is a decreasing null sequence of real numbers. Show that $\sum_{n=1}^{\infty} b_n \sin nt$ converges for every $t \in \mathbf{R}$. Give examples to show that the sequence $(b_n)_{n=1}^{\infty}$ need not be the Fourier sine series of an element of $\mathbf{V}$.

9.4 Convolutions, and Parseval's equation

Suppose that $f \in \mathbf{V}$ and that $\delta \in \mathbf{R}$. We set $T_\delta(f)(t) = f(t - \delta)$. $T_\delta(f)$ is a *translate* of f. Note that

$$\widehat{T_\delta(f)}_n = \frac{1}{2\pi} \int_{-\pi}^{\pi} f(t - \delta)e^{-int}\, dt = \frac{1}{2\pi} \int_{-\pi}^{\pi} f(t)e^{-in(t+\delta)}\, dt = e^{-in\delta}\hat{f}_n$$

We set
$$K_\delta(f) = \|f - T_\delta(f)\|_1 = \frac{1}{2\pi} \int_{-\pi}^{\pi} |f(t) - f(t - \delta)|\, dt$$

$K_\delta(.)$ is a semi-norm on $\mathbf{V}$, and $K_\delta(f) \leq 2\,\|f\|_1$.

Proposition 9.4.1 *If $f \in \mathbf{V}$ then $K_\delta(f) \to 0$ as $\delta \to 0$.*

Proof A little thought shows that if f is the indicator function of a proper subinterval of $[-\pi, \pi]$ then $K_\delta(f) = \delta/\pi$ for small enough values of δ, and so the result holds for f. It then follows from the semi-norm properties of K_δ that the result holds for step-functions. If $f \in \mathbf{V}$ and $\epsilon > 0$, there exists a step function g with $\|f - g\|_1 < \epsilon/3$. Then

$$K_\delta(f) \leq K_\delta(f - g) + K_\delta(g) \leq 2\epsilon/3 + K_\delta(g) < \epsilon,$$

if $|\delta|$ is sufficiently small. □

If $f, g \in V$ we define the *convolution product* $f \star g$ to be the function

$$f \star g(t) = \frac{1}{2\pi} \int_{-\pi}^{\pi} f(t-s)g(s)\, ds.$$

Example 9.4.2 If $f \in V$ then $f \star \gamma_n = \hat{f}_n \gamma_n$. In particular,

$$\gamma_m \star \gamma_n = \begin{cases} \gamma_n & \text{if } m = n, \\ 0 & \text{otherwise.} \end{cases}$$

For

$$f \star \gamma_n(t) = \frac{1}{2\pi} \int_{-\pi}^{\pi} e^{in(t-s)} f(s)\, ds = \frac{1}{2\pi} \int_{-\pi}^{\pi} e^{int} e^{-ins} f(s)\, ds = e^{int} \hat{f}_n.$$

Here are some of the properties of convolutions.

Proposition 9.4.3 *Suppose that $f, f_1, f_2, g \in V$ and $\alpha_1, \alpha_2 \in \mathbf{C}$.*
(i) $f \star g$ *is a continuous function.*
(ii) $f \star g = g \star f$.
(iii) $T_\delta(f) \star g = T_\delta(f \star g)$.
(iv) $(\alpha_1 f_1 + \alpha_2 f_2) \star g = \alpha_1(f_1 \star g) + \alpha_2(f_2 \star g)$.

Proof (i) The function $f \star g$ is certainly 2π-periodic. If $t, \delta \in \mathbf{R}$ then

$$|(f \star g)(t + \delta) - (f \star g)(t)| = \left| \frac{1}{2\pi} \int_{-\pi}^{\pi} (f(t - \delta - s) - f(t-s))g(s)\, ds \right|$$

$$= \left| \int_{-\pi+\delta}^{-\pi+\delta} f(t-s)(g(s+\delta) - g(s))\, ds \right|$$

$$= \left| \frac{1}{2\pi} \int_{-\pi}^{\pi} f(t-s)(g(s+\delta) - g(s))\, ds \right|$$

$$\leq \|f\|_\infty . K_\delta(g),$$

and so the result follows from the preceding proposition.

(ii) Making the change of variables $u = t - s$, it follows that

$$(g \star f)(t) = \frac{1}{2\pi} \int_{-\pi}^{\pi} g(t-u)f(u)du = \frac{1}{2\pi} \int_{t-\pi}^{t+\pi} f(t-s)g(s)\, ds$$

$$= \frac{1}{2\pi} \int_{-\pi}^{\pi} f(t-s)g(s)\, ds = (f \star g)(t).$$

(iii) For

$$(T_\delta(f) \star g)(t) = \frac{1}{2\pi} \int_{-\pi}^{\pi} f(t - \delta - s)g(s)\, ds = (f \star g)(t - \delta) = T_\delta(f \star g)(t).$$

(iv) This follows directly from the definition of convolution. $\qquad \square$

Convolution is an essential element of Fourier analysis.

Theorem 9.4.4   *If $f, g \in V$ and $n \in \mathbf{Z}$ then $(\widehat{f \star g})_n = \hat{f}_n \hat{g}_n$.*

Proof   Here is a quick and easy proof. Changing the order of integration,

$$(\widehat{f \star g})_n = \frac{1}{2\pi} \int_{-\pi}^{\pi} \left(\frac{1}{2\pi} \int_{-\pi}^{\pi} f(t-s) g(s) \, ds \right) e^{-int} \, dt$$

$$= \frac{1}{2\pi} \int_{-\pi}^{\pi} \left(\frac{1}{2\pi} \int_{-\pi}^{\pi} f(t-s) e^{-int} \, dt \right) g(s) \, ds$$

$$= \frac{1}{2\pi} \int_{-\pi}^{\pi} \left(\frac{1}{2\pi} \int_{-\pi}^{\pi} f(u) e^{-in(s+u)} \, du \right) g(s) \, ds$$

$$= \hat{f}_n \frac{1}{2\pi} \int_{-\pi}^{\pi} e^{-ins} g(s) \, ds = \hat{f}_n \hat{g}_n.$$

Unfortunately, we need to justify the change of order of integration. We do this for continuous functions in Volume II, and, more generally, in Volume III.

Instead, we proceed as follows. Note that $I_\delta \star I_\delta = J_\delta$, and that $((I_\delta)_n)^2 = (J_\delta)_n$, where I_δ and J_δ are the functions of Examples 9.3.9 and 9.3.10. Thus the result holds when $f = g = I_\delta$. It now follows from Proposition 9.4.3 that if $\mathcal{D}$ is a dissection of $[-\pi, \pi]$ into intervals of equal length and if f, g are step functions that are constant on the intervals of $\mathcal{D}$ then $(\widehat{f \star g})_n = \hat{f}_n \hat{g}_n$.

We now use a standard approximation argument. If $f, g \in V$ and $\epsilon > 0$, there exist step functions h and j, of this form described above, with $\|f - h\|_\infty < \epsilon$ and $\|g - j\|_\infty < \epsilon$. If $n \in \mathbf{N}$ then

$$f \star g = h \star j + (f - h) \star g + h \star (g - j)$$
$$\text{and } \hat{f}_n \hat{g}_n = \hat{h}_n \hat{j}_n + (\hat{f}_n - \hat{h}_n) \hat{g}_n + \hat{h}_n (\hat{f}_n - \hat{h}_n).$$

Hence

$$\left| (\widehat{f \star g})_n - (\widehat{h \star j})_n \right| = \leq |((f - h) \widehat{\star} g)_n| + |(h \star \widehat{(g - j)})_n|$$
$$\leq \epsilon(\|g\|_\infty + \|h\|_\infty) \leq \epsilon(\|g\|_\infty + \|f\|_\infty + \epsilon).$$

Similarly,

$$|\hat{f}_n \hat{g}_n - \hat{h}_n \hat{j}_n| \leq |(\hat{f}_n - \hat{h}_n)| \cdot |\hat{g}_n| + |\hat{h}_n| \cdot |\hat{g}_n - \hat{j}_n|$$
$$\leq \epsilon(\|g\|_\infty + \|h\|_\infty) \leq \epsilon(\|g\|_\infty + \|f\|_\infty + \epsilon).$$

Since $(\widehat{h \star j})_n = \hat{h}_n \hat{j}_n$, it follows that

$$|(\widehat{f \star g})_n - \hat{f}_n \hat{g}_n| \leq 2\epsilon(\|g\|_\infty + \|f\|_\infty + \epsilon).$$

Since ϵ is arbitrary, it follows that $\widehat{(f \star g)}_n = \hat{f}_n \hat{g}_n$.   □

Corollary 9.4.5 $\sum_{n=-\infty}^{\infty} \hat{f}_n \hat{g}_n e^{int}$ *converges uniformly to* $(f \star g)(t)$.

Proof By the Cauchy–Schwarz inequality and Bessel's inequality,

$$
\sum_{n=-\infty}^{\infty} |\hat{f}_n \hat{g}_n| \le \left(\sum_{n=-\infty}^{\infty} |\hat{f}_n|^2 \right)^{\frac{1}{2}} \cdot \left(\sum_{n=-\infty}^{\infty} |\hat{g}_n|^2 \right)^{\frac{1}{2}}
$$

$$
\le \left(\frac{1}{2\pi} \int_{-\pi}^{\pi} |f(t)|^2 \, dt \right)^{\frac{1}{2}} \cdot \left(\frac{1}{2\pi} \int_{-\pi}^{\pi} |g(t)|^2 \, dt \right)^{\frac{1}{2}} < \infty,
$$

and so the result follows from Theorem 9.3.6.   □

Corollary 9.4.6   *If* $f, g, h \in V$ *then* $(f \star g) \star h = f \star (g \star h)$.

Proof For each has Fourier series $\sum_{n=-\infty}^{\infty} \hat{f}_n \cdot \hat{g}_n \cdot \hat{h}_n e^{int}$.   □

We can therefore write $f \star g \star h$ for the common value.

Corollary 9.4.7 (Parseval's equation)

$$
\frac{1}{2\pi} \int_{-\pi}^{\pi} f(t)\overline{g(t)} \, dt = \sum_{n=-\infty}^{\infty} \hat{f}_n \overline{\hat{g}_n}.
$$

In particular, $\frac{1}{2\pi} \int_{-\pi}^{\pi} |f(t)|^2 \, dt = \sum_{n=-\infty}^{\infty} |\hat{f}_n|^2$.

Proof As in the previous section, let $S(g)(t) = \overline{g(-t)}$. Then

$$
(S(g) \star f)(0) = \frac{1}{2\pi} \int_{-\pi}^{\pi} f(t)\overline{g(t)} \, dt \text{ and } \widehat{(S(g) \star f)}_n = \hat{f}_n \overline{\hat{g}_n}.
$$ □

Example 9.4.8

$$
1 - \frac{1}{3^3} + \frac{1}{5^3} - \frac{1}{7^3} + \cdots = \frac{\pi^3}{32}.
$$

Let $f = I_{\pi/2} \star J_{\pi/2}$. Then $\hat{f}_0 = 1$; if $n > 0$ then

$$
\hat{f}_n = \begin{cases} 0 & \text{if } n \text{ is even,} \\ 8(-1)^k/\pi^3(2k+1)^3 & \text{if } n = 2k+1 \text{ is odd,} \end{cases}
$$

and $\hat{f}_n = \hat{f}_{-n}$ if $n < 0$. Also $f(0) = 3/2$, and so

$$\frac{3}{2} = 1 + 16 \sum_{k=0}^{\infty} \frac{(-1)^k}{\pi^3 (2k+1)^3},$$

which gives the result.

Exercises

9.4.1 By applying Parseval's equation to $J_{\pi/2}$, show that

$$\sum_{n=1}^{\infty} 1/n^4 = \pi^4/90.$$

9.4.2 Calculate the function $I_{\pi/2} \star J_{\pi/2}$, and deduce that

$$\sum_{n=1}^{\infty} 1/n^6 = \pi^6/945.$$

9.4.3 Calculate the function $J_{\pi/2} \star J_{\pi/2}$, and deduce that

$$\sum_{n=1}^{\infty} 1/n^8 = \pi^8/9450.$$

9.5 An example

Things can go wrong! We now give an example of an even continuous periodic function whose Fourier series fails to converge at 0. First, let $f_j(t) = \sin 2j|t|$, for $j \in \mathbf{N}$. Then f_j is an even function, and

$$a_0(f_j) = \frac{2}{\pi} \int_0^{\pi} \sin 2jt \, dt = 0.$$

If $n \in \mathbf{N}$ then

$$a_n(f_j) = \frac{2}{\pi} \int_0^{\pi} \sin 2jt \cos nt \, dt$$

$$= \frac{2}{\pi} \int_0^{\pi} \sin(2j+n)t + \sin(2j-n)t \, dt$$

$$= \frac{2}{\pi(2j+n)} + \frac{2}{\pi(2j-n)} \quad \text{if } n \text{ is odd,}$$

and $a_n(f_j) = 0$ if n is even. Note that $a_n(f_j)$ is non-negative if $n \leq 2j$ and is negative if n is odd and greater than $2j$. We now set

$$s_{2r}(f_j) = \sum_{j=1}^{2r} a_n(f_j) \cos 0 = \sum_{j=1}^{2r-1} a_n(f_j).$$

Note that $s_{2r}(f_j)$ increases for $r \leq j$, and then decreases. The maximum value is

$$s_{2j}(f_{2j}) = \frac{2}{\pi} \sum_{k=1}^{j} \frac{1}{2k-1} \geq \frac{\log j}{\pi}.$$

On the other hand, if $r > j$ then

$$s_{2r}(f_j) = \sum_{k=1}^{r} \frac{1}{2j + (2k-1)} + \sum_{k=1}^{r} \frac{1}{2j - (2k-1)}$$

$$= \sum_{k=j+1}^{r+j} \frac{1}{2j-1} + \left(\sum_{k=1}^{j} \frac{1}{2j-1} - \sum_{k=1}^{r-j} \frac{1}{2j-1} \right)$$

$$= \sum_{k=r-j+1}^{r+j} \frac{1}{2j-1} \geq 0.$$

Now let $N_j = 2^{j^3+1}$, let $g_j = f_{N_j}/j(j+1)$, let $h_k = \sum_{j=1}^{k} g_j$. and let $h = \sum_{j=1}^{\infty} g_j$. Since $|f_j(t)| \leq 1$ for all t and all j, it follows from Weierstrass' uniform M test that $h_k \to h$ uniformly, and so h is continuous. Further, $a_n(h) = \lim_{k \to \infty} a_n(h_k)$, and $s_{2r}(h) = \lim_{k \to \infty} s_{2r}(h_k)$. Since all the summands are non-negative,

$$s_{N_j}(h) \geq s_{N_j}(h_j) \geq \frac{s_{N_j}(f_{N_j})}{j(j+1)} \geq \frac{\log N_j}{\pi j(j+1)} = \frac{j^3 \log 2}{\pi j(j+1)} \geq j/5.$$

Thus the sequence $(s_{2r}(h))_{r=1}^{\infty}$ is unbounded, and so it does not converge.

9.6 The Dirichlet kernel

Suppose that $f \in \mathbf{V}$. Let us look more closely at the partial sum $s_n(f)(t) = \sum_{j=-n}^{n} \hat{f}_j e^{ijt}$. Since $\hat{f}_j = \frac{1}{2\pi} \int_{-\pi}^{\pi} f(s) e^{-ins} \, ds$, it follows that

$$s_n(f)(t) = \frac{1}{2\pi} \int_{-\pi}^{\pi} f(s) \left(\sum_{j=-n}^{n} e^{ij(t-s)} \right) ds = (D_n \star f)(t),$$

where $D_n(0) = \sum_{j=-n}^n 1 = 2n + 1$ and

$$D_n(t) = \sum_{j=-n}^n e^{ijt} = \frac{\sin(n+\frac{1}{2})t}{\sin\frac{1}{2}t} = \frac{\sin nt}{\tan\frac{1}{2}t} + \cos nt$$

for $0 < |t| \leq \pi$. The function D_n is called the n-th *Dirichlet kernel*.

Here are the principal properties of the Dirichlet kernel.

Theorem 9.6.1   *Let $t_j = 2\pi j/(2n+1)$ and let $I_j = (1/2\pi) \int_{t_{j-1}}^{t_j} |D_n(t)|\, dt$, for $1 \leq j \leq n$.*

(i) D_n *is an even continuous function in* **V**.

(ii) D_n *is a decreasing function on* $[0, t_1]$.

(iii) $D_n(t_j) = 0$, $D_n(t) > 0$ *if* $t_{j-1} < t < t_j$, *where j is odd, and $D_n(t) < 0$ if $t_{j-1} < t < t_j$, where j is even, for $1 \leq j \leq n$.*

(iv) $I_1 < 1$ *and* $I_1 > I_2 > \cdots > I_n > 0$.

(v) *If* $-\pi \leq a < b \leq \pi$ *then* $|(1/2\pi) \int_a^b D_n(t)\, dt| < 2$.

(vi) $I_j(t) \geq 2/\pi^2 j$, *and* $\frac{1}{2\pi} \int_{-\pi}^\pi |D_n(t)|\, dt \geq (4/\pi^2) \log n$.

Proof (i) and (ii) Each of the summands in the definition is continuous, even, and decreasing on $[0, t_1]$.

(iii) Since $\sin\frac{1}{2}t > 0$ for $0 < t \leq \pi$, this follows from the corresponding properties of the function $\sin(n+\frac{1}{2})t$.

(iv) Since $0 < D_n(t) < 2n + 1$ for $t \in (0, t_1]$, it follows that $0 < I_1 < (2n+1)t_1/2\pi = 1$. Further, if $1 \leq j < n$ then

$$I_{j+1} = \frac{1}{2\pi} \int_{t_j}^{t_{j+1}} \frac{|\sin(n+\frac{1}{2})t|}{\sin\frac{1}{2}t}\, dt = \frac{1}{2\pi} \int_0^{t_1} \frac{|\sin(n+\frac{1}{2})t|}{\sin\frac{1}{2}(t+t_j)}\, dt$$

$$< \frac{1}{2\pi} \int_0^{t_1} \frac{|\sin(n+\frac{1}{2})t|}{\sin\frac{1}{2}(t+t_{j-1})}\, dt = I_j.$$

(v) It is enough to show that $|(1/2\pi) \int_0^b D_n(t)\, dt| < 1$, for $0 < b \leq \pi$, since D_n is an even function. This follows because the integral is the sum of terms which decrease in absolute value, and alternate in sign.

(vi) If $t_{j-1} \leq t \leq t_j$ then

$$|D_n(t)| \geq \frac{|\sin(n+\frac{1}{2})t|}{\sin\frac{1}{2}t_j} \geq \frac{2|\sin(n+\frac{1}{2})t|}{t_j},$$

so that

$$I_j \geq \frac{1}{2\pi t_j} \int_{t_{j-1}}^{t_j} |\sin(n+\frac{1}{2})t|\, dt = \frac{1}{\pi t_j} \int_0^{t_1} \sin(n+\frac{1}{2})t\, dt = \frac{1}{\pi t_j} \cdot \frac{2t_1}{\pi} = \frac{2}{\pi^2 j}.$$

Thus

$$\frac{1}{2\pi} \int_{-\pi}^{\pi} |D_n(t)| \, dt \geq \frac{4}{\pi^2} \sum_{j=1}^{n} \frac{1}{j} \geq \frac{4}{\pi^2} \log n. \qquad \square$$

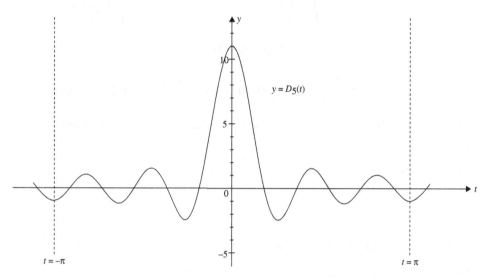

Figure 9.6. The Dirichlet kernel D_5.

The Dirichlet kernel is not very well behaved. First, it takes both positive and negative values. Secondly, $\frac{1}{2\pi} \int_{-\pi}^{\pi} |D_n(t)| \, dt \to \infty$ as $n \to \infty$. This last property underlies the fact that Fourier series of continuous functions in $\mathbf{V}$ need not converge point-wise.

Nevertheless, the Dirichlet kernel can be used to provide useful information about the convergence of Fourier series. We need to impose conditions that are generally stronger than continuity. Suppose that $f \in \mathbf{V}$, that $t \in [-\pi, \pi]$ and that we want to investigate the convergence of the Fourier series of f at t. We set

$$\phi_t(f)(s) = \tfrac{1}{2}(f(t+s) + f(t-s)) - f(t) \text{ for } |s| \leq \pi,$$

$$\theta_t(f)(s) = \begin{cases} \phi_t(s)/\tan s/2 & \text{for } 0 < |s| \leq \pi, \\ 0 & \text{for } s = 0, \end{cases}$$

and extend by periodicity.

Note that $\phi_t(f)$ is an even function and that $\theta_t(f)$ is an odd function. The function $\phi_t(f)$ is in $\mathbf{V}$, and $\|\phi_t(f)\|_1 \leq 2\|f\|_1$.

The function $\theta_t(f)$ can behave badly near 0 (although the function $\theta_t(f)(s)\sin ns$ is bounded on $[-\pi, \pi]$). If f is differentiable at t then $\theta_t(f)(s) \to 0$ as $s \to 0$. Since the Dirichlet kernel is an even function, and $\frac{1}{2\pi}\int_{-\pi}^{\pi} D_n(t)\, dt = 1$,

$$s_n(f)(t) - f(t) = \frac{1}{2\pi}\int_{-\pi}^{\pi}\left(\frac{f(t+s)+f(t-s)}{2} - f(t)\right)D_n(s)\, ds$$

$$= \frac{1}{\pi}\int_0^\pi \phi_t(f)(s)D_n(s)\, ds$$

$$= \frac{1}{\pi}\int_0^\pi \theta_t(f)(s)\sin ns\, ds + \frac{1}{\pi}\int_0^\pi \phi_t(f)(s)\cos ns\, ds.$$

We use this to give a criterion for the Fourier series of f to converge to $f(t)$ at t.

Theorem 9.6.2 (Dini's test)   *If the improper integral*

$$I = \frac{1}{\pi}\int_0^\pi |\theta_t(f)(s)|\, ds = \lim_{\eta \searrow 0}\frac{1}{\pi}\int_\eta^\pi |\theta_t(f)(s)|\, ds$$

is finite, then $s_n(f)(t) \to f(t)$ *as* $n \to \infty$.

Proof   If $\epsilon > 0$ there exists $0 < \eta < \pi$ such that

$$\frac{1}{\pi}\int_0^\eta |\theta_t(f)(s)|\, ds = I - \frac{1}{\pi}\int_\eta^\pi |\theta_t(f)(s)|\, ds < \epsilon/2.$$

Let

$$g(s) = \begin{cases} 0 & \text{if } |s - t| < \eta, \\ f(s) & \text{if } \eta \le |s - t| \le \pi, \end{cases}$$

and extend g by periodicity to obtain a function in $\mathbf{V}$. Then $\theta_t(g)$ is an odd function in $\mathbf{V}$ which vanishes in $(-\eta, \eta)$, and

$$s_n(g)(t) = \frac{1}{\pi}\int_0^\pi \theta_t(g)(s)\sin ns\, ds + \frac{1}{\pi}\int_0^\pi \phi_t(g)(s)\cos ns\, ds = \widehat{\theta_t(g)}_n + \widehat{\phi_t(g)}_n.$$

Thus $s_n(g)(t) \to 0$ as $n \to \infty$, and so there exists n_0 such that $|s_n(g)(t)| < \epsilon/2$ for $n \ge n_0$.

On the other hand,

$$|s_n(g)(t) - (s_n(f)(t) - f(t))| = \left|\frac{1}{\pi}\int_0^\eta \theta_t(f)(s)\sin ns\, ds\right|$$

$$\le \frac{1}{\pi}\int_0^\eta |\theta_t(f)(s)|\, ds < \epsilon/2,$$

and so $|s_n(f)(t) - f(t)| < \epsilon$ for $n \ge n_0$. $\quad\square$

Note that the condition in Dini's test is equivalent to the requirement that the improper integral $(1/\pi) \int_0^\pi |\phi_t(f)(s)|/s\, ds$ should be finite.

We can say more, if f vanishes on an interval.

Theorem 9.6.3 (Riemann's localization theorem) *Suppose that $f \in V$, that $[a, b] \subseteq [\pi, \pi]$ and that $f(t) = 0$ for $t \in [a, b]$. Suppose that $0 < \delta < (b - a)/2$. Then $s_n(f)(t) \to 0$ as $n \to \infty$ uniformly on $[a + \delta, b - \delta]$.*

Proof We need two lemmas, of interest in themselves.

Lemma 9.6.4 *If $f \in V$ and $n \in \mathbf{Z} \setminus \{0\}$ then $|\hat{f}_n| \le K_{\pi/n}(f)/2$.*

Proof Since $e^{-in(t+\pi/n)} = -e^{-int}$,

$$|\hat{f}_n| = \frac{1}{2} \left| \frac{1}{2\pi} \int_{-\pi}^\pi f(t)(e^{-int} - e^{-in(t+\pi/n)})\, dt \right|$$

$$= \frac{1}{2} \left| \frac{1}{2\pi} \int_{-\pi}^\pi (f(t) - f(t + \pi/n))e^{-int})\, dt \right| \le \frac{K_{\pi/n}(f)}{2}. \qquad \square$$

Corollary 9.6.5 *If $n \in \mathbf{N}$ then*

$$|a_n(f)| \le K_{\pi/n}(f)/2 \text{ and } |b_n(f)| \le K_{\pi/n}(f)/2.$$

Lemma 9.6.6 *Suppose that $f, g \in V$, that $t \in \mathbf{R}$ and that $\delta > 0$. Let $h_t(s) = f(t - s)g(s)$. Then*

$$K_\delta(h_t) \le \|g\|_\infty K_\delta(f) + \|f\|_\infty K_\delta(g).$$

Proof

$$K_\delta(h_t) = \frac{1}{2\pi} \int_{-\pi}^\pi |f(t - s + \delta)g(s + \delta) - f(t - s)g(s)|\, ds$$

$$\le \frac{1}{2\pi} \int_{-\pi}^\pi |(f(t - s + \delta) - f(t - s))g(s + \delta)|\, ds +$$

$$\frac{1}{2\pi} \int_{-\pi}^\pi |f(t - s)(g(s + \delta) - g(s))|\, ds$$

$$\le \|g\|_\infty \frac{1}{2\pi} \int_{-\pi}^\pi |f(t - s + \delta) - f(t - s)|\, ds +$$

$$\|f\|_\infty \frac{1}{2\pi} \int_{-\pi}^\pi |(g(s + \delta) - g(s))|\, ds$$

$$= \|g\|_\infty K_\delta(f) + \|f\|_\infty K_\delta(g). \qquad \square$$

The importance of this lemma is that the right-hand side of the inequality does not involve t.

We now prove the theorem. Let

$$g(t) = \begin{cases} 0 & \text{if } |t| < \delta \text{ or } |t| = \pi, \\ 1/\tan\frac{1}{2}t & \text{if } \delta \le |t| < \pi, \end{cases}$$

and extend by periodicity. Then $g \in \mathbf{V}$. If $t \in [a + \delta, b - \delta]$ then

$$s_n(f)(t) = \frac{1}{2\pi}\int_{-\pi}^{\pi} f(t-s)g(s)\sin ns\, ds + \frac{1}{2\pi}\int_{-\pi}^{\pi} f(t-s)\cos ns\, ds,$$

so that

$$|s_n(f)(t)| \le \tfrac{1}{2}(\|g\|_\infty K_{\pi/n}(f) + \|f\|_\infty K_{\pi/n}(g) + K_{\pi/n}(f)).$$

The right-hand side of this inequality does not involve t, and tends to 0 as $n \to \infty$. $\square$

Suppose that $f \in \mathbf{V}$, that $t_0 \in (-\pi, \pi]$ and that $0 < \eta < \pi$. Let $g(t) = f(t)$ if $|t - t_0| < \eta$, let $g(t) = 0$ if $\eta < |t - t_0| \le \pi$, and extend by periodicity. Since $f - g = 0$ on $(t_0 - \eta, t_0 + \eta)$, Riemann's localization theorem says that the Fourier series for f converges at t_0 (or at any point in $(t_0 - \eta, t_0 + \eta)$) if and only if the same holds for g: convergence is a local property, depending only on the values of f near t_0.

Let us apply these results to the function I_δ of Example 9.3.9. It follows that

$$s_n(f)(t) \to \begin{cases} \pi/\delta & \text{uniformly in } |t| < \delta - \eta, \text{ for } 0 < \eta < \delta \\ \pi/2\delta & \text{if } t = \delta \text{ or } t = -\delta \\ 0 & \text{uniformly in } \delta + \eta < |t| \le \pi, \text{ for } 0 < \eta < \pi - \delta. \end{cases}$$

In particular, if we set $\delta = \pi/2$ and $t = 0$, it follows that

$$1 - \frac{1}{3} + \frac{1}{5} - \cdots = \frac{\pi}{4}.$$

We now show that if f is monotonic in an open interval, then the Fourier series converges at each point of the interval. Recall that if f is monotonic in an open interval I and that $t \in I$ then $f(t+) = \inf\{f(s) : s \in I, s > t\}$ and that $f(t-) = \sup\{f(s) : s \in I, s < t\}$.

Theorem 9.6.7 (Jordan's theorem)   *Suppose that $f \in \mathbf{V}$ and that f is monotonic in an open interval I. If $t \in I$ then the Fourier series for f converges at t to $\frac{1}{2}(f(t+) + f(t-))$.*

Proof We make several simplifications: we remove the jump, and we localize. We can clearly suppose that f is increasing in I. By a change of variables, we can suppose that $t = 0$, so that we need to show that $\sum_{j=-n}^{n} \hat{f}_n = \frac{1}{2}(f(0+) + f(0-))$. We can also suppose that $f(0) = \frac{1}{2}(f(0+) + f(0-))$. Let $j(0) = j(\pi) = 0$, let $j(s) = 1$ for $0 < s \leq \pi$ and $j(s) = -1$ for $-\pi < s < 0$, and extend j by periodicity. Then j is an odd function, and so $\sum_{j=-n}^{n} \hat{j}_n = 0$ for all $n \in \mathbf{N}$. Now let

$$g(s) = f(s) - \tfrac{1}{2}(f(0+) - f(0-))j(s) - f(0).$$

Then

$$\sum_{k=-n}^{n} \hat{g}_k = \sum_{k=-n}^{n} \hat{f}_k - f(0),$$

and so we need to show that $\sum_{j=-n}^{n} \hat{g}_n = 0$.

From the construction, $g(0) = 0$ and g is continuous at 0. Suppose that $\epsilon > 0$. There exists $\delta > 0$ such that $(-\delta, \delta) \subseteq I$ and such that $-\epsilon/5 < g(-\delta) \leq g(\delta) < \epsilon/5$. Now set $h(s) = g(s)$ if $|s| \leq \delta$ and set $h(s) = 0$ if $\delta < |s| \leq \pi$, and extend by periodicity. Since $g(s) - h(s) = 0$ on $[-\delta, \delta]$. $\sum_{j=-n}^{n} \hat{h}_j - \sum_{j=-n}^{n} \hat{g}_j \to 0$ as $n \to \infty$, by Riemann's localization theorem. Thus there exists n_0 such that

$$\Big| \sum_{j=-n}^{n} \hat{h}_j - \sum_{j=-n}^{n} \hat{g}_j \Big| < \epsilon/5 \text{ for } n \geq n_0.$$

By Du Bois–Reymond's mean-value theorem (Corollary 8.6.4), there exists $-\delta < c < \delta$ such that

$$\sum_{j=-n}^{n} \hat{h}_j = \frac{1}{2\pi} \int_{-\delta}^{\delta} D_n(s) h(s) \, ds$$

$$= \frac{h(-\delta)}{2\pi} \int_{-\delta}^{c} D_n(s) \, ds + \frac{h(\delta)}{2\pi} \int_{c}^{-\delta} D_n(s) \, ds.$$

Using Theorem 9.6.1 (v), it follows that

$$\Big| \sum_{j=-n}^{n} \hat{h}_j \Big| \leq \frac{\epsilon}{5} \left(\Big| \frac{1}{2\pi} \int_{-\delta}^{c} D_n(s) \, ds \Big| + \Big| \frac{1}{2\pi} \int_{c}^{\delta} D_n(s) \, ds \Big| \right) \leq \frac{4\epsilon}{5}.$$

Thus $\big| \sum_{j=-n}^{n} \hat{g}_j \big| < \epsilon$ for $n \geq n_0$. $\qquad \square$

Exercise

9.6.1 Suppose that $f \in \mathbf{V}$ and that f is monotonic and continuous in an open interval I. Show that the Fourier series for f converges uniformly to f in any closed subinterval of I.

9.7 The Fejér kernel and the Poisson kernel

If f is a continuous function in $\mathbf{V}$, it is an easy matter to calculate its harmonics. On the other hand, the example of Section 9.5 shows that the partial sums $s_n(f)(t) = \sum_{j=-n}^{n} \hat{f}_j e^{ijt}$ need not converge to $f(t)$. Can we use the harmonics to reconstruct f? This is the problem of *harmonic synthesis*. We give two important examples of harmonic synthesis.

The first was given by Lipót Fejér, at the age of nineteen. It is based on the idea that the average of terms in a sequence can behave better than the terms themselves. If $f \in \mathbf{V}$, we set

$$\sigma_n(f) = \frac{1}{n+1} \sum_{j=0}^{n} s_j(f) = \frac{1}{n+1} \sum_{j=0}^{n} (D_j \star f) = F_n \star f,$$

where $F_n = \left(\sum_{j=0}^{n} D_j\right)/(n+1)$ is the *Fejér kernel*. Using the formulae

$$2\sin(j + \tfrac{1}{2})t \sin \tfrac{1}{2}t = \cos jt - \cos(j+1)t \text{ and } 1 - \cos 2\alpha t = 2\sin^2 \alpha t,$$

we see that $f_n(0) = n + 1$ and that

$$F_n(t) = \frac{1}{n+1} \sum_{j=0}^{n} \frac{\sin(j + \tfrac{1}{2})t \sin \tfrac{1}{2}t}{2\sin^2 \tfrac{1}{2}t}$$

$$= \frac{1}{n+1}\left(\frac{1 - \cos(n+1)t}{2\sin^2 \tfrac{1}{2}t}\right) = \frac{1}{n+1}\left(\frac{\sin^2((n+1)t/2)}{\sin^2 \tfrac{1}{2}t}\right),$$

for $0 < |t| \le \pi$.

The Fejér kernel has three important properties.

- $F_n(t)$ is a non-negative function.
- $\frac{1}{2\pi}\int_{-\pi}^{\pi} F_n(t)\,dt = \left(\sum_{j=0}^{n} \frac{1}{2\pi}\int_{-\pi}^{\pi} D_j(t)\,dt\right)/(n+1) = 1$.
- If $0 < \delta < \pi$ then $F_n(t) \to 0$ uniformly on $\{t : \delta < |t| \le \pi\}$.

A sequence of functions in $\mathbf{V}$ with these properties is called an *approximate identity*.

Theorem 9.7.1 *If $(\phi_n)_{n=0}^{\infty}$ is an approximate identity in $\mathbf{V}$ and $f \in \mathbf{V}$ is continuous at t_0 then $(\phi_n \star f)(t_0) \to f(t_0)$ as $n \to \infty$. If f is continuous*

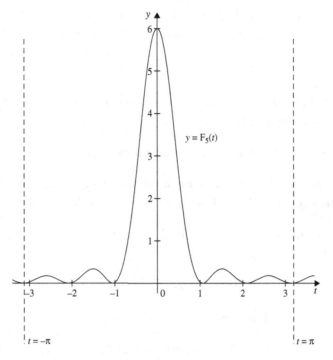

Figure 9.7a. The Fejér kernel.

on a closed interval $[a, b]$ *and* $0 < \eta < (b - a)/2$, *then* $(\phi_n \star f)(t) \to f(t)$, *as* $n \to \infty$, *uniformly in* $[a + \eta, b - \eta]$.

Proof Suppose that $\epsilon > 0$. There exists $0 < \delta < \pi$ such that if $|s - t_0| < \delta$ then $|f(s) - f(t_0)| < \epsilon/3$. Then

$$|(\phi_n \star f)(t_0) - f(t_0)| = |\frac{1}{2\pi} \int_{-\pi}^{\pi} (f(t_0 - s) - f(t_0))\phi_n(s)\, ds| \leq I_1 + I_2 + I_3,$$

where

$$I_1 = \frac{1}{2\pi} \int_{-\pi}^{-\delta} (|f(t_0 - s)| + |f(s)|)\phi_n(s)\, ds$$

$$\leq \|f\|_\infty \sup\{\phi_n(t) : -\pi \leq t \leq -\delta\},$$

$$I_2 = \frac{1}{2\pi} \int_{-\delta}^{\delta} (|f(t_0 - s) - f(s)|)\phi_n(s)\, ds < \epsilon/3,$$

$$I_3 = \frac{1}{2\pi} \int_{-\pi}^{-\delta} (|f(t_0 - s)| + |f(s)|)\phi_n(s)\, ds$$

$$\leq \|f\|_\infty \sup\{\phi_n(t) : \delta \leq t \leq \pi\},$$

so that there exists n_0 such that $|(\phi_n \star f)(t_0) - f(t_0)| < \epsilon$ for $n \geq n_0$.

If f is continuous on $[a, b]$ then it is uniformly continuous, and there exists $0 < \delta < \pi$ such that if $s, t \in [a, b]$ and $|s - t| < \delta$ then $|f(s) - f(t)| < \epsilon/3$. Thus, choosing $\delta < \eta$ in the preceding argument, n_0 can be chosen so that $|(\phi_n \star f)(t) - f(t)| < \epsilon$ for all $t \in [a, b]$, for $n \geq n_0$. □

Corollary 9.7.2 *If f is a continuous function in* **V** *then $\sigma_n(f)(t) \to f(t)$ as $n \to \infty$, uniformly in t.*

We have a version of Riemann's localization theorem.

Corollary 9.7.3 *If $f(t) = 0$ on a closed interval $[a, b]$ and $0 < \eta < (b - a)/2$ then $(\phi_n \star f)(t) \to 0$, as $n \to \infty$, uniformly in $[a + \eta, b - \eta]$.*

Proposition 9.7.4 *If $f \in$ **V** and if $s_n(f)(t) \to l$ as $n \to \infty$ then $\sigma_n(f)(t) \to l$ as $n \to \infty$.*

Proof This is part of Exercise 4.6.2. □

This has the following important consequence.

Corollary 9.7.5 *If $f \in$ **V** is continuous at t_0 and if $s_n(f)(t_0)$ converges as $n \to \infty$, then it converges to $f(t_0)$.*

Proof For if $s_n(f)(t_0) \to l$ as $n \to \infty$, then $\sigma_n(f)(t_0) \to l$ as $n \to \infty$, and so $l = f(t_0)$. □

Corollary 9.7.2 shows that a continuous function in **V** can be uniformly approximated by trigonometric polynomials: we use this to show that a continuous function on $[0, 1]$ can be uniformly approximated by polynomials.

Theorem 9.7.6 *If f is a continuous function on $[0, 1]$ and $\epsilon > 0$ there exists a polynomial p such that $|f(x) - p(x)| < \epsilon$ for all $x \in [0, 1]$.*

Proof We need a lemma.

Lemma 9.7.7 *For each $n \in \mathbf{Z}^+$ there exists a polynomial T_n such that $\cos nt = T_n(\cos t)$ for all $t \in \mathbf{R}$.*

Proof The proof is by induction on n. The result is true for $n = 1$ and $n = 2$. Suppose that it is true for all $m \leq n$, where $n \geq 1$. Then

$$\cos(n + 1)t = 2\cos nt \cos t - \cos(n - 1)t = 2T_n(\cos t)\cos t - T_{n-1}(\cos t). \quad \square$$

The polynomial T_n is called the n-th *Chebyshev polynomial*.

We now prove the theorem. Let $g(t) = f(\cos t)$. Then g is a continuous even function in $\mathbf{V}$, and so there exists $n \in \mathbf{N}$ such that $|\sigma_n(g)(t) - g(t)| < \epsilon$ for all t. But $\sigma_n(g)$ is an even trigonometric polynomial, and so

$$\sigma_n(g)(t) = \sum_{j=0}^{n} c_j \cos jt = \sum_{j=0}^{n} c_j T_j(\cos t)$$

for some constants $c_0, \ldots, c_n$. Thus if $x = \cos t \in [0, 1]$ and $p = \sum_{j=0}^{n} c_j T_j$ then

$$|f(x) - p(x)| = |f(\cos t) - p(\cos t)| = |g(t) - \sigma_n(g)(t)| < \epsilon. \qquad \square$$

The second example of harmonic synthesis is obtained by damping the contributions for large values of $|n|$. Suppose that $f \in \mathbf{V}$ and that $0 \le r < 1$. Then we set

$$P_r(f)(t) = \sum_{n=-\infty}^{\infty} r^{|n|} \hat{f}_n e^{int}.$$

Since $|\hat{f}_n| \le \|f\|_1$, the series converges absolutely, and converges uniformly in t. Thus $P_r(f)$ is a continuous function. Let us set

$$P_{r,n} = \sum_{j=-n}^{n} r^{|j|} \gamma_j \text{ and } P_r = \sum_{j=-\infty}^{\infty} r^{|j|} \gamma_j.$$

Then $P_{r,n} \to P_r$ as $n \to \infty$, uniformly in t, and so

$$P_r(f)(t) = \lim_{n \to \infty} \frac{1}{2\pi} \int_{-\pi}^{\pi} P_{r,n}(t - s) f(s) \, ds$$

$$= \lim_{n \to \infty} (P_{r,n} \star f)(t) = (P_r \star f)(t).$$

The function $(r, t) \to P_r(t)$ is the *Poisson kernel*. Now

$$P_r(t) = \sum_{j=0}^{\infty} r^j e^{ijt} + \sum_{j=0}^{\infty} r^j e^{-ijt} - 1$$

$$= \frac{1}{1 - re^{it}} + \frac{1}{1 - re^{-it}} - 1 = \frac{1 - r^2}{1 - 2r \cos t + r^2}.$$

Note that $P_r(0) = (1 + r)/(1 - r)$, that $P_r(t) \ge 0$ and that P_r is an even function. Further,

$$\frac{1}{2\pi} \int_{-\pi}^{\pi} P_r(t) \, dt = \sum_{j=-\infty}^{\infty} \left(\frac{r^{|j|}}{2\pi} \int_{-\pi}^{\pi} e^{ijt} \, dt \right) = 1.$$

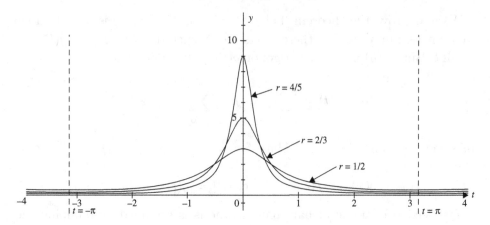

Figure 9.7b. The Poisson kernel.

Since

$$1 - 2r\cos t + r^2 = (1-r)^2 + 2r(1-\cos t) \geq 2r(1-\cos t) = 4r\sin^2 \tfrac{1}{2}t,$$

$P_r(t) \to 0$ uniformly on $\{t : \delta \leq |t| \leq \pi\}$ as $r \nearrow 1$, for $0 < \delta < \pi$.

Thus the Poisson kernel is an approximate identity (though here the parameter r is in $[0,1)$, and we are concerned with limits as r increases to 1). Thus we have the following.

Theorem 9.7.8   *If $f \in \mathbf{V}$ is continuous at t_0 then $P_r(f)(t_0) \to f(t_0)$ as $r \nearrow 1$. If f is a continuous function, then $P_r(f)(t) \to f(t)$, uniformly in t, as $r \nearrow 1$.*

Which of these two methods is more powerful? In order to answer this, we need a stronger version of Abel's theorem (Theorem 6.6.5).

Theorem 9.7.9   *Suppose that $(a_n)_{n=0}^\infty$ is a real or complex sequence. Let $s_n = \sum_{j=0}^n a_j$, and let $\sigma_n = (\sum_{j=0}^n s_j)/(n+1)$, for $n \in \mathbf{Z}^+$. Suppose that $\sigma_n \to \sigma$ as $n \to \infty$. Then $\sum_{n=0}^\infty a_n x^n \to \sigma$ as $x \nearrow 1$.*

Proof   The proof is very similar to the proof of Theorem 6.6.5. By replacing a_0 by $a_0 - \sigma$, we can suppose that $\sigma = 0$. Let $s_n = \sum_{j=0}^n a_j$, for $n \in \mathbf{Z}^+$. Suppose that $0 \leq x < 1$. Let $f(x) = \sum_{n=0}^\infty a_n x^n$. Recall that $1/1 - x)^2 = \sum_{n=0}^\infty (n+1)x^n$. Each of the series $\sum_{n=0}^\infty a_n x^n$ and $\sum_{n=0}^\infty (n+1)x^n$ converges absolutely, and so by Proposition 4.6.1, the convolution product $\sum_{n=0}^\infty c_n x^n$ converges absolutely to $f(x)/(1-x)^2$. But

$$c_n = \sum_{j=0}^n a_j(n+1-j) = (n+1)\sigma_n,$$

and so $f(x) = (1-x)^2 \sum_{n=0}^{\infty}(n+1)\sigma_n x^n$.

Suppose that $0 < \epsilon < 1$. There exists n_0 such that $|\sigma_n| < \epsilon/2$ for $n \geq n_0$, and so

$$|(1-x)^2 \sum_{n=n_0}^{\infty} s_n x^n| \leq \epsilon(1-x)^2(\sum_{n=n_0}^{\infty}(n+1)x^n)/2$$

$$\leq \epsilon(1-x)^2(\sum_{n=0}^{\infty}(n+1)x^n)/2 = \epsilon/2.$$

On the other hand, the sequence $(\sigma_n)_{n=0}^{\infty}$ is bounded: let $M = \sup\{|\sigma_n| : n \in \mathbf{Z}^+\}$. Then

$$|(1-x)^2 \sum_{n=0}^{n_0-1}(n+1)\sigma_n x^n| \leq (1-x)^2 M n_0(n_0+1)/2.$$

Let $\eta = (\epsilon/((M+1)n_0(n_0+1))^{\frac{1}{2}}$. If $1 - \eta < x < 1$ then

$$|(1-x)^2 \sum_{n=0}^{n_0-1}(n+1)\sigma_n x^n| < \epsilon/2,$$

and so

$$|\sum_{n=0}^{\infty} a_n x^n| = |f(x)| = |(1-x)^2 \sum_{n=0}^{\infty}(n+1)\sigma_n x^n| < \epsilon. \qquad \square$$

It follows that the Poisson kernel is more powerful than the Fejér kernel.

Exercise

9.7.1 Suppose that $(a_n)_{n=0}^{\infty}$ is a real or complex sequence. Let $s_n = \sum_{j=0}^{n} a_j$, and let $\sigma_n = (\sum_{j=0}^{n} s_j)/(n+1)$, for $n \in \mathbf{Z}^+$. Suppose that $\sigma_n \to \sigma$ as $n \to \infty$. Suppose that $K > 0$. Let $W_K = \{z : |1-z| \leq K(1-|z|)\}$. Show that $\sum_{n=0}^{\infty} a_n z^n \to \sigma$ as $z \to 1$ in W_K.

10

Some applications

The theory that we have developed is meant to be used. In this chapter, we consider various applications of the results that we have established, and in particular will introduce some of the important special functions of analysis. Some details are omitted; you should provide them. We shall return later to some of the topics considered here in Volumes II and III.

10.1 Infinite products

Suppose that $(a_j)_{j=0}^{\infty}$ is a sequence of real numbers, and that $a_j \neq -1$ for all $j \in \mathbf{Z}^{+}$. Let $p_n = \prod_{j=0}^{n}(1+a_j)$. We say that the *infinite product* $\prod_{j=0}^{\infty}(1+a_j)$ converges to p if $p \neq 0$ and $p_n \to p$ as $n \to \infty$. If $p_n \to 0$ as $n \to \infty$ we say that the product *diverges to* 0. If the product converges, then $a_n = (p_n - p_{n-1})/p_{n-1} \to 0$ as $n \to \infty$: this means that we can restrict attention to products for which $a_j > -1$ for all $j \in \mathbf{N}^{+}$, so that $1 + a_j > 0$ for all $j \in \mathbf{N}$, and $p_n > 0$ for all $n \in \mathbf{N}$. The logarithmic function then enables us to reduce the problem of convergence of the product to the convergence of a sum. The function log is a continuous bijection of $(0, \infty)$ onto $(-\infty, \infty)$, with continuous inverse exp. Hence $p_n \to p$ (with $p \neq 0$) as $n \to \infty$ if and only if

$$\sum_{j=0}^{n} \log(1 + a_j) = \log p_n \to \log p \text{ as } n \to \infty.$$

The general principle of convergence takes the following form.

Proposition 10.1.1 *Suppose that $(a_j)_{j=0}^{\infty}$ is a sequence of real numbers, and that $a_j > -1$ for all $j \in \mathbf{Z}^{+}$. Then the product $\prod_{j=0}^{\infty}(1 + a_j)$ converges* *if and only if, given $\epsilon > 0$, there exists $n_0 \in \mathbf{Z}^+$ such that*

$$\left| \prod_{j=n+1}^{m} (1 + a_j) - 1 \right| = \left| \frac{p_m - p_n}{p_n} \right| < \epsilon$$

for $m > n \geq n_0$.

Proof Let us prove this directly. Suppose that $p_n \to p$, and that $\epsilon > 0$. Then $p/p_n \to 1$, and so there exists n_0 such that $|p_m - p_n| < \epsilon/2p$ and $p/p_n < 2$, for $m > n \geq n_0$. Then

$$\left| \prod_{j=n+1}^{m} (1 + a_j) - 1 \right| = \left| \frac{p_m - p_n}{p_n} \right| = |\frac{p_m - p_n}{p}| . \frac{p}{p_n} < \epsilon$$

for $m > n \geq n_0$.

Conversely suppose that the condition is satisfied. Then there exists n_0 such that

$$\left| \frac{p_m}{p_n} - 1 \right| = \left| \prod_{j=n+1}^{m} (1 + a_j) - 1 \right| < 1/2,$$

for $m > n \geq n_0$, and so $p_{n_0}/2 \leq p_n \leq 2p_{n_0}$ for $n \geq n_0$. Given $\epsilon > 0$, there exists $n_1 \geq n_0$ such that $|(p_m/p_n) - 1| < \epsilon/2p_{n_0}$ for $m > n \geq n_1$. If $m > n \geq n_1$ then

$$|p_m - p_n| = \left| \frac{p_m}{p_n} - 1 \right| . p_n < \epsilon,$$

so that, by the general principle of convergence, $(p_n)_{n=0}^{\infty}$ converges, to p say. Further, since $p_n \geq p_{n_0}/2$ for $n \geq n_0$, $p \geq p_{n_0}/2 > 0$. $\square$

If the infinite product $\prod_{j=0}^{\infty}(1 + a_j)$ converges, then $a_j \to 0$ as $j \to \infty$. If $\sum_{j=0}^{\infty} a_j^2 < \infty$, we can say much more.

Theorem 10.1.2 *Suppose that $(a_j)_{j=0}^{\infty}$ is a sequence of real numbers, that $a_j > -1$ for all $j \in \mathbf{Z}^+$ and that $\sum_{j=0}^{\infty} a_j^2 < \infty$. Then the infinite product $\prod_{j=0}^{\infty}(1 + a_j)$ converges if and only if $\sum_{j=0}^{\infty} a_j$ converges.*

Proof We use the fact that if $|h| < 1/2$ then, by Taylor's theorem,

$$\log(1 + h) - h = -\frac{h^2}{2(1 + \theta h)^2}, \quad \text{for some } 0 < \theta < 1,$$

so that

$$h - 2h^2 \leq \log(1 + h) \leq h.$$

There exists n_0 such that $|a_j| < 1/2$ for $j \geq n_0$. If $m > n \geq n_0$ then

$$\sum_{j=n+1}^{m} \log(1 + a_j) \leq \sum_{j=m+1}^{n} a_j \leq \sum_{j=n+1}^{m} \log(1 + a_j) + 2 \sum_{j=n+1}^{m} a_j^2,$$

and it follows easily from this that $(\sum_{j=0}^{n} \log(1+a_j))_{n=0}^{\infty}$ is a Cauchy sequence if and only if $(\sum_{j=0}^{n} a_j)_{n=0}^{\infty}$ is a Cauchy sequence. The result therefore follows from the general principle of convergence. □

Next let us consider the cases when the terms a_j are all positive.

Proposition 10.1.3 *Suppose that $(a_j)_{j=0}^{\infty}$ is a sequence of positive numbers, and that $a_j < 1$ for all $j \in \mathbf{N}^+$. The following statements are equivalent.*

(i) $\sum_{j=0}^{\infty} a_j$ converges.
(ii) $\prod_{j=0}^{\infty}(1 + a_j)$ converges.
(iii) $\prod_{j=0}^{\infty}(1 - a_j)$ converges.
(iv) If $|b_j| \leq 1$ and $a_j b_j \neq -1$ for $j \in \mathbf{Z}^+$ then $\prod_{j=0}^{\infty}(1 + b_j a_j)$ converges.

Proof If (i) holds and $|b_j| \leq 1$ for $j \in \mathbf{Z}^+$ then $\sum_{j=0}^{\infty}(a_j b_j)^2 < \infty$, and so (iv) holds, by Theorem 10.1.2. Clearly, (iv) implies (iii). Suppose that $\prod_{j=0}^{\infty}(1 - a_j)$ converges, to q, say. Then since $1 + a_j < 1/(1 - a_j)$,

$$\prod_{j=0}^{n}(1 + a_j) \leq \left(\prod_{j=0}^{n}(1 - a_j) \right)^{-1} \leq 1/q,$$

and so the increasing sequence $(\prod_{j=0}^{n}(1 + a_j))_{n=0}^{\infty}$ converges: (ii) holds.

Suppose that (ii) holds. Let $p_n = \prod_{j=0}^{n}(1 + a_j)$ and let $p = \prod_{j=0}^{\infty}(1 + a_j)$. Since $a_j \to 0$ as $j \to \infty$, there exists n_0 such that $a_j < 1$ for $j \geq n_0$. But, by the mean-value theorem, $\log(1 + x) = x/(1 + \theta x)$ for some $0 < \theta < 1$ and so $\log(1 + x) \geq x/2$, for $0 \leq x \leq 1$. Therefore

$$\sum_{j=0}^{n} a_j - \sum_{j=0}^{n_0} a_j \leq 2 \sum_{j=n_0+1}^{n} \log(1 + a_j) = 2\log(p_n/p_{n_0}) \leq 2\log(p/p_{n_0})$$

for $n \geq n_0$, so that $\sum_{j=0}^{\infty} a_j$ converges. Thus (i) holds. □

Corollary 10.1.4 *If $(a_j)_{j=0}^{\infty}$ is a real sequence, none of whose terms takes the value -1, and if $\sum_{j=0}^{\infty} |a_j| < \infty$, then the infinite product $\prod_{j=0}^{\infty}(1 + a_j)$ converges. Further, if σ is a permutation of $\mathbf{N}^+$ then $\prod_{j=0}^{\infty}(1 + a_{\sigma(j)})$ converges, to the same value.*

Such a product is said to be *absolutely convergent*.

Exercises

10.1.1 Use the logarithmic function to deduce Proposition 10.1.1 from Theorem 3.6.2.

10.1.2 Let $a_{2k} = 1/\sqrt{k+1}$ and let $a_{2k+1} = -1/\sqrt{k+1}$. Show that $\sum_{j=0}^{\infty} a_j$ converges, whereas $\prod_{j=1}^{\infty}(1 + a_j)$ diverges to 0.

10.1.3 Let $a_0 = 0$, let $a_{2k-1} = 1/\sqrt{k}$ and let $a_{2k} = -1/\sqrt{k} + 1/k$, for $k \in \mathbf{N}$. Show that $\sum_{j=0}^{\infty} a_j$ diverges, whereas $\prod_{j=1}^{\infty}(1 + a_j)$ converges.

10.1.4 Why do these examples not contradict Theorem 10.1.2?

10.1.5 Let $p_1 < p_2 < \cdots$ be an enumeration of the primes. By considering products of the form $\prod_{j=1}^{n}(1 - 1/p_j)^{-1}$, or otherwise, show that $\prod_{j=1}^{\infty}(1 - 1/p_j)$ diverges to 0. Deduce that $\sum_{j=1}^{\infty}(1/p_j) = \infty$.

10.2 The Taylor series of logarithmic functions

Integrating the identity

$$\frac{1}{1+x} = 1 - x + \cdots + (-x)^{n-1} + \frac{(-x)^n}{1+x}$$

we see that

$$\log(1 + x) = x - \frac{x^2}{2} + \cdots - \frac{(-x)^n}{n} + \int_0^x \frac{(-t)^n}{1+t}\,dt,$$

for $x > -1$. Suppose that $-1 < x < 1$. Then the remainder term tends to 0 as $n \to \infty$, and so

$$\log(1 + x) = x - \frac{x^2}{2} + \cdots - \frac{(-x)^n}{n} + \cdots = \sum_{n=1}^{\infty} \frac{(-1)^{n+1}x^n}{n}.$$

Since

$$\log(1 - x) = -x - \frac{x^2}{2} - \cdots - \frac{x^n}{n} + \cdots,$$

it follows that

$$\log\left(\frac{1+x}{1-x}\right) = \log(1 + x) - \log(1 - x) = 2\left(x + \frac{x^3}{3} + \frac{x^5}{5} + \cdots\right),$$

for $-1 < x < 1$. We shall use this formula when we establish Stirling's formula.

10.2.1 Show that

$$\frac{1}{n}\left(1+\frac{1}{n}\right)^{n\log n} \to 1 \text{ as } n \to \infty.$$

10.3 The beta function

The *beta function* $B(x, y)$ is defined for $x > 0$ and $y > 0$ as

$$B(x, y) = \int_0^1 t^{x-1}(1 - t)^{y-1}\, dt.$$

Note that if $x < 1$ or $y < 1$ then this is an improper integral. Note also that $B(x, 1) = 1/x$. The change of variables $s = 1 - t$ shows that $B(x, y) = B(y, x)$. If we make the change of variables $t = \sin^2 \theta$ then $1 - t = \cos^2 \theta$ and $dt/d\theta = 2\sin\theta\cos\theta$, so that

$$B(x, y) = 2\int_0^{\pi/2} \sin^{2x-1}\theta \cos^{2y-1}\theta\, d\theta.$$

Proposition 10.3.1   *If $x > 0$ and $y > 0$ then*

$$B(x, y + 1) = \frac{yB(x, y)}{x + y}.$$

Proof   Integrating by parts,

$$B(x, y + 1) = \int_0^1 t^{x-1}(1 - t)^y\, dt$$

$$= \int_0^1 t^{x+y-1}\left(\frac{1}{t} - 1\right)^y\, dt$$

$$= \left[\frac{t^{x+y}}{x+y}\left(\frac{1}{t} - 1\right)^y\right]_0^1 + \frac{y}{x+y}\int_0^1 t^{x+y-2}\left(\frac{1}{t} - 1\right)^{y-1}\, dt$$

$$= \frac{y}{x+y}\int_0^1 t^{x-1}(1 - t)^{y-1}\, dt = \frac{y}{x+y}B(x, y). \qquad \square$$

This means that we can calculate the value of $B(x, y)$ for all positive x and y if we can calculate it for $0 < x \le 1$ and $0 < y \le 1$. Let

$$I_s = \int_0^{\pi/2} \sin^s \theta\, d\theta \text{ for } s > -1.$$

Corollary 10.3.2 (*i*) $B(1/2, 1/2) = \pi$.
(*ii*) $B(x/2, 1/2) = I_{x-1}$.
(*iii*) $sI_s = (s-1)I_{s-2}$, *for* $s > 1$.
(*iv*) *If* $k \in \mathbf{N}$ *then*

$$I_{2k} = \int_0^{\pi/2} \sin^{2k}\theta \, d\theta = \frac{(2k-1)(2k-3)\ldots 1}{(2k)(2k-2)\ldots 2} I_0$$

$$= \frac{(2k)!}{2^{2k}(k!)^2} \cdot \frac{\pi}{2} = \frac{1}{2^{2k}} \cdot \binom{2k}{k} \cdot \frac{\pi}{2}$$

and

$$I_{2k+1} = \int_0^{\pi/2} \sin^{2k+1}\theta \, d\theta = \frac{(2k)(2k-2)\ldots 2}{(2k+1)(2k-1)\ldots 3} I_1 = \frac{2^{2k}(k!)^2}{(2k+1)!}.$$

Proof These results all follow easily from the equation

$$B(x, y) = 2 \int_0^{\pi/2} \sin^{2x-1}\cos^{2y-1}\theta \, d\theta. \qquad \square$$

Now

$$1 \geq \frac{I_{2k+1}}{I_{2k}} = \frac{2k}{2k+1} \cdot \frac{I_{2k-1}}{I_{2k}} \geq \frac{2k}{2k+1},$$

so that $I_{2k+1}/I_{2k} \to 1$ as $k \to \infty$. Thus we obtain *Wallis' formula*

$$\frac{2^{4k+1}(k!)^4}{(2k+1)!(2k)!} \to \pi \text{ as } k \to \infty.$$

Let us establish a corresponding result for an infinite product. Since

$$\frac{2^k.k!}{(2k)!} = \frac{2.2.\ldots.2k}{1.2.\ldots.2k} = \frac{1}{1.3.5.\ldots.(2k-1)} \quad \text{and} \quad \frac{2^k.k!}{(2k+1)!} = \frac{1}{3.5.\ldots.(2k+1)},$$

it follows that

$$\frac{2^{4k}(k!)^4}{(2k)!(2k+1)!} = \frac{2.2.4.4.\ldots.(2k).(2k)}{1.3.3.5.\ldots.(2k-1)(2k+1)}$$

$$= \prod_{j=1}^{k} \left(\frac{(2j)^2}{(2j-1)(2j+1)} \right) = \prod_{j=1}^{k} \left(1 + \frac{1}{4j^2 - 1} \right).$$

Thus it follows from Wallis' formula that

$$\prod_{j=1}^{\infty} \left(1 + \frac{1}{4j^2 - 1} \right) = \frac{\pi}{2}.$$

If $0 < x < 1$ then

$$B(x, 1-x) = 2 \int_0^{\pi/2} \tan^{2x-1} \theta \, d\theta.$$

Our main concern now is to find an expression for this integral. Making the change of variables $v = t/(1-t)$, we find that

$$B(x, y) = \int_0^\infty \frac{v^{x-1}}{1+v^{x+y}} \, dv, \text{ so that } B(x, 1-x) = \int_0^\infty \frac{v^{x-1}}{1+v} \, dv.$$

Now

$$\int_0^1 \frac{v^{x-1}}{1+v} \, dv = \int_0^1 v^{x-1}(1 - v + \cdots + (-1)^n v^n) \, dv + (-1)^{n+1} \int_0^1 \frac{v^{x-1}v^n}{1+v} \, dv.$$

The second term on the right-hand side tends to 0 as $n \to \infty$ (why?), and so

$$\int_0^1 \frac{v^{x-1}}{1+v} \, dv = \frac{1}{x} + \sum_{n=1}^\infty (-1)^n \frac{1}{x+n}.$$

Similarly, making the change of variables $v = 1/w$, we find that

$$\int_1^\infty \frac{v^{x-1}}{1+v} \, dv = \int_0^1 \frac{1}{w^x(1+w)} \, dw$$

$$= \int_0^1 w^{-x} - w^{1-x} + \cdots + (-1)^n w^{n-x} \, dw + (-1)^{n-1} \int_0^1 \frac{w^{n+1-x}}{1+w} \, dw.$$

Again, the second term on the right-hand side tends to 0 as $n \to \infty$, and so

$$\int_1^\infty \frac{v^{x-1}}{1+v} \, dv = \sum_{n=1}^\infty (-1)^{n-1} \frac{1}{n-x}.$$

Adding the two integrals, we find that

$$B(x, 1-x) = 2 \int_0^{\pi/2} \tan^{2x-1} \theta \, d\theta = \frac{1}{x} + \sum_{n=1}^\infty (-1)^n \left(\frac{1}{n+x} - \frac{1}{n-x} \right)$$

$$= \frac{1}{x} - 2 \sum_{n=1}^\infty (-1)^n \left(\frac{1}{n^2 - x^2} \right).$$

We shall see in Volume III that this sum can be evaluated; its value is $\pi/\sin \pi x$.

The beta function is logarithmically convex: if x_0, x_1, y_0 and y_1 are positive and $0 < \theta < 1$ then, putting $x_\theta = (1-\theta)x_0 + \theta x_1$, $y_\theta = (1-\theta)y_0 + \theta y_1$ and using Hölder's inequality with indices $1/(1-\theta)$ and $1/\theta$, we see that

$$B(x_\theta, y_\theta) = \int_0^1 t^{x_\theta}(1-t)^{y_\theta}\, dt$$

$$= \int_0^1 (t^{x_0})^{1-\theta}(t^{x_1})^\theta ((1-t)^{y_0})^{1-\theta}((1-t)^{y_1})^\theta\, dt$$

$$= \int_0^1 (t_0^x(1-t)^{y_0})^{1-\theta}(t^{x_1}(1-t)^{y_1})^\theta\, dt$$

$$\leq \left(\int_0^1 (t_0^x(1-t)^{y_0})\, dt\right)^{1-\theta} \cdot \left(\int_0^1 (t_1^x(1-t)^{y_1})\, dt\right)^\theta$$

$$= B(x_0, y_0)^{1-\theta} B(x_1, y_1)^\theta.$$

Exercises

10.3.1 Show that $xB(x, y+1) = yB(x+1, y)$.

10.3.2 Show that $B(x, y) = B(x+1, y) + (B(x, y+1)$.

10.3.3 Show that

$$\prod_{j=1}^n \left(1 - \frac{1}{4j^2}\right) \to \frac{2}{\pi} \text{ as } n \to \infty.$$

10.3.4 Show that $n(\int_0^{\pi/2} \sin^n t\, dt)^2 \to 2\pi$ as $n \to \infty$. [Consider the cases n odd and n even separately.]

10.4 Stirling's formula

We wish to estimate the size of $n!$, as n becomes large. Let

$$a_n = \frac{n!e^n}{n^{n+1/2}}, \text{ and let } b_n = \log a_n = \log(n!) - (n+1/2)\log n + n.$$

Set $s = 1/(2n+1)$ (so that $(n+1)/n = (1+s)/(1-s)$). Using the result about logarithmic functions established in Section 10.2,

$$b_n - b_{n+1} = \log(n+1) - (n+1/2)\log n + (n+3/2)\log(n+1) - 1$$

$$= (n+1/2)\log\left(\frac{n+1}{n}\right) - 1$$

$$= \frac{1}{2s}\log\left(\frac{1+s}{1-s}\right) - 1 = \frac{s^2}{3} + \frac{s^4}{5} + \cdots$$

Thus

$$b_n - b_{n+1} \le \frac{s^2}{3}\left(\frac{1}{1-s^2}\right) = \frac{1}{12n(n+1)} = \frac{1}{12n} - \frac{1}{12(n+1)},$$

and $b_n - b_{n+1} \ge \frac{s^2}{3} = \frac{1}{3(2n+1)^2} > \frac{1}{12(n+1)} - \frac{1}{12(n+2)}.$

These inequalities show that the sequences

$$(b_n)_{n=1}^\infty \text{ and } \left(b_n - \frac{1}{12(n+1)}\right)_{n=1}^\infty$$

are both decreasing sequences, and that the sequence $(b_n - 1/12n)_{n=1}^\infty$ is an increasing sequence. Thus all three sequences tend to a common limit b. Consequently $(a_n)_{n=1}^\infty \to a$ as $n \to \infty$, where $a = e^b$, so that $n! \sim a n^{n+1/2} e^{-n}$.

It remains to determine a. We use Wallis' formula.

$$\frac{4^{n+1}(n!)^4}{(2n)!(2n+1)!} \sim \frac{a^4 4^{n+1} n^{4n+2} e^{-4n}}{a^2 (2n)^{4n+1} e^{-4n}(2n+1)} = \frac{a^2 n}{2n+1} \sim a^2/2.$$

But $4^{n+1}(n!)^4/((2n)!(2n+1)!) \to \pi$ as $n \to \infty$, by Wallis' formula, and so $a = \sqrt{2\pi}$. Thus we obtain *Stirling's formula*

$$n! \sim \sqrt{2\pi}.n^{n+\frac{1}{2}} e^{-n}.$$

More precisely,

$$e^{1/12(n+1)}\sqrt{2\pi n}\left(\frac{n}{e}\right)^n \le n! \le e^{1/12n}\sqrt{2\pi n}\left(\frac{n}{e}\right)^n.$$

10.5 The gamma function

Suppose that $a > 0$. The exponential functions e^{ax} grows faster than any polynomial, as $x \to +\infty$. On the other hand, $n!$ grows faster than e^{an}, as $x \to \infty$, for any $a \in \mathbf{R}$. Can we find a continuous function f of a natural kind on $[0, \infty)$ such that $f(n) = n!$ for $n \in \mathbf{N}$? Perhaps surprisingly, the answer is 'yes'; in fact, the function that we shall construct, the *gamma function* Γ, satisfies $\Gamma(n+1) = n!$.

We want to define the improper integral $\int_0^\infty t^{x-1}e^{-t}\,dt$. We consider the intervals $[0, 1]$ and $[1, \infty]$ separately.

First, suppose that $0 < x < 1$. Then $t^{x-1}e^{-t} \to \infty$ as $x \searrow 0$. But $t^{x-1}e^{-t} \le t^{x-1}$, and $\int_\epsilon^1 t^{x-1}\,dt = (1 - \epsilon^x)/x \to 1/x$ as $\epsilon \searrow 0$. Thus $I(\epsilon) = \int_\epsilon^1 t^{x-1}e^{-t}\,dt$ is a decreasing function on $(0, 1]$ which is bounded above by $1/x$, and so the improper integral $\int_0^1 t^{x-1}e^{-t}\,dt$ exists. If $x \geq 1$, the function $t^{x-1}e^{-t}$ is a continuous function on $[0,1]$, and so the Riemann integral $\int_0^1 t^{x-1}e^{-t}\,dt$ exists.

The function $t^{x-1}e^{-t}$ is continuous on $[1,\infty)$. If $n \in \mathbf{N}$ and $n > x$ then $e^t \geq t^n/n!$, so that $t^{x-1}e^{-t} \leq n!t^{x-n-1}$. Thus

$$\int_1^T t^{x-1}e^{-t}\,dt \leq n! \int_1^T t^{x-n-1}\,dt = \frac{n!(1 - T^{x-n})}{n - x} \leq \frac{n!}{n - x}.$$

Consequently, the improper integral $\int_0^\infty t^{x-1}e^{-t}\,dt$ exists.

We can therefore define the gamma function for $0 < t < \infty$ as

$$\Gamma(x) = \int_0^\infty t^{x-1}e^{-t}\,dt;$$

it exists as an improper integral.

Note that $\Gamma(1) = \int_0^\infty e^{-t}\,dt = 1$.

Proposition 10.5.1 *If $x > 0$ then $x\Gamma(x) = \Gamma(x+1)$.*

Proof Integrating by parts,

$$\int_\epsilon^X t^{x-1}e^{-t}\,dt = \left[\frac{t^x e^{-t}}{x}\right]_\epsilon^X + \frac{1}{x}\int_\epsilon^X t^x e^{-t}\,dt.$$

Now $t^x e^{-t} \to 0$ as $t \to 0$ and as $t \to \infty$, and so it follows that $x\Gamma(x) = \Gamma(x+1)$. $\qquad\square$

Corollary 10.5.2 *If $n \in \mathbf{Z}$ then $\Gamma(n+1) = n!$.*

Proof The result holds for $n = 1$, since $\Gamma(2) = \Gamma(1) = 1$. The result then follows by induction. $\qquad\square$

Proposition 10.5.3 *Γ is a continuous function on $(0,\infty)$.*

Proof Suppose that $x \in (0,\infty)$ and that $0 < a < x < b < \infty$. Suppose that $\epsilon > 0$. There exist $0 < \eta < 1 < R < \infty$ such that

$$\int_0^\eta t^{a-1}e^{-t}\,dt < \epsilon/5 \text{ and } \int_R^\infty t^{b-1}e^{-t}\,dt < \epsilon/5.$$

There then exists $0 < \delta < \min(x - a, b - x)$ such that if $|x - y| < \delta$ then $|t^{y-1} - t^{x-1}| < \epsilon/5(R - \eta)$. If $|x - y| < \delta$ then

$$\Gamma(y) - \Gamma(x) = I_0 + (I_1 - I_2) + (I_3 - I_4),$$

where

$$I_0 = \int_\eta^R (t^{y-1} - t^{x-1})e^{-t}\, dt,$$

$$I_1 = \int_0^\eta t^{y-1}e^{-t}\, dt, \qquad I_2 = \int_0^\eta t^{x-1}e^{-t}\, dt,$$

$$I_3 = \int_R^\infty t^{y-1}e^{-t}\, dt, \qquad I_4 = \int_R^\infty t^{x-1}e^{-t}\, dt.$$

The modulus of each integral is less that $\epsilon/5$, so that $|\Gamma(y) - \Gamma(x)| < \epsilon$. ☐

The beta and gamma functions are closely related.

Proposition 10.5.4 *If $x > 0$ and $y > 0$ then*

$$\Gamma(x)\Gamma(y) = B(x, y)\Gamma(x + y).$$

Proof Changing variables by setting $t = su$, exchanging the order of integration (this is justified in Volume II), and setting $w = s(1 + u)$, we find that

$$\Gamma(x)\Gamma(y) = \left(\int_0^\infty s^{x-1}e^{-s}\, ds \right) \left(\int_0^\infty t^{y-1}e^{-t}\, dt \right)$$

$$= \int_0^\infty s^{x-1}e^{-s} \left(\int_0^\infty s^y u^{y-1}e^{-su}\, du \right) ds$$

$$= \int_0^\infty u^{y-1} \left(\int_0^\infty s^{x+y-1}e^{-s(1+u)}\, ds \right) du$$

$$= \int_0^\infty u^{y-1} \left(\int_0^\infty \frac{w^{x+y-1}e^{-w}}{(1+u)^{x+y}}\, dw \right) du$$

$$= \left(\int_0^\infty \frac{u^{y-1}}{(1+u)^{x+y}}\, du \right) \Gamma(x + y).$$

Setting $v = u/(1 + u)$, we find that

$$\int_0^\infty \frac{u^{y-1}}{(1+u)^{x+y}}\, du = \int_0^1 v^{y-1}(1 - v)^{x-1}\, dv = B(x, y). \qquad ☐$$

Exercises

10.5.1 Show that

$$\Gamma(x) = \int_1^\infty \frac{(\log y)^{x-1}}{y^2}\, dy = \int_0^1 \left(\log \frac{1}{u} \right)^{x-1} dx.$$

10.5.2 Show that the gamma function is a logarithmically convex function on $(0, \infty)$.

10.5.3 Show that if $0 < x < 1$ then

$$\frac{1}{ex} \le \Gamma(x) \le 1 + \frac{1}{x}.$$

10.6 Riemann's zeta function

It follows from the integral test that $\sum_{j=1}^{\infty}(1/j^s)$ diverges if $s \le 1$, and converges if $s > 1$. If $s > 1$, we set $\zeta(s) = \sum_{j=1}^{\infty}(1/j^s)$. The function ζ is called *Riemann's zeta function*; it is a decreasing function of s.

It follows from the integral test that

$$\frac{1}{s-1} = \int_1^\infty \frac{dx}{x^s}\, ds < \zeta(s) = 1 + \sum_{j=2}^{\infty}(1/j^s)$$

$$\le 1 + \int_1^\infty \frac{dx}{x^s}\, ds = 1 + \frac{1}{s-1} = \frac{s}{s-1},$$

so that $\zeta(s) \to \infty$ as $s \searrow 1$ and $\zeta(s) \to 1$ as $s \to \infty$.

It was Euler who first considered the sum as a function of the real variable s. Later, Riemann considered ζ as a function of a complex variable; he introduced the notation ζ for the function and s for the variable.

Euler recognized the importance of the zeta function for number theory, and initiated the study of analytic number theory. Let $2 = p_1 < p_2 < \ldots$ be the sequence of primes, in increasing order. We shall use Theorem 2.6.6, the fundamental theorem of arithmetic: every $n \ge 2$ can be written uniquely in the form $n = p_1^{a_1} \ldots p_k^{a_k}$, where $a_1, \ldots, a_k \in \mathbf{Z}^+$ and $a_k \ne 0$.

We set

$$P_n(s) = \prod_{j=1}^n \left(\frac{1}{1 - 1/p_j^s}\right) = \prod_{j=1}^n \left(1 + \frac{1}{p_j^s - 1}\right).$$

Thus $P_n(s)$ is a continuous decreasing function on $(0, \infty)$.

Since

$$\frac{1}{1 - 1/p_j^s} = 1 + \frac{1}{p_j^s} + \frac{1}{p_j^{2s}} + \cdots,$$

and since all the terms are non-negative, we can expand the products, to obtain

$$P_n(s) = \sum \{\frac{1}{(p_1^{k_1} \ldots p_n^{k_n})^s} : k_1, \ldots, k_n \in \mathbf{Z}^+\}.$$

This provides an analytic proof of the fact that there are infinitely many primes. For if there are only finitely many primes $p_1, \ldots, p_n$ then every positive integer can be written in the form $p_1^{k_1} \ldots p_n^{k_n}$, so that $P_n(1) = \sum_{j=1}^{\infty} 1/j = \infty$, giving a contradiction. We can say more.

If $s > 1$ then it follows that $P_n(s) \leq \sum_{n=1}^{\infty} 1/n^s = \zeta(s)$. The sequence $(P_n(s))_{n=1}^{\infty}$ is increasing, and so $P_n(s)$ converges to a limit $P(s)$, with $P(s) \leq \zeta(s)$. On the other hand, suppose that $N \in \mathbf{N}$, and let $\{p_1, \ldots p_k\}$ be the set of primes less than N. Then every $j < N$ can be written as a product of powers of $p_1, \ldots p_k$, and so

$$P(s) \geq P_k(s) \geq \sum_{j=1}^{N-1} \frac{1}{j^s}.$$

Since this holds for all N, $P(s) \geq \zeta(s)$, and so we have the following.

Proposition 10.6.1   *If $1 < s < \infty$ then*

$$\zeta(s) = \prod_{j=1}^{\infty} \left(\frac{1}{1 - 1/p_j^s} \right) = \prod_{j=1}^{\infty} \left(1 + \frac{1}{p_j^s - 1} \right),$$

where $(p_1 < p_2 < \ldots)$ is the sequence of primes, arranged in increasing order.

Corollary 10.6.2 $\sum_{n=1}^{\infty} (1/p_n) = \infty$.

Proof   If not, then, by Proposition 10.1.3,

$$\prod_{j=1}^{\infty} \left(1 - \frac{1}{p_j} \right)$$

is convergent to a non-zero limit P, say. But

$$1/\zeta(s) = \prod_{j=1}^{\infty} \left(1 - \frac{1}{p_j^s} \right) \geq P$$

for $s > 1$, so that $1/P \geq \zeta(s)$. Since $\zeta(s) \to \infty$ as $s \searrow 1$, this gives a contradiction.   □

10.7 Chebyshev's prime number theorem

Again, let $2 = p_1 < p_2 < \cdots$ denote the sequence of primes, in increasing order. The fact that $\sum_{j=1}^{\infty} 1/p_j = \infty$ shows not only that there are infinitely many primes, but also that they occur fairly frequently; for example, if $(a_j)_{j=1}^{\infty}$ is a sequence of positive numbers for which $\sum_{j=1}^{\infty} a_j < \infty$ then $\liminf_{j \to \infty} a_j p_j = 0$. This raises the question; how are the prime numbers distributed? If $x > 0$, let $\pi(x)$ be the number of primes not greater than x.

In 1792, Gauss, at the age of fifteen, conjectured that $\pi(x) \sim x/\log x$: that is, $\pi(x)\log x/x \to 1$ as $x \to \infty$. In 1850, Chebyshev showed, by elementary real analysis, that this was the right rate of growth.

First, let us introduce some notation. Suppose that f is a real-valued function defined on $\mathbf{N}$, and that $x > 0$. We set

$$\sum_{p \le x} f(p) = \sum \{f(p) : p \text{ a prime}, p \le x\}$$

$$\sum_{y < p \le x} f(p) = \sum \{f(p) : p \text{ a prime}, y < p \le x\}$$

and $$\sum_{p^m \le x} f(p^m) = \sum \{f(p^m) : p \text{ a prime}, m \in \mathbf{N}, p^m \le x\},$$

and use similar notations for products.

Chebyshev introduced two auxiliary functions:

$$\theta(x) = \sum_{p \le x} \log p \text{ and } \psi(x) = \sum_{p^m \le x} \log p.$$

He proved the following.

Theorem 10.7.1 (Chebyshev's prime number theorem)

(i) $\liminf_{x \to \infty} \frac{\pi(x)\log x}{x} = \liminf_{x \to \infty} \frac{\theta(x)}{x} = \liminf_{x \to \infty} \frac{\psi(x)}{x}$,

$\limsup_{x \to \infty} \frac{\pi(x)\log x}{x} = \limsup_{x \to \infty} \frac{\theta(x)}{x} = \limsup_{x \to \infty} \frac{\psi(x)}{x}$.

(ii) $\limsup_{x \to \infty} \frac{\theta(x)}{x} < \lg 4 < 1.387$.

(iii) $\liminf_{x \to \infty} \frac{\psi(x)}{x} > \frac{1}{2}\log 2 > 0.346$.

Proof of (i) Clearly $\theta(x) \le \psi(x)$. Let $c_p(x) = \sup\{m : p^m \le x\}$. Then

$$\psi(x) = \sum_{p \le x} c_p(x)\log p = \sum_{p \le \sqrt{x}} c_p(x)\log x + \sum_{\sqrt{x} < p \le x} c_p(x)\log x$$

$$\le \sqrt{x}\log x + \theta(x),$$

since $c_p(x) = 1$ for $\sqrt{x} < p \le x$. Thus $\psi(x)/x - \theta(x)/x \to 0$ as $x \to \infty$. Consequently

$$\liminf_{x \to \infty} \frac{\theta(x)}{x} = \liminf_{x \to \infty} \frac{\psi(x)}{x} \text{ and } \limsup_{x \to \infty} \frac{\theta(x)}{x} = \limsup_{x \to \infty} \frac{\psi(x)}{x}.$$

Since $c_p(x)$ is the largest integer such that $p^m \le x$,

$$c_p(x) \log p \le \log x < (c_p(x) + 1) \log p,$$

so that $c_p(x)$ is the integral part of $\log x / \log p$. Consequently,

$$\psi(x) = \sum_{p \le x} \lfloor \frac{\log x}{\log p} \rfloor \cdot \log p \le \pi(x) \log x.$$

Thus

$$\liminf_{x \to \infty} \frac{\psi(x)}{x} \le \liminf_{x \to \infty} \frac{\pi(x) \log x}{x},$$

$$\limsup_{x \to \infty} \frac{\psi(x)}{x} \le \limsup_{x \to \infty} \frac{\pi(x) \log x}{x}.$$

Suppose that $0 < \alpha < 1$. Then

$$\theta(x) \ge \sum_{x^\alpha < p \le x} \log p \ge \sum_{x^\alpha < p \le x} \log x^\alpha$$

$$= (\alpha \log x)(\pi(x) - \pi(x^\alpha)) \ge (\alpha \log x)(\pi(x) - x^\alpha),$$

so that

$$\frac{\theta(x)}{x} \ge \alpha \left(\frac{\pi(x) \log x}{x} - x^{\alpha-1} \log x \right).$$

Since $x^{\alpha-1} \log x \to 0$ as $x \to \infty$,

$$\liminf_{x \to \infty} \frac{\theta(x)}{x} \ge \alpha \liminf_{x \to \infty} \frac{\pi(x) \log x}{x},$$

$$\limsup_{x \to \infty} \frac{\theta(x)}{x} \ge \alpha \limsup_{x \to \infty} \frac{\pi(x) \log x}{x}.$$

Since this holds for all $0 < \alpha < 1$,

$$\liminf_{x \to \infty} \frac{\theta(x)}{x} \ge \liminf_{x \to \infty} \frac{\pi(x) \log x}{x},$$

$$\limsup_{x \to \infty} \frac{\theta(x)}{x} \ge \limsup_{x \to \infty} \frac{\pi(x) \log x}{x}.$$

Proof of (ii) It is sufficient to show that $\theta(n)/n < \log 4$, for $n \in \mathbf{N}$, with $n \ge 2$. We prove this by induction. Certainly $\theta(2) = \log 2 < 2 \log 4$, $\theta(3) = \log 6 \le 3 \log 4$ and $\theta(4) = \log 6 \le 4 \log 4$. Suppose that the result holds for $2 \le j \le 2n$. Then, since $2n + 2$ is not prime,

$$\theta(2n + 2) - \theta(n + 1) = \theta(2n + 1) - \theta(n + 1) = \sum_{n+1 < p \le 2n+1} \log p = \log P,$$

where $P = \prod_{n+1<p\leq 2n+1} p$. Now if p is a prime and $n+1 < p \leq 2n+1$ then p divides $(2n+1)!$ and does not divide $(n+1)!$, and so p divides $\binom{2n+1}{n+1} = \binom{2n+1}{n}$. Thus P divides $\binom{2n+1}{n+1} = \binom{2n+1}{n}$. But

$$\binom{2n+1}{n+1} = \frac{1}{2}\left(\binom{2n+1}{n} + \binom{2n+1}{n+1}\right) \leq \frac{1}{2}\sum_{j=0}^{2n+1}\binom{2n+1}{j} = 2^{2n} = 4^n,$$

so that $P \leq 4^n$, $\log P \leq n \log 4$, and

$$\theta(2n+2) = \theta(2n+1) \leq \theta(n+1) + n\log 4 \leq (2n+1)\log 4.$$

Proof of (iii) It is sufficient to show that $\psi(2n)/2n \geq \frac{1}{2}\log 2$, for $n \in \mathbf{N}$. By the fundamental theorem of arithmetic, if $n \in \mathbf{N}$ we can write n uniquely as $\prod_{p\leq n} p^{v_p(n)}$. The quantity $v_p(n)$ is the *p-adic valuation* of n. Since $\{1,\dots,n\}$ contains $\lfloor n/p \rfloor$ multiples of p, $\lfloor n/p^2 \rfloor$ multiples of p^2, and so on,

$$v_p(n!) = \sum_{j=1}^{\infty} \lfloor n/p^j \rfloor.$$

(Of course, this is a finite sum.) Now

$$\binom{2n}{n} = \frac{(2n)!}{(n!)^2} = \prod_{p\leq 2n} p^{g_p(n)},$$

where

$$g_p(n) = v_p((2n)!) - 2v_p(n!) = \sum_{j=1}^{\infty}\lfloor 2n/p^j \rfloor - 2\sum_{j=1}^{\infty}\lfloor n/p^j \rfloor$$

$$= \sum_{j=1}^{\infty}\left(\lfloor 2n/p^j \rfloor - 2\lfloor n/p^j \rfloor\right).$$

Now if $\lfloor 2n/p^j \rfloor$ is even, there are as many numbers of the form ap^j in $\{1,\dots,n\}$ as there are in $\{n+1,\dots,2n\}$, and if $\lfloor 2n/p^j \rfloor$ is odd, there is one less number of the form ap^j in $\{1,\dots,n\}$ as there are in $\{n+1,\dots,2n\}$ (why is this?). Thus

$$\lfloor 2n/p^j \rfloor - 2\lfloor n/p^j \rfloor = 0 \text{ if } \lfloor 2n/p^j \rfloor \text{ is even,}$$
$$= 1 \text{ if } \lfloor 2n/p^j \rfloor \text{ is odd,}$$

so that $0 \leq g_p(n) \leq c_p(2n)$. Consequently

$$\log \binom{2n}{n} = \sum_{p\leq 2n} g_p(n)\log p \leq \sum_{p\leq 2n} c_p(2n)\log p = \psi(2n).$$

But

$$\binom{2n}{n} = \frac{(n+1)(n+2)\ldots(2n)}{1.2.\ldots.n} \geq 2^n,$$

so that $\log \binom{2n}{n} \geq n \log 2$, and $\psi(2n)/2n \geq \frac{1}{2} \log 2$. □

Chebyshev also showed that

$$\liminf_{x \to \infty} \frac{\pi(x) \log x}{x} \leq 1 \leq \limsup_{x \to \infty} \frac{\pi(x) \log x}{x},$$

but the real difficulty is to show that $\pi(x) \log x/x$ tends to a limit as $x \to \infty$.

Gauss' conjecture was eventually proved independently by Hadamard and de la Vallée Poussin in 1896.

Theorem 10.7.2 (The prime number theorem) $\pi(x) \log x/x \to 1$ *as* $x \to \infty$.

Their proofs are difficult, and use the theory of functions of a complex variable.

10.8 Evaluating $\zeta(2)$

An outstanding problem at the beginning of the eighteenth century was the evaluation of

$$\zeta(2) = 1 + \frac{1}{4} + \frac{1}{9} + \cdots + \frac{1}{n^2} + \cdots .$$

One of Euler's early triumphs was to show that $\zeta(2) = \pi^2/6$. We gave a proof using Fourier series in Example 9.3.10: here we present one given by Euler. We start by applying the binomial theorem. Suppose that $0 < x < 1$. Now

$$\frac{d}{dx}(\sin^{-1} x)^2 = 2 \frac{\sin^{-1}(x)}{\sqrt{1 - x^2}}$$

$$= \frac{2x}{\sqrt{1 - x^2}} + 2 \sum_{j=1}^{\infty} \left(\frac{(2j)!}{(2j+1)2^{2j}(j!)^2} \frac{x^{2j+1}}{\sqrt{1 - x^2}} \right),$$

by Corollary 8.5.5. Integrating term by term (justify this carefully),

$$(\sin^{-1} x)^2 = 2\int_0^x \frac{t}{\sqrt{1-t^2}}\, dt + 2\sum_{j=1}^\infty \left(\frac{(2j)!}{(2j+1)2^{2j}(j!)^2}\int_0^x \frac{t^{2j+1}}{\sqrt{1-t^2}}\, dt\right),$$

for $0 \le x < 1$. Since all the terms are non-negative increasing functions of x, it follows from Theorem 6.3.10 that the formula also holds if we set $x = 1$:

$$\frac{\pi^2}{8} = \int_0^1 \frac{t}{\sqrt{1-t^2}}\, dt + \sum_{j=1}^\infty \left(\frac{(2j)!}{(2j+1)2^{2j}(j!)^2}\int_0^1 \frac{t^{2j+1}}{\sqrt{1-t^2}}\, dt\right).$$

Setting $t = \sin\theta$, and applying Corollary 10.3.2,

$$\int_0^1 \frac{t^{2j+1}}{\sqrt{1-t^2}}\, dt = \int_0^{\pi/2} \sin^{2j+1}\theta\, d\theta = \frac{2^{2j}(j!)^2}{(2j+1)!}.$$

Substituting, we see that

$$\frac{\pi^2}{8} = 1 + \sum_{j=1}^\infty \frac{1}{(2j+1)^2} = 1 + \frac{1}{9} + \frac{1}{25} + \cdots$$

$$= \left(1 + \frac{1}{4} + \frac{1}{9} + \cdots\right) - \frac{1}{4}\left(1 + \frac{1}{4} + \frac{1}{9} + \cdots\right)$$

$$= \frac{3}{4}\left(1 + \frac{1}{4} + \frac{1}{9} + \cdots\right),$$

giving the result.

10.9 The irrationality of e^r

We have seen in Section 4.2 that e is irrational; we now show that e^r is irrational for all non-zero rational r. If $r = p/q$ and e^r is rational, then $e^p = (e^r)^q$ is rational, and so is $e^{-p} = 1/e^p$. It is therefore enough to show that e^{-k} is irrational, for each positive integer k. Suppose, if possible, that $e^{-k} = p/q$, where p and q are integers.

We use the fact that if f is a differentiable function on $[a, b]$ then

$$\frac{d}{dx}(-e^{-kx}f(x)) = e^{-kx}(kf(x) - f'(x)),$$

which suggests the possibility of cancellation.

Suppose that f is an $(m+1)$-times differentiable on $[a, b]$; set

$$g(x) = -e^{-kx}(k^m f(x) + k^{m-1}f'(x) + \cdots + f^{(m)}(x)).$$

Then $g'(x) = e^{-kx}(k^{m+1}f(x) - f^{(m+1)}(x))$, so that

$$g(b) - g(a) = \int_a^b e^{-kx}(k^{m+1}f(x) - f^{(m+1)}(x))\, dx.$$

We apply this to the polynomial function

$$\beta_n(x) = \frac{x^n(1-x)^n}{n!} = \frac{1}{n!}\sum_{j=0}^n \binom{n}{j}(-1)^j x^{n+j},$$

on the interval $[0,1]$. We choose this function, because $\beta_n(0) = 0$ and $\beta_n^{(h)}(0) = 0$ for $1 \le h < n$ and for $h > 2n$. If $h = n+j$, where $0 \le j \le n$, then

$$\beta_n^{(h)}(0) = (-1)^j \frac{(n+j)!}{n!}\binom{n}{j} = (-1)^j j!\binom{n+j}{j}\binom{n}{j},$$

which is an integer. Since $\beta_n(x) = \beta_n(1-x)$, similar phenomena occur when $x = 1$. We now take $m = 2n$, and set

$$g_n(x) = -e^{-kx}(k^{2n}\beta_n(x) + k^{2n-1}\beta_n'(x) + \cdots + \beta_n^{(2n)}(x)).$$

Then $g_n(1) = e^{-k}r = pr/q$ and $g(0) = s$, where r and s are integers. Further, since $\beta_n^{(2n+1)}(x) = 0$ for all x,

$$g_n(1) - g_n(0) = \int_0^1 e^{-kx}(k^{2n+1}\beta_n(x) - \beta_n^{(2n+1)})\, dx = \int_0^1 e^{-kx}k^{2n+1}\beta_n(x)\, dx.$$

Thus

$$q\left(\int_0^1 e^{-kx}k^{2n+1}\beta_n(x)\, dx\right) = pr - qs, \text{ an integer.}$$

But $0 < \beta_n(x) < 1/n!$ and $e^{-kx} \le 1$ for $0 < x < 1$, and so

$$0 < q(\int_0^1 e^{-kx}k^{2n+1}\beta_n(x)\, dx) \le q\left(\frac{k^{2n+1}}{n!}\right).$$

Since $k^{2n+1}/n! \to 0$ as $n \to \infty$, it follows that

$$0 < q\left(\int_0^1 e^{-kx}k^{2n+1}\beta_n(x)\, dx\right) < 1$$

for large enough n, giving the required contradiction.

10.10 The irrationality of π

We use the idea of the previous example to show that π is irrational. In fact, we do rather more, and show that π^2 is irrational. Suppose that $\pi^2 = p/q$, where p and q are integers.

Suppose that u is a twice differentiable function on an interval $[a, b]$. Let

$$g(x) = u(x) \cos \pi x,$$
$$h(x) = u'(x) \sin \pi x.$$

Then

$$g'(x) = u'(x) \cos \pi x - \pi u(x) \sin \pi x,$$
$$h'(x) = \pi u'(x) \cos \pi x + u''(x) \sin \pi x,$$

so that

$$(h - \pi g)'(x) = (\pi^2 u(x) + u''(x)) \sin \pi x.$$

This again suggests the possibility of cancellation.

Suppose that f is a $2n + 2$-times differentiable function; set

$$u(x) = \pi^{2n+2} f(x) - \pi^{2n} f''(x) + \cdots + (-1)^n \pi^2 f^{(2n)}(x).$$

Then

$$(h/\pi - g)'(x) = \left(\pi^{2n+1} f(x) + (-1)^n \pi f^{(2n+2)}(x) \right) \sin \pi x.$$

Since $h(1) = h(0) = 0$, it follow that

$$g(0) - g(1) = \int_0^1 \left(\pi^{2n+1} f(x) + (-1)^n \pi f^{(2n+2)}(x) \right) \sin \pi x \, dx.$$

Let us take $f(x) = \beta_n(x)$, as before. Recall that $\beta_n(0) = \beta_n(1) = 0$ and that $\beta_n^{(h)}(0)$ and $\beta_n^{(h)}(1)$ are integers for all $h \in \mathbf{N}$. Thus $q^{n+1} g(0) = q^{n+1} u(0)$ and $q^{n+1} g(1) = -q^{n+1} u(1)$ are both integers. Since $\beta_n^{(2n+2)}(x) = 0$ for all x,

$$g(0) - g(1) = \int_0^1 \pi^{2n+1} \beta_n(x) \sin \pi x \, dx,$$

so that

$$q^{n+1} g(0) - q^{n+1} g(1) = \frac{p^{n+1}}{\pi} \int_0^1 \beta_n(x) \sin \pi x \, dx$$

is an integer. But

$$0 < \frac{p^{n+1}}{\pi} \int_0^1 \beta_n(x) \sin \pi x \, dx \le \frac{p^{n+1}}{\pi . n!},$$

and $p^{n+1}/\pi.n! \to 0$ as $n \to \infty$, so that

$$0 < \frac{p^{n+1}}{\pi} \int_0^1 \beta_n(x) \sin \pi x \, dx < 1$$

for large enough n, giving the required contradiction.

Appendix A

Zorn's lemma and the well-ordering principle

A.1 Zorn's lemma

We show that Zorn's lemma is a consequence of the axiom of choice.

Theorem A.1.1 *Assume the axiom of choice. Suppose that $(X, \leq)$ is a non-empty partially ordered set with the property that every non-empty chain (totally ordered subset) of X has an upper bound. Then there exists a maximal element in X.*

Proof[1] We need a few more definitions. Suppose that A is a subset of a partially ordered set $(X, \leq)$ and that $x \in X$. x is a *strict upper bound* for A if $a < x$ for all $a \in A$. A totally ordered set $(S, \leq)$ is *well-ordered* if every non-empty subset of S has a least element. A subset D of a totally ordered set $(S, \leq)$ is an *initial segment* of S if whenever $x \in S$, $d \in D$ and $x \leq d$ then $x \in D$.

We break the proof into a sequence of lemmas and corollaries. If A is a subset of X, let A' be the set of strict upper bounds of A. If $A = \emptyset$, then $A' = X$.

Let $\mathcal{C}$ be the set of chains in X.

Lemma A.1.2 *Suppose that there exists $C \in \mathcal{C}$ for which $C' = \emptyset$. Then C has a unique upper bound, which is a maximal element of X.*

Proof C has an upper bound c. Then $c \in C$, and is the unique upper bound for C, since C is a chain. If $x \geq c$, then x is an upper bound for C, and so is equal to c. Thus c is a maximal element of X. □

We must therefore find a chain C for which $C' = \emptyset$. Let s be a choice function on $P(X) \setminus \emptyset$: if A is a non-empty subset of X then $s(A) \in A$. We

---

[1] Thanks to Peter Johnstone for showing me how to simplify the proof.

use the choice function to define a successor function, and use this to find a large chain in $\mathcal{C}$.

Let $\mathcal{T}$ be the set of chains C in X with the property that if D is an initial segment of C and $D \neq C$ then $s(D')$ is the least element of $C \setminus D$.

Lemma A.1.3 $\mathcal{T} \neq \emptyset$.

Proof The set $C = \{s(X)\}$ is a non-empty chain in $\mathcal{T}$. Suppose that D is an initial segment in C. If $D \neq C$ then $D = \emptyset$, and $s(X)$ is the least element of $C \setminus D$. Thus $C \in \mathcal{T}$. $\square$

Lemma A.1.4 *If $C \in \mathcal{T}$ and $C' \neq \emptyset$ then $C^+ = C \cup \{s(C')\} \in \mathcal{T}$.*

Proof C^+ is certainly a chain, with greatest element $s(C')$. Suppose that D is an initial segment of C^+ and that $D \neq C^+$. There are two possibilities. First, D is a proper subset of C. Then $s(D')$ is the least element of $D' \cap C$, and is therefore the least element of $D' \cap C^+$. Secondly, $D = C$. Then $s(D') = s(C')$ is the least element of $C^+ \setminus D$. $\square$

Corollary A.1.5 *We order $\mathcal{T}$ by inclusion. It is enough to show that $\mathcal{T}$ has a greatest element M.*

Proof For if $M' \neq \emptyset$, then $M^+ \in \mathcal{T}$, contradicting the maximality of M. $\square$

We show that $\mathcal{T}$ is totally ordered by inclusion.

Lemma A.1.6 *Suppose that $C, D \in \mathcal{T}$, and that C is not contained in D. Then D is an initial segment of C.*

Proof Let

$$E = \{x \in C \cap D : \text{if } y \leq x \text{ then } y \in C \text{ if and only if } y \in D\}.$$

Then E is an initial segment of both C and D. Since $E \subseteq D$, $E \neq C$, and $s(E')$ is the least element of $C \setminus E$. Suppose that $E \neq D$. Then $s(E')$ is the least element of $D \setminus E$. But this implies that $s(E') \in E$, giving a contradiction. Thus $D = E$, and so D is an initial segment of C. $\square$

Lemma A.1.7 *Let $M = \cup_{C \in \mathcal{T}} C$. Then $M \in \mathcal{T}$.*

Proof M is certainly a chain in X. Suppose that D is an initial segment in M, and that $D \neq M$. Then $D^+ \subseteq M$, and so $s(D') \in M$. If $x \in M \setminus D$ then $x \in C \setminus D$ for some $C \in \mathcal{T}$, and $s(D')$ is the least element of $C \setminus D$. Thus $s(D') \leq x$, and so $s(D')$ is the least element of $M \setminus D$. Thus $M \in \mathcal{T}$. $\square$

Corollary A.1.8   *M is the greatest element of T.*

Proof   For if $C \in T$, then $C \subseteq M$, by the definition of M.   □

This completes the proof of Theorem A.1.1.   □

A.2 The well-ordering principle

Zorn's lemma implies that any non-empty set can be well-ordered.

Theorem A.2.1 (The well-ordering principle)   *If S is a non-empty set, Zorn's lemma implies that there is a total order on S under which S is well-ordered.*

Proof   We sketch the proof, and leave it to the reader to supply all the details. Let T be the set of all pairs $(A, \leq_A)$, where A is a non-empty subset of S and $\leq_A$ is a well-ordered total order on A. T is not empty (consider singleton sets). We define a partial order on T by setting $(A, \leq_A) \leq (B, \leq_B)$ if A is an initial segment of B, and the two partial orders agree on A: $x \leq_A y$ if and only if $x \leq_B y$. We argue as in Theorem A.1.1. If C is a chain in T, set $M = \cup\{A : (A, \leq_A) \in C\}$. If $x, y \in M$, there exists $(A, \leq_A) \in C$ such that $x, y \in A$. Set $x \leq_M y$ if $x \leq_A y$. This is well defined, and defines a total order on M. This total order is a well-ordering of M, so that $(M, \leq_M) \in T$. If $(A, \leq_A) \in C$ then A is an initial segment of $(M, \leq_M)$. Hence $(M, \leq_M)$ is an upper bound for C. We can apply Zorn's lemma: there exists a maximal element $(N, \leq_N)$ of T. We claim that $N = X$. If not, there exists $z \in X \setminus N$. Let $N' = N \cup \{z\}$. Define a partial order $\leq_{N'}$ on N' by setting $x \leq_{N'} y$ if $x, y \in N$ and $x \leq_N y$, and by setting $x \leq'_N z$ for all $x \in N'$. Then $N' \in T$, and $(N, \leq_N)$ is strictly less than $(N', \leq_{N'})$, contradicting the maximality of $(N, \leq_N)$.   □

There are circumstances in which it may be more convenient to prove results using the well-ordering principle, rather than the axiom of choice. The well-ordering principle is used to develop the theory of ordinals.

It is easy to deduce the axiom of choice from the well-ordering principle.

Theorem A.2.2   *The well-ordering principle implies the axiom of choice.*

Proof   Suppose that $\{B_\alpha\}_{\alpha \in A}$ is a non-empty family of non-empty sets. Let $X = \cup_{\alpha \in A} B_\alpha$, and let $\leq$ be a well-ordered total order on X. If $\alpha \in A$, let $c(\alpha)$ be the least element of B_α. Then c is a choice function.   □

Exercises

A.2.1 Give the details of the proof of Theorem A.2.1.

A.2.2 Use the well-ordering principle to prove Theorem 1.9.2.

A.2.3 Modify the proof of Theorem A.1.1 in the following way to deduce the well-ordering principle directly from the axiom of choice.

Let s be a choice function on the non-empty subsets of a non-empty set X. Let $\mathcal{T}$ be the set of pairs $(C, \leq_C)$, where C is a subset of X and $\leq_C$ is a well-ordered total order on C with the property that if D is an initial segment of C and $C \setminus D$ is not empty then $s(X \setminus D)$ is the least element of $C \setminus D$.

(i) Suppose that $(C \leq_C)$ and $(D \leq_D)$ are elements of X, and that C is not contained in D. Show that D is an initial segment of C, and that the two total orders $\leq_C$ and $\leq_D$ agree on D.

(ii) Let $M = \{C : (C \leq_C) \in \mathcal{T}\}$. Define a total order $\leq_M$ on M, verifying that it is well-defined, and show that $(M, \leq_M) \in \mathcal{T}$.

(iii) Show that $M = X$.

Index